CHEMICALS VIA HIGHER PLANT BIOENGINEERING

ADVANCES IN EXPERIMENTAL MEDICINE AND BIOLOGY

Recent Volumes in this Series

Volume 455
RHEUMADERM: Current Issues in Rheumatology and Dermatology
Edited by Carmel Mallia and Jouni Uitto

Volume 456
RESOLVING THE ANTIBIOTIC PARADOX:
Progress in Understanding Drug Resistance and Development of New Antibiotics
Edited by Barry P. Rosen and Shahriar Mobashery

Volume 457
DRUG RESISTANCE IN LEUKEMIA AND LYMPHOMA III
Edited by G. J. L. Kaspers, R. Pieters, and A. J. P. Veerman

Volume 458
ANTIVIRAL CHEMOTHERAPY 5: New Directions for Clinical Application and Research
Edited by John Mills, Paul A. Volberding, and Lawrence Corey

Volume 459
IMPACT OF PROCESSING ON FOOD SAFETY
Edited by Lauren S. Jackson, Mark G. Knize, and Jeffrey N. Morgan

Volume 460
MELATONIN AFTER FOUR DECADES
Edited by James Olcese

Volume 461
CYTOKINES, STRESS, AND DEPRESSION
Edited by Robert Dantzer, Emmanuele Wollman, and Raz Yirmiya

Volume 462
ADVANCES IN BLADDER RESEARCH
Edited by Laurence S. Baskin and Simon W. Hayward

Volume 463
ENZYMOLOGY AND MOLECULAR BIOLOGY OF CARBONYL METABOLISM 7
Edited by Henry Weiner, Edmund Maser, David W. Crabb, and Ronald Lindahl

Volume 464
CHEMICALS VIA HIGHER PLANT BIOENGINEERING
Edited by Fereidoon Shahidi, Paul Kolodziejczyk, John R. Whitaker, Agustin Lopez Munguia, and Glenn Fuller

---

A Continuation Order Plan is available for this series. A continuation order will bring delivery of each new volume immediately upon publication. Volumes are billed only upon actual shipment. For further information please contact the publisher.

CHEMICALS VIA HIGHER PLANT BIOENGINEERING

Edited by

Fereidoon Shahidi
Memorial University of Newfoundland
St. John, Newfoundland, Canada

Paul Kolodziejczyk
POS Pilot Plant Corporation
Saskatoon, Saskatchewan, Canada

John R. Whitaker
University of California
Davis, California

Agustin Lopez Munguia
Institute of Biotechnology-UNAM
Cuernavaca, Morelos, Mexico

and

Glenn Fuller
USDA/ARS
Albany, California

Kluwer Academic/Plenum Publishers
New York, Boston, Dordrecht, London, Moscow

Based on proceedings of a symposium on Chemicals via Higher Plant Bioengineering,
held at the 5th North American Chemical Congress, November 11–15, 1997, in Cancun, Mexico

ISBN 0-306-46117-X

233 Spring Street, New York, N.Y. 10013

10 9 8 7 6 5 4 3 2 1

A C.I.P. record for this book is available from the Library of Congress

Printed in the United States of America

PREFACE

Food and raw material for its production was generally produced via the traditional agriculture. On the other hand, novel chemicals were manufactured in the laboratory or extracted from plant and animal sources. However, as the world population is steadily increasing, there is a decrease in traditional agriculture productivity and concerns are also expressed over the damage inflicted to the environment and restrictions that might be enforced in food production. At the same time, there is an increasing demand for high quality agricultural products as well as for food ingredients related to both the traditional or newly discovered nutrients or phytochemicals.

Trends and developments in the area of plant biotechnology and bioengineering has allowed manipulation of genes and/or insertion of new genes, thus production of transgenic plants. Starting from the introduction of agronomic traits, particularly stress resistance to diverse environmental factors, process and sensory characteristics, food quality and production of novel varieties of plant-based products through genetic engineering, biotechnology is changing the agriculture and the concept of production of plant-based raw materials. Increasing attention is being paid on research for production of plants that can provide a wide array of food and non-food products. Perhaps the first non-food product that plant biotechnology would achieve is production of large scale custom-designed industrial oils, but the list of chemicals is long, ranging from oils and specific triacylglycerols to biopolymers, enzymes, blood components, among others.

This monograph assembles the latest developments in plant biotechnology as presented by lead scientists during the Fifth Chemical Congress of North America held in November 1997 in Cancun or subsequently solicited by the editors. It emphasizes the production of chemicals from higher plants, but also includes fundamental aspects of plant biochemistry. We are grateful to all contributors for their outstanding efforts that made the production of this state-of-the-art monograph possible.

Fereidoon Shahidi
Paul P. Kolodziejczyk
John R. Whitaker
Augustin Lopez Munguia
Glenn Fuller

CONTENTS

1

NOVEL CHEMICALS FROM PLANTS VIA BIOENGINEERING

An Overview

Paul P. Kolodziejczyk[1] and Fereidoon Shahidi[2]

[1]POS Pilot Plant Corporation
118 Veterinary Road
Saskatoon, SK, S7N 2R4, Canada
[2]Department of Biochemistry
Memorial University of Newfoundland
St. John's, NF, A1B 3X9, Canada

ABSTRACT

Novel chemicals were traditionally extracted from medicinal plants or produced synthetically. However, new development in the field of bioengineering has allowed production of novel products from plants such as edible and industrial oils as well as specific chemicals which could be used as foods with remedial effects.

INTRODUCTION

Traditional production of novel chemicals took place in the laboratory and processing plants. However, developments in the production of genetically-modified plants as well as transgenic animals and microorgenisms have gained popularity in recent years. As an example, production of canola, a genetically altered rapeseed allowed reduction of erucic acid from 25–50% to less than 1% and glucosinolates to minute quantities (Kott et al., 1990; Shahidi et al., 1997). As a result, canola has become the third most important world source of vegetable oil after soybean and palm.

Introduction of new genes to canola has allowed production of an array of specialty oils and even re-introduction of erucic acid to the oils for selected products (Figure 1, Murphy, 1994). Thus, use of plants as vehicles for preparation of novel chemicals, including edible and industrial oils, is important.

Chemicals via Higher Plant Bioengineering, edited by Shahidi *et al.*
Kluwer Academic / Plenum Publishers, New York, 1999.

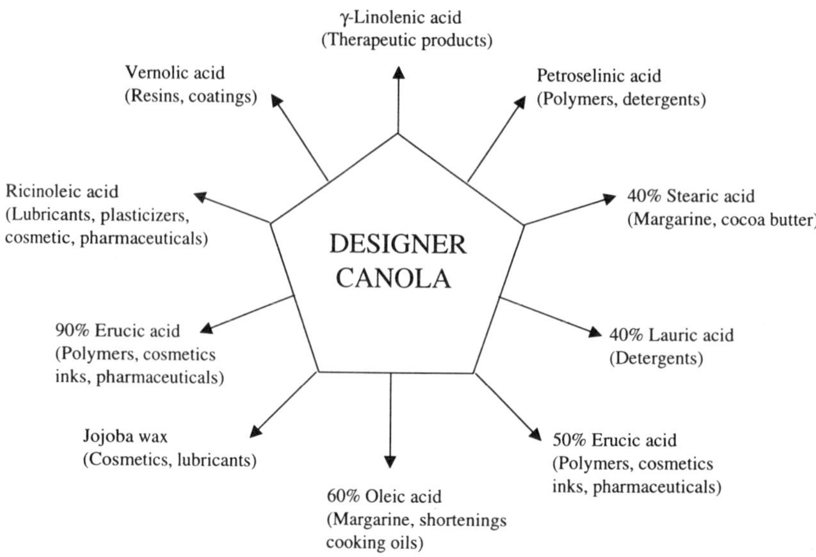

Figure 1. Designer canola varieties produced or under development. Adapted from Murphy, 1994.

MOLECULAR FARMING AND PLANT BIOREACTORS

Plant bioreactors are plant or plant cell cultures that have been genetically engineered to produce novel chemicals. The horizons of potential applications of biotechnology reaches far beyond the traditional meaning of agriculture. Proprietary technologies may position companies as producers and suppliers of bulk products to the biopharmaceutical and enzyme industries. Current development is focused not only on high-value therapeutic proteins, edible vaccines and antibodies for human immunotherapy, but also on industrial applications like the production of enzymes for food and feed additives, engineering plants in order to produce specially designed seed which can be used as a delivery vehicle for industrial enzymes or specialty oils (Kolodziejczyk, 1998).

As a result of genetic engineering, plants can be used as versatile, renewable and low-cost source of molecules previously not available from plant sources. The number of compounds that have been successfully produced in transgenic crops is steadily increasing as novel genes involved in specific biosynthetic pathways are identified and inserted into commercially important crops. Transgenic plants offer the potential to be one of the most economical systems for large scale production of proteins intended for industrial, pharmaceutical, veterinary and agriculture use. A number of mammalian proteins with therapeutic potential have been expressed in transgenic plants.

Using antibodies as drugs is not new, however, these antibodies are usually manufactured in cloned animal cells. There are two serious risks related to using animal cell originated antibodies. First is the possibility of allergic reactions due to the presence of traces of foreign antibodies (animal not human) in the drug; second, is the possible contamination of therapeutics by animal pathogens, viruses and proteins such as Bovine Spongiform Encephalopathy (BSE) commonly known as the mad cow disease, which has recently become more evident. This is an important safety issue since plants do not serve as hosts for human pathogens (Ma, 1995).

The cost of production is also an important factor. Cloned animal cells are very inefficient production facilities. The cost of operating the fermentors is also extremely high. On the contrary, production of protein in plants is inexpensive. However, efficient extraction and purification processes need to be developed to produce large quantities of recombinant proteins from field grown transgenic plants. Typically, a recombinant protein is expressed in plant seed which is indistinguishable from that in normal seeds. When needed, the seed is processed and the recombinant protein is enzymatically or chemically cleaved from the plant derived protein. Usually, the recombinant protein might be stored in the seed for prolonged periods of time without any deterioration. This is another advantage over animal cell production which requires instant processing and purification.

Environmental concerns might also drive the development of new technologies. The impact of production and disposal of millions of tons of non-degradable material is taxing. Promising opportunities are provided by the breakthroughs in transgenic technologies, allowing production of very large volumes of biotechnology driven products at modest costs and low emission of pollutants. Traditional chemical technology of plastics production is based mainly on non-renewable petrochemicals and high energy input, accompanied by the undesirable emission of carbon dioxide and other gases. Now specialized agriculture offers new exciting opportunities, plants are considered a renewable source of input material and they do not emit environmental pollutants. On the contrary, plants in their biosynthesis use the energy from sun and absorb carbon dioxide from the environment.

Lately, microbial polyesters (the polyhydroxyalkanoates) have been developed. These materials might be produced as a spin-off in oil crops. When appropriate polymerization, molecular weight control and economically sound recovery procedures are developed, the biodegradable plastics from agriculture might become cost competitive with petrochemical polymers.

The population explosion is an obvious fact of our time. World population will grow from 2.5 billion in 1950 to 9.8 billion in the year 2050. At the same time the cultivated area per person will diminish from 0.22 to 0.05 ha. It is obvious that new highly efficient plants for both human nutrition and animal feed are needed. Recently, new GMO (Genetically Modified Organisms) or transgenic microorganisms, plants and animals are getting a lot of publicity and are appearing on the covers of major magazines. They create real excitement in the mass media, as their impact reaches far beyond the scientific or technological aspects. The public safety, ethical and environmental issues, and international trade barriers linked to the GMO are subjects of hot debates, violent demonstrations and governmental actions.

The legislation related to the use of biotechnology in agriculture, especially in food production and processing, differs from country to country. Several countries are currently reviewing their biosafety legislation and attempts are made to harmonize such legislation globally.

It is obvious that *populus* of countries experiencing overproduction of agriculture products, such as those in Europe and North America, will differ in their perception of bioengineered plants from those in the developing countries where food is in short supply and lives of millions of people are threatened by malnutrition and starvation. Therefore, it is not surprising that large multinational chemical companies have recently entered into biotech and seed market. The agricultural technology business is undergoing an unprecedented wave of consolidation. A few big players, all large chemical and pharmaceutical firms, are buying-up not only seed companies, but also the high quality agricultural biotech companies. Megacorporations in agricultural biotechnology will not only yield significant economic power, but political power as well through control of food supplies to vast and highly populated regions of the world (Kolodziejczyk, 1998).

Will bioengineering prove to be a beneficial and safe tool? What will happen to agricultural biotechnology? With evident erosion, shortage of irrigation water and about 80 million people added to the world's population each year the food production in many areas will likely decline. Does biotechnology have the potential to alleviate many of these problems? The world food problems are daunting. The FAO (Food and Agricultural Organization) estimates that at this time 800 million people do not have sufficient food. New plant varieties may make it possible to improve the food supply without increasing of usage of farmland, water or other resources.

It must be stressed, however, that new technologies alone are not sufficient. Other key issues are governments regulations, international trade and corporate profits, sharing research and technology, and legally binding international biosafety protocols.

As we enter the 21st century, will this be the golden age of biotechnology? It is hoped so, and we believe that this view is shared by other authors of this book, who by their contribution to advancement of theoretical and applied biotechnology are charting a new course for plant bioengineering. However, since we are entering a new and unknown territory, we should be cognizant of our responsibilities as scientists; put forward our best efforts and use our critical judgment when new tools are constructed and utilized.

REFERENCES

Ma, J.K. Generation and assembly of secretory antibodies in plants. Science 1995, 268, 716–719.

Murphy, D.J. Transgenic plants – A future source of novel edible and industrial oils. Lipid Technol. 1994, 6(4), 84–91.

Kolodziejczyk, P.P. Chemists in the new world of biotechnology. Can. Chem. News 1998, 50 (4), 16–20.

Kott, L.S.; Erickson, L.R.; Beversdorf, W.D. The role of biotechnology in canola/rapeseed research. In Canola and Rapeseed: Production, Chemistry, Nutrition and Processing Technology. Shahidi, F., Ed. Van Nostrand Reinhold, New York. 1990. pp. 47–78.

Shahidi, F.; Daun, J.K.; De Clerq, D.R. Glucosinolates in Brassica oilseeds: Processing effects and extraction. In Antinutrients and Phytochemicals in Food. Shahidi, F., Ed. ACS Symposium Series 662, American Chemical Society, Washington, D.C., pp. 152–170.

2

RECENT PROGRESS IN AGRICULTURAL BIOTECHNOLOGY AND OPPORTUNITIES FOR CONTRACT RESEARCH AND DEVELOPMENT

Paul P. Kolodziejczyk and Paul Fedec

POS Pilot Plant Corporation
118 Veterinary Road
Saskatoon, SK, S7N 2R4, Canada

ABSTRACT

The global market for agriculture products and agriculture-based value-added products is undergoing change as the top players in agriculture and agricultural biotechnology face increased consolidation and ultimately form alliances in development, production and marketing. Transgenic plants for human consumption and industrial applications are entering the marketplace. Novel, genetically engineered, plant-based organisms (GMO) designed for resistance to herbicides, pesticides and environmental stress or for the production of valuable chemicals, pharmaceuticals and vaccines are available. A growing demand for bioprocessing, test production, scale-up or providing data for registration has created new opportunities for contract research and development (CR&D) firms.

INTRODUCTION

Trends currently evident in agricultural biotechnology may be divided into several categories. The major trends are provided below.

a. The growth potential in the agricultural biotechnology industry continues to be very promising. A recent study shows that amongst US biotechnology product sales, the agricultural sector will experience the fastest growth (Table 1), even though this sector remains only third in sales volume, after human therapeutics and diagnostics.
b. The global market for agriculture products and agriculture-based value-added products is undergoing change as the top players in agriculture and agricultural

Chemicals via Higher Plant Bioengineering, edited by Shahidi *et al.*
Kluwer Academic / Plenum Publishers, New York, 1999.

Table 1. USA biotechnology product sales forecast (millions of 1996 US dollars)

Sector	Base year 1996	Forecast 2001	Forecast 2006	Avg. annual growth rate (%)
Human therapeutics	7,555	13,935	25,545	13
Human diagnostics	1,760	2,705	4,050	9
Agriculture	285	740	1,740	20
Specialties	275	690	1,600	19
Non-medical diagnostics	225	330	465	8
Totals	10,100	18,400	32,400	12

Source: Consulting Resources Corp.

biotechnology face increased consolidation and ultimately form alliances in development, production and marketing.

c. Transgenic plants for human consumption and industrial applications are entering the marketplace. Novel, genetically engineered, plant based organisms (GMO) designed for resistance to herbicides, pesticides and environmental stress or for the production of valuable pharmaceuticals and vaccines are available. The increased power of plant genetics can be utilized to enhance crop production and quality, but also allows plants to be exploited as sophisticated bioreactors, thereby producing a wide array of industrial chemicals and intermediates, for example, plastics, flavors and colorants, pharmaceuticals, antibodies, vaccines, blood proteins, and enzymes for food, feed and industrial uses.
d. Green factories, i.e. plant bioreactors, will not only facilitate production of chemicals in an environmentally friendly manner by utilizing carbon dioxide and the sun's energy in their photosynthesis, but they will also produce a new generation of safer agrochemicals. Spurred by environmental pressures and customer demand, the move towards low-dose active ingredients is becoming a very important trend. The worldwide agrochemical industry was valued at about $ 32.5 billion at the end-user level in 1996. North America accounts for the largest segment of the global market with a 30.3% share, followed by East Asia at 26.5 %, Western Europe and Latin America both at 10.9%, Eastern Europe at 2.9% and rest of the world accounting for 4.1 percent.
e. In the pharmaceutical market, the chiral or single isomer version of a drug makes a dramatically safer product. The side effects of many drugs are eliminated or minimized. Single isomers of many drugs offer the ability to deliver effective therapy. It is estimated, that by 2005, some 60% of all new drugs released will be in the form of a single isomer. The pharmaceutical industry is already undergoing 'chirality cleansing' and the amount of chirally pure drugs on the market is expected to grow at a fast pace with plant bioreactors expected to play an important role in this process. Plants, with their ability to produce chiral drugs, might be more efficient than traditional chemical synthesis. The revolution in chiral agrochemicals and chiral drugs produced in plants is about to commence. Approximately 140 active ingredients, or 25 percent of current commercial agrochemicals, have a chiral center. However, with only 10 and 15 percent currently offered in the chirally pure form, there is a market opportunity for the development of single enantiomer compounds. The amount of herbicides and pesticides currently used in agriculture could be reduced through the use of

biologically active single isomers along with corresponding reduction of non-active ballast. These chemicals could be applied at one-half the rate and achieve the same results. That translates into a 50% reduction in herbicide or pesticide load to the environment. Plants with their ability to produce chiral drugs might be more efficient than traditional chemical synthesis.

f. A growing demand for bioprocessing, test production, scale-up or providing data for registration and regulatory agencies has created new opportunities for contract research and development (CR&D) firms. Biotechnology companies are supplementing their internal R&D with external resources to satisfy competing pressures. Contract R&D laboratories offer technical expertise, laboratory space and specialized, state-of-the-art laboratory equipment that are not available to many biotechnology companies. Knowledge of regulatory agencies expectations that are not always resident with biotechnology companies (especially for companies planning to register both inside and outside a particular country), the ability to meet extremely demanding timelines, that in many instances could not be met with the client's in-house capabilities; and compliance with GLP, GMP and ISO in testing and processing are additional factors favoring outsourcing. There is visible trend - one of going from "bricks to brains", indicative of a change from high asset base, low R&D investment business to low asset base, high R&D investment operations. Tax incentives in some countries have created a favorable climate for this shift. In other countries, where labour is expensive and employment rights are strongly protected, it can be more attractive to buy the time and services of a contract company rather than to hire new or additional staff. Many companies are achieving savings in personnel cost by using CR&D organizations because fixed costs now become variable costs.

g. Globalisation of trade and the formation of several strong trading blocks: EEC, NAFTA or Mercosur, allows swift capital flows between the block countries, taking advantage of lower labour costs and may contribute in the future to an even faster growth of contract research and development organizations.

PLANTS AS BIOREACTORS

Humans have always been dependent on plants for the production of food, clothing and building materials. For millennia, generations of humans have used chemical substances produced in plants, substances which nowadays we classify as primary or secondary metabolites. This purely artificial classification describes primary metabolites as substances essential to the physiology of the plant and are biosynthesized by plants in large quantities. Primary metabolites very often act as structural or storage material. Traditionally, this classification includes proteins, carbohydrates (lignins and starches), and seed oils among others. These substances are important commercial commodities which are usually produced in large quantities and have a rather low price per weight unit (usually less then $ 1 per 1 kg). Examples of current production volumes for several typical commodities that may be classified as primary metabolites are given in Table 2.

In the 50's the term secondary plant metabolites was coined to supplement the classification. As Bonner and Galston (1952) described :

> But individual species of higher plants are able to make not only many chemical materials characteristic of, and essential to, the functioning of the plant, but also particular chemical

substances which are not essential to the economy of the plant and which have no recognizable role in metabolism. It is characteristic of such compounds that they often occur only in scattered species distributed at random throughout the plant kingdom. Among these secondary products of metabolic byways are the alkaloids, the terpenes, rubbers, sterols and steroids, the tannins, and many of the other plant materials which contribute to the welfare of mankind.

Even without basic scientific knowledge and long before the dawn of modern science, humans were using plants and plant derived substances as medicine, flavorants, spices, natural colorants and dyes, substances of abuse or poison.

The folk medicine of all civilizations provides us with many examples of drugs derived from plants. The oldest documented evidence referring to the use of plant medicines comes from China and India and is over 3500 years old. Cultures of Mesopotamia, Egypt, Greece and Rome had a share in the advancement of our knowledge about substances of plant origin. This information was recorded by Hippocrates in his works *Corpus Hippocraticum* from the 4th century B.C. and is recognized as the beginning of medicine and pharmacy.

Pepper, cinnamon, cloves and several other spices were used as food additives and/or stabilizers. Their economic importance prompted Columbus to undertake a venture which resulted in the discovery of America.

The application of plant dyes was known to ancient civilizations. Indigo, the dye extracted from the leguminous plant *Indigofera indicum,* is mentioned in Brahman texts *Atharvaveda* and in *Samyutta Nikaya* originating four thousand years ago. In Egyptian papyrus from 1500 B.C., found in the Denderah temple, the following recipe was recorded for the dyeing process (Figure 1).

Today, we know that plant-derived indican can undergo an enzymatic hydrolysis to indoxyl and glucose. The colorless indoxyl, in the presence of oxygen, is oxidized to indigo (blue) as shown in Figure 2.

A tremendous effort by organic chemists working with natural plant products was made in the middle of this century which brought about the development of new physical and physiochemical methods of separation, purification, and structure elucidation. Simultaneously, the interest of the pharmaceutical industry in these compounds increased. Some of the most useful drugs in today's medical practice include: aspirin, morphine, digitalis and Vinca alkaloids, taxol and camptothecin. Spectacular, and extremely profitable, was the development of contraceptive pills. The natural steroid product from Mexican yam, diosgenin, was used as a building block for human female and male hormones (Fig. 3) as well as for hormonal antagonists used in birth control pills (Djerassi, 1992).

It is estimated that about 25% of the current drugs in the market have a component which is a direct extract or a chemically-modified substance from plants. The total value

Table 2. Commodity prices and levels of production

Commodity	World production [million T]
Vegetable oils & fats	94,257,300
Sugar	121,775,300
Flour/meal of oil seeds	101,265
Cotton lint	19,560,150
Flax fiber	534,059

Source: FAO-STAT 1997.

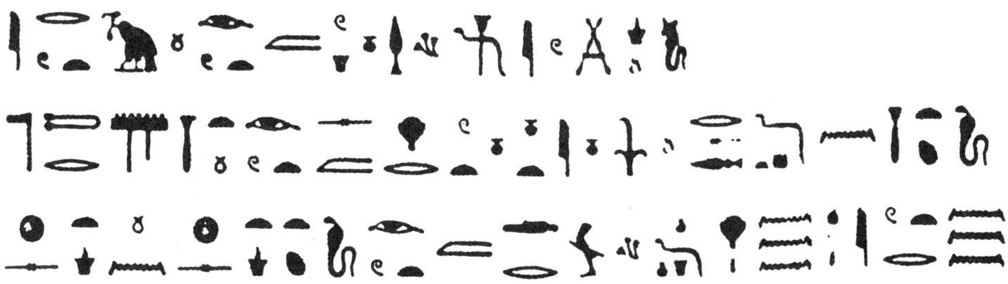

Figure 1. Hieroglyphs describing a dyeing process (Chassinat, 1921). Translation: "The light blue fabric made with fresh 'wan' to wrap Hathor in. The divine green fabric made with 'herouit' from ... to ornate Ouaddjit body. The blue fabric of the Blue Goddess made with indigo ground in the river's water"(Cardon and Chassinet, 1990).

of plant metabolites used in the pharmaceutical industry today is US $22.6 billion and will grow to US $30.7 by the year 2002 (Genetic Engeenering News, 1998). On the other hand, the plant kingdom provides several drugs of abuse like opium, heroin (which is chemically modified morphine) and cocaine.

Secondary metabolites often play poorly understood roles in plant physiology. Some secondary metabolites act as regulatory plant hormones or signal transmitters (gibberellic acid), others as repellent of predators (glucosinolates, cyanogenics), attractants of insects crucial for pollination and still others as protective agents against environmental stress like drought, salinity or heavy metals.

In contrast to primary metabolites, secondary metabolites are rather scarce, their levels in plants vary and are dependent on several factors like tissue type, developmental stage of the plant, environmental conditions, etc. This feature is also directly related to cell specialization. There is a substantial difference in the production of secondary metabolites by unicellular organisms (major producers of antibiotics for the pharmaceutical in-

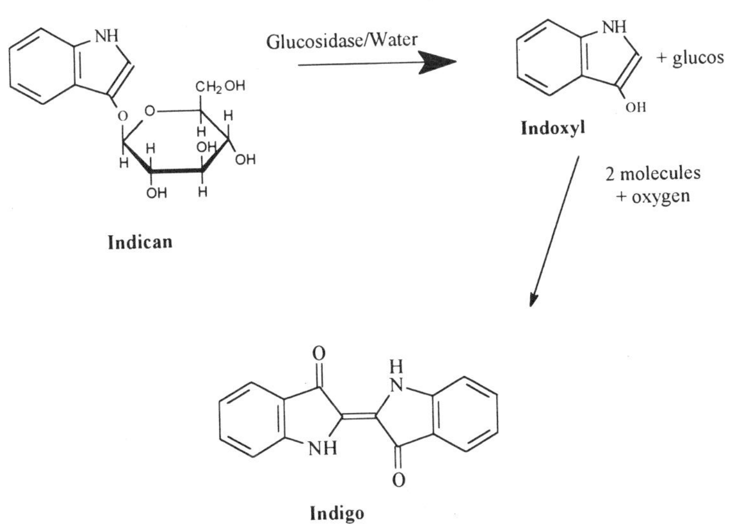

Figure 2. Formation of indigo dye from the indican present in the *Indigofera indicum* plant. The enzyme glucosidase is plant specific.

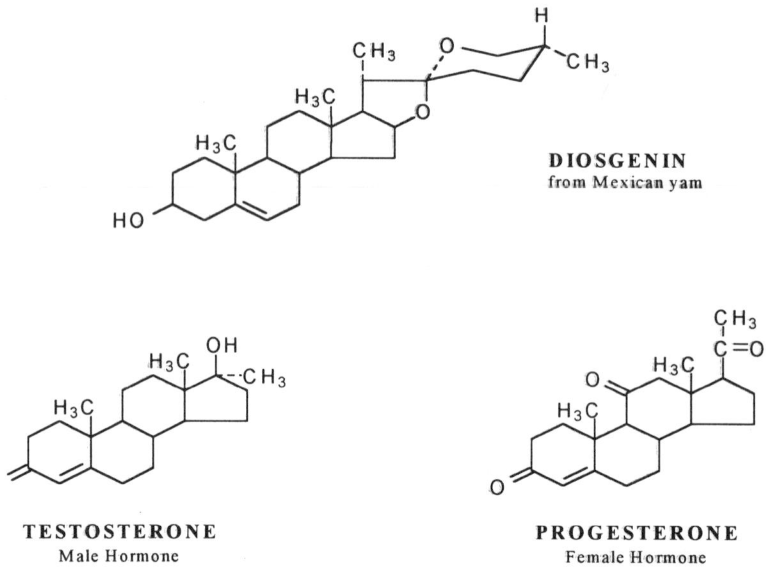

Figure 3. Structures of diosgenin from Mexican yam and male and female hormones produced by chemical modifications. Diosgenin is used as a starting material for production of contraceptive pill and hormonal drugs.

dustry) and plants. In unicellular organisms, e.g. bacteria, cells following a phase of rapid division may pass into a state of specialization during which secondary metabolites are formed.

In multicellular organisms such as plants, formation of secondary metabolites is expressed as a specific feature of certain organs or tissues during restricted periods of their development and specialization.

The synthesis of secondary metabolites in most instances requires multistep reactions. Their structure is often very complex, as such compounds play a very specific physiological function in plants. Some of these metabolites have proven to be very important weapons in medicine's arsenal. However, due to their complex structure, there is usually no economically viable method to produce them on a commercial scale by chemical synthesis (Fig.4).

On the other hand, the pharmaceutical industry needs a steady and controlable supply of drugs, or raw material for drugs production. Some of the plants used for extraction of valuable drugs grow in areas which were difficult to access for geographic or political reasons. Other plants like the Pacific yew, producer of taxol, are rare and have an extremely slow growth cycle. Harvesting of existing trees is not especially welcomed by public opinion.

As a result of these factors, several attempts were undertaken to produce drugs of major importance, including vinblastine and taxol, through the use of cell or tissue cultures. The most often used approach includes 'hairy root' culture or suspended or immobilized plant or tissue culture.

However, in reality, plant cell cultures were very difficult to control, the stability of cell lines and cell aggregation often created serious problems. Even if some experiments succeeded in producing the desired substance in amounts greater than those present in the respective plants (Kutney et al., 1983); so far, there has been no large scale application of

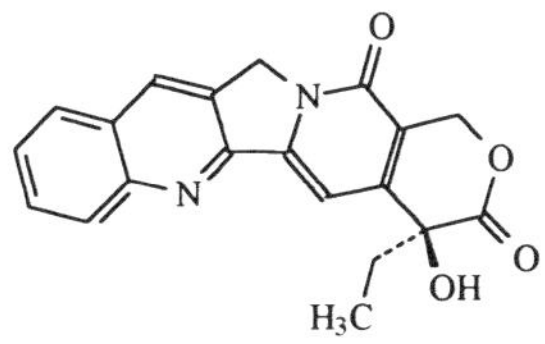

Campththecin
(from *Camptotheca acuminata*)

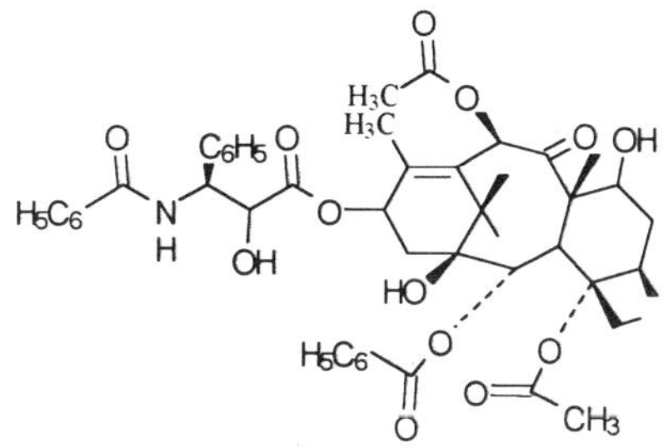

Taxol (Paciltaxel)
(from Pacific Yew *Taxus brevifolia*)

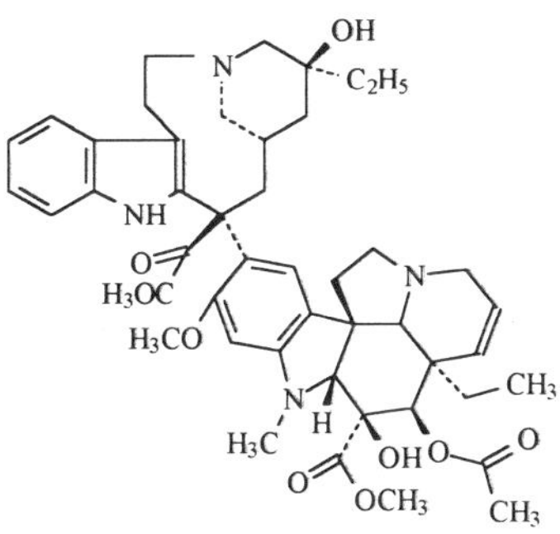

Vinblastine
(from *Vinca rosea*)

Figure 4. Chemical structures of several very important anticancer drugs isolated from plants.

this technology. Although industry remains very interested in plant cell culture, owing to its potential for the production of rare compounds, only a few species are currently cultured commercially for pharmaceutical use: *Lithospermum erythrorhizon* (shikonin production), *Coleus blumei* (rosmaric acid), *Coptis japonica* (berberine) and *Panax ginseng* (ginsenosides). Even for truly high value substances, like vinblastine and vincristine, the economics of the process favor extraction of the desired material from plant sources.

Recent developments in 'design' of living organisms has created the possibility of using plants as bioreactors; a kind of sophisticated 'vessel' for controlled production of desired chemical compounds. Interestingly, science has come full circle: from the discovery, following by the chemical synthesis of the first 'organic' chemical compound - urea (Wöhler, 1828) to the production of designer chemicals by targeted engineering of organisms. Up to Wöhler's time, scientists thought that compounds present in living plants and animals could not be synthesized in the laboratory from inorganic molecules, the '*vis vitalis*' - force of life - was supposed to do the trick. Today, we can manipulate most living organisms and use them, within limitations, to produce desired products.

Biotechnology is now changing this picture in a very dramatic way by using recombinant DNA technology to modify the chemical composition of existing crops or to produce biological macromolecules in crop plants for pharmaceutical applications. There are also possibilities open for using genes from different species and combining them into a gene construct which is expressed in the most effective vehicle for a particular product.

The most recent example is the biosynthesis of prenenolone and progesterone, human hormones, in yeast. By reconstituting *Saccharomyces cerevisiae* with six different mammalian and plant genes (from *Arabidopsis thaliana*), researches have created a self-sufficient microorganism that is able to synthesize progesterone using galactose as a carbon source (Duport et al., 1998).

MODIFICATION OF EXISTING CROPS

In recent years, we have seen unprecedented progress in the modification of oil profiles of oil bearing crops like rapeseed (canola), flax and soybean. An account of the accomplishments, problems and future perspectives in this field is given in another chapter of this book (Murphy, *this book*). The rationale behind such modifications is the possibility for production of chemical substances, both for human consumption and for industrial application, using plants as bioreactors. There is interest in, for example, modifying the seed oil profiles in order to secure renewable input materials for industrial markets. Canada's own canola is an excellent example of how focused research efforts can result in creation of a versatile plant which might be manipulated to provide desired products both for edible and industrial purposes.

Canola belongs to the *Brassica* species (rapeseed) which, in the Western world was used mainly for industrial applications. Traditional rapeseed contained between 25–50% erucic acid (cis-13-docosenoic, C 22:1) and glucosinolates (thioglucosides), some of which had goitrogenic or other antinutritional properties. The situation changed after plant breeding programs altered the chemical composition of rape oil by lowering the amount of erucic acid in the oil and glucosinolate content in the meal (Stefansson, 1983; Downey, 1990).

Largely due to major seed improvements, quality Brassica oilseed crops have become the third most important world source of vegetable oil after soybean and palm. Today's canola oil is virtually free of erucic acid and has gained an attractiveness as a well balanced, low saturates edible oil. Canola oil competes well with other vegetable oils due to its favorable fatty acid profile. In Canada, canola oil accounts for over 60% of the domestic production of refined vegetable oil. Canola meal, the residue after oil extraction, has a well balanced amino acid composition and is used in the animal feed industry, and might further serve as a useful vehicle for plastics or modified protein production in transgenic plants.

There is a bit of irony in the fact that biotechnology companies are now trying to produce canola with high levels of erucic acid for use in industrial processes. Several other canolas with modified fatty acid profiles are already in production and the wide variations in fatty acid profile can be easily visualized in Fig. 5. Some are suitable in niche markets for both edible and industrial applications. The latest interesting example is high lauric canola. This transgenic species was developed by Calgene for the detergent market. The gene producing high amounts of lauric acid (C12:0; 40%), originated from the California bay tree and was transferred into canola.

The use of biotechnology for modification of canola seed seems be limitless, every year we learn about new genes being expressed in canola. There are several reasons for this; one is the fact that canola is one of the best researched species and one of the easiest to manipulate. Several new canola varieties, like high γ-linolenic, or high carotene are soon to be released.

The fatty acid profile in flax has been successfully altered to lower the amount of α-linolenic acid from greater than 50% in traditional flaxseed to about 2% in the Linola® variety. This new variety has penetrated the market and is used as an alternative source of

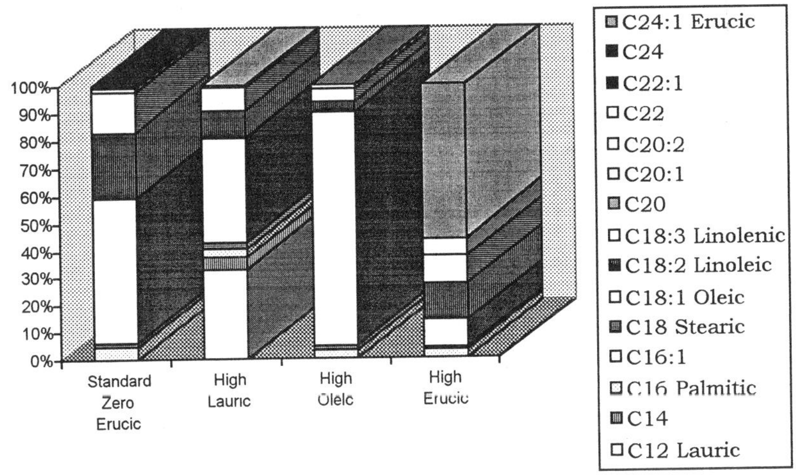

Figure 5. Fatty acid profiles of new canola varieties.

oil for human consumption and margarine production. Traditionally, flaxseed oil was considered as an industrial oil only because it readily oxidized making it suitable as a base for paint production (drying oil) (Kolodziejczyk and Fedec, 1995). The change in fatty acid composition for the two varieties is given in Fig. 6.

Plastics, traditionally a product of the chemical industry, developed and mastered in this century, present exceptional performance features but are also a well recognized burden to the environment. Environmental concerns from the impact of production and disposal of millions of tons of non-degradable material might soon force plastics to move to the agriculture sector. These opportunities are provided by the breakthroughs in transgenic technologies, allowing production of very large volumes at modest costs and low emission of pollutants. Traditional chemical technology of plastics production is based mainly on non-renewable petrochemicals and high energy input, accompanied by undesirable carbon dioxide and other emissions. Here specialized agriculture offers a new and exciting opportunity, plants are considered renewable sources of input material and they do not emit environmental pollutants. To the contrary, plants through their biosynthesis use sun energy and absorb carbon dioxide from the environment.

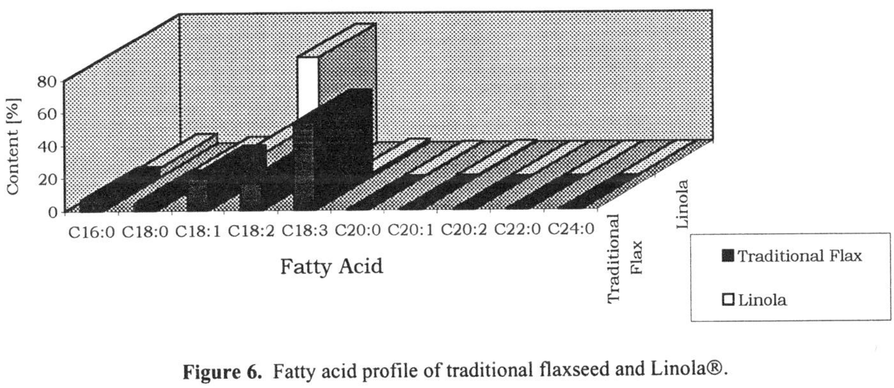

Figure 6. Fatty acid profile of traditional flaxseed and Linola®.

Table 3. Pharmaceutical applications of plant biotechnology

Examples	Reference
Vaccines	Moffat (1995), Dalsgaard (1997), Haq (1995),
Enzymes	Ma (*this book*)
Hormones	Pen (1996), Hood (*this book*)
Plasma proteins	Duport (1998)
Antibodies	Sijmons (1990), Theisen (1997) Whitelam (1996), Ma (1995)

Lately, microbial polyesters, the polyhydroxyalkanoates have been developed. These materials might also be produced in oil crops as a by-product of oil production. When appropriate polymerization, molecular weight control and economically sound recovery procedures are developed, the biodegradable plastics from agriculture might become cost competitive with petrochemical polymers (Poirier et al., 1995).

PRODUCTION OF BIOLOGICALLY ACTIVE MOLECULES IN CROP PLANTS

Recombinat DNA technology has proven to be a very effective tool for production of biologically active macromolecules in crop plants. Some of the possible pharmaceutical applications of plant biotechnology which have been tested already are summarized in Table 3. The potential medicinal uses of plant-derived recombinant products are unlimited, ranging from chirally pure pharmaceuticals, antibodies and edible oral vaccines to therapeutic peptides and proteins.

There are two basic approaches used to produce recombinant protein in host plants. The first one is based on the insertion of a stable transgene into a plant genome. The second approach uses plants as hosts for transient expression by using plant viruses as vectors. As expected, there are advantages and disadvantages to each approach. Creation of transgenic plants is time consuming and costly. The RNA viral vectors have been shown to be very cost effective. They provide a 'shortcut' for production of peptides, proteins and vaccines in a plant. The viral expression systems offer rapid production, high expression levels and make possible post-translational modification. The viral vectors might be another vehicle to express foreign gene products in plants. Several examples of this approach have already been tested in the production of therapeutic substances.

Agriculture is facing inevitable change and is poised to benefit from the recently accumulated knowledge of plant genomics. The primary sequences of major plant genes will be determined over the next few years. We can expect that this information will have a tremendous impact on crop production and quality.

For the pharmaceutical industry, understanding of plant genome function at the molecular level may provide a new tool for controlled drug design in plants. Historically, the medical and pharmaceutical industries used to benefit from the discoveries of natural products stemming from plant sources. These bioactive compounds have been used directly, or after chemical modification, to treat human disease. Today we can only imagine to what extent biotechnology will change the whole world of drug development and production.

In recent years much effort has been devoted to characterizing genes and respective proteins forming natural products. We have learned a great deal about enzymes and reaction sequences leading to biosynthesis of complex molecules. Now we are at the beginning of the path leading to redirection of enzyme systems to generate new compounds. This opens up the possibility of exploiting nature's own combinatorial devices, making use of 'programmed' chemistry that also opens the door to the production of 'natural' compounds by biotechnological means.

NEW INDUSTRY ALLIANCES

The agricultural technology business is undergoing an unprecedented wave of consolidation. A few big players, all large chemical and pharmaceutical firms, are buying up not only seed companies, but also high quality agricultural biotech companies. Monsanto has been the most acquisitive. The chemical giant now owns the biotech firms Calgene (Davis, CA) and Agracetus (Middleton, WI) out right, and has stakes in Ecogen (Langhorne, PA) and Dekalb Genetics (Dekalb, IL.). In addition to Holden's Foundation Seeds, they have acquired Corn States International, Asgrow Agronomics and a Brazilian firm, Monsoy. Novartis (E.Hanover; NJ) owns the seed companies Northrup King, S&G Seeds, Hilleshog, as well as Ciba Seeds from the old Ciba-Geigy dynasty. Dow Elanco, now known as Dow Agro Sciences and spawned from a cooperative venture between Dow Chemical and Eli Lilly (Indianapolis, IN), has acquired a majority interest in Mycogen (San Diego), which has its own captive seed business. DNA Plant Technology (Oakland, CA) is now majority owned by Empresa La Moderna, a Mexican agricultural biotechnology and tobacco company .

The well known German based company Hoechst (Frankfurt) in a joint venture with Schering (Berlin), created AgrEvo in 1994, a worldwide subsidiary who in Canada is promoting the herbicide resistant canola varieties. In very a symptomatic trend, AgrEvo acquired another biotechnology company Plant Genetic Systems (Belgium) in 1996.

OUT-SOURCING AS A TREND IN BIOTECHNOLOGY INDUSTRY: FASHION OR NECESSITY?

New chemical substances produced in transgenic plants often require novel processing technologies. Biotechnology companies are forced to contract out their R&D and often the preliminary stages of manufacturing, due to limited resources and their focus on technology rather than production. Bioprocessing services is a growing business which has recently come of age.

In the past, companies were pursuing their own end products and had operational R& D facilities that were often vastly underutilized. Most of these facilities had a high level of specialization geared to a particular product or process. Retro-fitting such facilities usually became complex and costly. Furthermore, construction of new facilities in a complex field generally requires a lengthy lead time and the opportunity may fade before the ground work is laid.

Now that traditional financial resources for the biotechnology industry have dwindled, the management of research and development is undergoing significant change in response to growing pressures to accomplish more with fewer resources at a faster pace. The faster pace is necessary in order to react to the competition seeking to exploit an identified opportunity. Often these opportunities are short lived and the "me too's" in the business don't realize a financial return after the band wagon has passed.

An emerging trend, seen in nearly all biotechnology companies, is a far greater use of partnerships, collaborations and outsourcing as a substitute for in-house R&D. Contracting is now an inherent part of novel biotech business approaches, as in the virtual company model. Thus independent and confidential, project dedicated Contract R&D (CR&D) facilities appear to be the answer to meet industry needs. Crucial assets to CR&Ds are: the know-how in GMP processing, access to technology, scale-up expertise and on-hand suitable equipment, quality control with established record keeping procedures and full analytical support.

This trend is increasingly evident with recent divestitures, mergers, asset swaps and joint ventures in agricultural, seed and chemical industries. It has become very difficult to sustain and manage multi-faceted R&D within one company, comprised of drugs, pesticides, seeds and other products; selling to different customer groups; and using the technology that includes chemistry, biochemistry, biology, pharmacology, biotechnology and plant genetics. Often the R & D equipment and facilities cannot be readily adapted from one field to another within the company structure.

A goal for every biotechnology company is to move products from the research phase, through the development stage and into production in the shortest possible time. Under pressure for growth, industrial R&D must discover new technologies and develop new products. Working with tighter budgets and fewer staff, managers have no choice but to supplement internal resources with external sources of research and technology as effectively as possible. Outsourcing portions of the basic research like speciality analytical services, methods development and validation or scale-up of bioprocessing helps the organization to reduce the average cycle time for products moving through the developmental process. Contract laboratories geared to mass through-put coupled with lower costs and short turn-around time provide convenience and flexibility.

Specialized research technologies and services that are used infrequently may simply be too expensive to add in house. It forces consideration of outsourced research or services required for specific instrumentation or researchers with a specific knowledge and expertise.

Some contract organizations are establishing relationships in which they become the entire service (i.e. analytical) or R&D arm of a company. By providing a comprehensive program and long-term relationships, they may be able to offer special prices for such complete services and, provided confidentiality is respected, similar services can be offered to more than one company in a similar field. There is added benefit in the concept that the outsourced service provider is acting as an independent third-party analyst.

At the end of the product pipeline in the biotech industry are the contract manufacturing services. Companies that offer contract manufacturing services to the biotechnology industry encompass a broad range of business interests and manufacturing capabilities. In the early days of biotechnology, the science was outpacing engineering and the production gap was a major obstacle for fast transfer of the discovery from the laboratory to the marketplace. Some of the major bioprocessing challenges include the design of large scale continuous systems and the gentle but efficient product separation, purification and recovery.

A well-known principle in engineering is that costs increase exponentially from concept, to system study, to the laboratory model, to pilot plant and finally to the production facility. The scale-up laws and costs are not linear. The need to develop new types of flexible processing facilities serving the fast growing biotech industry is evident. One of the difficulties associated with a new emerging technology is a tendency to reinvent well established technologies rather than adapt well proven methods from neighbouring fields.

The converse is equally true: a custom made technology, for a specific product or process, can be stretched only so much to fit a novel application. A solution is to keep an open mind and facilitate continuous dialogue between scientists and engineers in order to develop new processes and demonstrate their technical and economical feasibility.

The 1996 world-wide market for contract manufacturing of biopharmaceutical products was estimated at $ 200–300 million. Based on the current rate of growth, the market size is projected to reach $ 1 billion shortly after the year 2000 (McKown, 1997). This is not a mere fashion trend, it is the reality of an industry under pressure to be creative in a fast moving and extremely competitive business opportunity environment.

WHAT TO LOOK FOR WHEN CHOOSING YOUR CONTRACT R&D PARTNER

Quality standards are a critical factor in outsourcing agreements. It is increasingly common for companies to audit the quality system of the contractor prior to awarding the contract. This allows companies to ensure that their own quality standards are matched by the testing laboratory or contract R & D organization. In other cases, when contracts are awarded by open bidding procedures, the competing laboratories may be required to demonstrate specific certification or compliance before entering the bidding process. Such an approach is used by the Western Canada Canola and Rapeseed Recommending Committee (WCC/RRC). The Canola Council of Canada administers the co-operative program for testing all new traditional and transgenic canola varieties registered in Canada. The quality analysis of canola seed which includes oil content, protein content, fatty acid composition, chlorophyll level and glucosinolate level is contracted out to laboratories through the bidding process. The contract is awarded on the basis of both price and performance in a specially designed testing quality performance program administrated by the Canadian Grain Commission.

OUTSOURCING REQUIRES NEW MANAGEMENT STRATEGIES

There is a trend in the biotechnology industry to form new business relationships to enhance the research and developmental process. Virtual companies are turning over entire research programs to contract organizations and simply coordinating the research elements as they come together for decision making. With outsourcing becoming a basic strategy in the agricultural biotechnology, new approaches to effective managing of complex and strategic partnerships needs to be mastered. As a result of the rapid emergence of new technologies, clients require custom-tailored services which results in the need for a close client/CR&D relationship. From the contracting company's perspective, the crucial issue is the makeup of the outsourcing team, which consists of specific expertise to make the partnership work effectively. It is extremely important to maintain efficient and effective communication channels with the CR&D company. This of course, can be readily facilitated by electronic communication where information, instruction and data can be easily exchanged, reviewed, commented upon and agreed to quickly and confidentially.

Some companies are merely seeking to achieve cost-driven benefits from collaborative partnerships, others are going beyond traditional subcontracting. Often strategic agreements are formed where sophisticated projects with predetermined relationships and financial terms and a more defined long term set of strategic objectives are defined for both partners. This implies sharing certain aspects of intellectual property with the contract R&D organization. This may be technology or methodology, but very often it is largely the harmonization of regulatory aspects. In contract manufacturing cases, the CR&D on-site evaluation and audit is a first step to ensure good manufacturing practices (GMP) and often Good Laboratory Practice (GLP) compliance and validation of the facilities and related processes. In most cases, the contracting out partner provides standard operating procedures (SOPs) to ensure familiarity with their GMP requirements.

The up-front definition of the contract is critical. This includes a proposal that clarifies the technological requirements, the services the company wants to buy and their extent, the overall structure of the project and a clear understanding of intellectual property ownership. Setting clear expectations before initiating the project is often even more critical. The proposal should also address each partner's responsibilities in the event of project failure or deviation.

The risk of project failure may be reduced through sufficient testing and cross-company dialogue. The contract R&D partner can take additional steps to reduce the risk and build confidence in developing the new partnership. A feasibility study or bench-top scale investigation will cost less money than reprocessing or modification in a full-scale pilot plant run. To ensure that its own interests are secure, the CR&D quality assurance department needs to review and approve the project before execution.

The contract R&D organization must have in place well defined project management coordinating policies and procedures. The reporting structure will generally include the Directors of Research and Quality Assurance and the Principle Investigator or Study Director (or Project Leader). The flow of technical information between the two parties will be on a need to know basis but will need to be disseminated to the technical and operations staff in a manner that allows the orderly planning, execution and review of the contracted research program. Typically the senior management of the contract R&D company is not party to the specifics of the technical program, but only administers of the reporting structures and the financial aspects. During the execution of the research and testing programs, data collection, documentation, deviations and corrective actions must be handled appropriately, recorded and initialized according to protocols agreed upon between the parties. Generally these involve the use of accepted procedures recognized internationally under GMP or GLP guidelines. More recently, ISO 9000 series quality assurance certification is being requested by international players. Auditing of the entire program by Quality Assurance at a level of about 10% insures that the deliverables are met and that the basal requirements for the contracting partner's future validation, verification and regulatory needs are in place. The final products of the research program may be product or ingredient samples, analytical data, original copies of all raw processing data and an approved for release final report. Subsequently, the parties may explore additional support through production of materials for functionality testing and market evaluation using the same protocols that guarantee the validity of process methods, analyses and supporting documentation. Upon the completion of the contracted program the relationship can be terminated, respecting the contractual conditions for confidentiality and dissolution.

FROM OUR OWN EXPERIENCE

The POS Pilot Plant Corporation has operated as a fee for service research and development facility for over 20 years in the area of processing and product development from grains and oilseeds. In particular, during the past few years we have seen a significant shift to the outsourcing of R&D by the biotechnology industry. The key ingredients that the CR&D partner must bring to the equation are confidentiality, a recognized expertise in both staff and facilities, a respected track record for delivery of programs, stability, an international presence and respect and strong adherence to quality principles. Sub-contracting offers numerous benefits to the contracting out organization. They can access resources quickly that are specialized and not available in their own organization, particularly those that may not be in the long term requirements of the company. Through this mutually beneficial arrangement, the contracting company is able to take advantage of the CR&D's strict adherence to total budgets for expense and time, a clearly defined set of objectives, little room for the program to go "off course" because of the "nice to know" aspects that creep in to delay the program and send it over budget, generation of the maximum amount of information for the minimum input and the availability of instant resources for troubleshooting deviations. As skill sets become increasingly specialized in agricultural biotechnology, as consolidation continues to occur and corporate objectives drive the economics into situations of tight budgets set against narrow opportunity timeframes or even short lifespans of the end product, outsourcing of research and development from well organized and established service providers will continue to be an opportunity for both parties to develop strategic alliances to maximize the gain from input resources.

REFERENCES

Bonner, J.; Galston, A.W. "Highways and Byways in Plant Metabolism". In *Principles of Plant Physiology*; p. 298–317; W.H. Freeman & Co., San Francisco, **1952**.

Cardon, D., du Chatenet, G. "Du pays de Cocagne au pays des cow-boys: Le 'blues' des plantes à indigo". In *Guide des Teintures naturelles*, Perret, D. (ed.), Delachaux et Niestlé, Paris, pp. 133–159, **1990**. Translation by: M.-A. de Larminat.

Chassinat, E. In *Papyrus médical Copte*, Mémoires de l'Institut français d'Archeologie orientale du Caire, t. XXXII, p.210, **1921.**

Dalsgaard, K. Plant-derived vaccine protects target animals against a viral disease. *Nature Biotechnology* **1997**, 15, 248–252.

Djerassi C., "Birth of the Pill". In *The Pill, Pygmy Chimps, and Degas' Horse*. Basic Books, pp. 49–65, **1992**.

Downey, R.K. Brassica oilseed breeding - achievements and opportunities. *Plant Breed. Abstracts*, **1990**, 60, 1165–1170.

Duport, C.; Spagnoli, R.; Degryse, E.; Pompon, D. Self-sufficient biosynthesis of pregnenolone and progesterone in engineered yeast. Nature Biotechnology **1998,** 16, 186–189.

Genetic Engeeneering News, **1998,** 18 (3), 1.

Haq, T.A., Mason, H.S., Clements, J.D., Arntzen, C.J. Oral immunization with a recombinant bacterial antigen produced in transgenic plants. *Science* **1995,** 268, 714–716.

Hood, E., E., Kunsnadi, A., Nikolov, Z., Howard, J. A., Molecular farming of industrial proteins from transgenic maize. In: *this book.*

Kolodziejczyk, P.P.; Fedec P. Processing Flaxseed for Human Nutrition. In "*Flax in Human Nutrition*",; Cunnane, S.C., Thompson, L.U., Eds.; AOCS Press, Champaign, Ill., pp. 261–280, **1995**.

Kutney, J.P.; Aweryn, B.; Choi, L.S.L.; Honda, T.; Kolodziejczyk, P.; Lewis, N.G.; Sato, T.; Sleigh, S.K.; Stuart, K.L.; Worth, B.R.; Kurtz, W.G.W.; Chatson, K.B.; Constabel, F. Studies in plant tissue culture. The synthesis and biosynthesis of indole alkaloids. Tetrahedron **1983,** 39, 3781–3795.

Ma, J.K.-C., Hiatt, A., Hein, M., Vine, N.D., Wang, F., Stabila, P., van Dolleweerd, C., Mostov, K., Lehner, T. Generation and assembly of secretory antibobies in plants. *Science* **1995,** 268: 716–719.

Ma, S.; Jevinkar, A.M. Autoantigens produced in plants for oral tolerance therapy of autoimmune diseases. In: *this book.*

McKown, R.L. Contract Manufacturing Sustains Growth as the Biotechnology Industry Matures. *Genetic Engineering News* **1997,** 17 (17), 10.

Moffat, A.S. Exploring transgenic plants as a new vaccine source. *Science* **1995**, 268, 658–660.

Murphy, D.J. Manipulation of plant oil composition for the production of valuable chemicals -progress, problems and prospects. - In: *this book*.

Pen, J. Transgenic seed: a novel enzyme product. *Agro-Food-Industry Hi-Tech* **1996,** January/February, 9- 13.

Poirier, Y., Nawrath, C., Somerville, C., Production of polyhydroxyalkanoates, a family of biodegradable plastics and elastomers, in bacteria and plants. *Bio/Technology* **1995,** 13, 142–150.

Sijmons, P.L. Production of correctly processed human serum albumin in transgenic plants. *Bio/Technology* **1990,** 8, 217–221.

Stefansson, B.R. In "*High and Low Erucic Acid Rapeseed Oils*", Kramer, J.K.G., Sauer, F.D., Pigden, W.J. (eds.), Academic Press, New York, pp. 143–160, **1983.**

Theisen, M. Les plantes comme bioréacteurs. *Biofutur* **1997,** 168, 47–51.

Whitelam, G.C., Cockburn, W. Antibody expression in transgenic plants. *Trends Plant Sci.* **1996,** 8, 268–272.

3

MANIPULATION OF PLANT OIL COMPOSITION FOR THE PRODUCTION OF VALUABLE CHEMICALS

Progress, Problems, and Prospects

Denis J. Murphy*

Department of Brassica and Oilseeds Research
John Innes Centre, Norwich Research Park, Colney Lane
Norwich, NR4 7UH, United Kingdom

ABSTRACT

The first transgenic crop with a modified seed oil composition was released for commercial use in 1995. Many additional transgenic rapeseed varieties expressing novel seed oils or proteins are now under development. Recent research advances in *Arabidopsis* molecular genetics and the emerging relatedness between the *Arabidopsis* and *Brassica* genomes will enable more radical manipulation of many key agronomic traits in rapeseed. These will range from greatly improved seed oil yield to the production of seed oils with defined fatty acid contents ranging from C8 to C24 chain lengths and including many desirable functionalities including double bonds, triple bonds, hydroxy and epoxide groups. Much of the technology currently being development in rapeseed and soybean oil modification will be directly applicable to other oil crops including high yield tropical perennials such as oil palm.

INTRODUCTION

Oil crops are one of the world's most valuable agricultural commodities and are probably worth well over $100 bn/yr. The current volume of traded vegetable oils (ie that

* Tel: +44 1603 452571, Fax: +44 1603 259882, E-mail: denis.murphy@bbsrc.ac.uk

Chemicals via Higher Plant Bioengineering, edited by Shahidi *et al.*
Kluwer Academic / Plenum Publishers, New York, 1999.

enter into national and international markets, rather than being consumed locally) is over 70 MT/yr and is predicted to increase to over 110 MT/yr by the year 2010 (Murphy, 1996). Almost three quarters of the globally traded vegetable oil production comes from four major crops, ie soybean, oil palm, rapeseed and sunflower. According to current trends, the domination by the "Big Four" oil crops is set to continue for at least another decade or more. Indeed, with the establishment of large new plantations of oil palm in South East Asia, particularly in Indonesia and Papua-New Guinea. Oil palm production is set to double in the next 15 years and it is predicted that this tropical perennial crop will then be the most important source of vegetable oils in the world.

Oil crops are sources of many industrial products and have the potential to provide the raw materials for many additional products, some of which may be of very high value. In addition, it should be remembered that, after cereals, oil crops are the second most important source of edible calories for human societies. The application of scientific plant breeding techniques during the twentieth century has led to enormous improvements in both yield and quality of agricultural products which has permitted the enormous growth in the human population of the planet. Nevertheless, conventional plant breeding techniques are limited to existing genetic variation within individual species, or sexually compatible related species.

The past decade has witnessed dramatic developments in the potential for further manipulation of our major agricultural crops using biotechnological methods, largely based on DNA manipulation and related technologies including genomics. These developments now allow for the transfer of commercially useful traits from widely related species into a crop plant of interest, using techniques such as protoplast fusion and embryo rescue in conjunction with powerful molecular marker-assisted selection methods. The advent of genetic engineering has now widened still further the scope for creating additional genetic variation so that genes can potentially be accessed from any organism or, indeed, made synthetically and then transferred to the crop species of interest. These new technologies will also allow us to domesticate entirely new crop species over a relatively short timescale in order to produce many novel edible and industrial products.

IMPROVING OIL QUALITY

The fatty acid composition of the major oil crops is relatively restricted, as shown in Table 1. In particular these oils are dominated by high levels of C18 unsaturated fatty acids such as oleic, linoleic and linolenic acids. To some extent this reflects the fact that 80% of all vegetable oils are used for edible purposes. Nevertheless, most oil crops are not used directly for edible use but must be processed, eg by hydrogenation and/or mixed with other vegetable oils before they are suitable for formulation in a specific product such as margarine or cooking oil. There is now the possibility to "tailor-make" oil crops to produce a very specific fatty acid profile for a particular end use. An example of this is the use of a gene knock-out strategy in order to virtually eliminate linolenic acid from soybeans, as has recently been achieved by Dupont (Kinney, 1997). Another potential strategy is to enhance the nutritional quality of a particular oil by transferring genes responsible for the production of long-chain n-3 polyunsaturated fatty acids such as docosahexaenoic acid (DHA) and eicosapentaenoic acid (EPA). It is also possible to enhance the production of important fat-soluble vitamins, such as α-tocopherol, in oil crops.

While there are many possibilities for using biotechnology to improve the edible quality of vegetable oils, the possibilities for improving and extending the range of industrial

Table 1. Fatty acid compositions of the "Big Four" oil crops (% w/w)

Fatty Acid	Soybean	Oilpalm (mesocarp)	Rapeseed	Sunflower
16:0	11	45	4.5	5.5
18:0	4	4.5	1.5	5
18:1	22	38	61	20
18:2	53	11	22	69
18:3	7.5	0.2	10	0.1

products are even more varied and exciting. Although there is relatively limited variation in the fatty acid content of the "Big Four" oil crops, other oilseeds (many of which are not domesticated) show an astonishing diversity in their fatty acid chain length and functionalities, as shown in Table 2. Since the formation of such fatty acids is under genetic control, the genes responsible can be isolated and transferred to one of the high-yielding major crop species. In principle, therefore, it is now possible to envisage the production on an agricultural scale of a vegetable oil containing any fatty acid chain length ranging from C6 to C24 with varying degrees of unsaturation, epoxidation, branching or hydroxylation, depending on the requirements of the customer (Murphy, 1996). Oilseeds can also be used as vehicles for the large scale production of a wide range of fat-soluble products, including vitamins, pharmaceuticals and agricultural control agents. More recently a novel technology has been developed for the use of the oil-binding proteins, oleosins, as carriers for the large-scale production of commercially valuable proteins ranging from industrial enzymes to high value pharmaceutical peptides such as hirudin. In this chapter, some of the technological and societal drivers for these developments will be examined, followed by a review of the progress to date, some of the problems associated with translating these scientific and technical developments into commercial reality and finally, the future prospects for oil crop biotechnology in the coming century will be discussed.

TECHNOLOGICAL DRIVERS

Oilseeds and particularly rapeseed and soybean have been amongst the first major agricultural crops to benefit from the use of modern biotechnology. Unlike many transgenic crop releases, which have involved traits such as herbicide tolerance for disease resistance, the majority of released transgenic rapeseed varieties contain genes which alter oil quality traits such as fatty acid chain length or unsaturation. One of the major reasons for the rapid uptake of gene transfer technology in rapeseed is the relative ease of manipulation of *Brassica* spp in tissue culture. This has allowed for the development of fairly straightforward transformation systems, both in rapeseed (Moloney et al, 1989) and a variety of related *Brassica* crop species. The development of tissue culture and gene transfer systems for soybean has been much more challenging and, while relatively routine transformation systems have been developed in companies such as Dupont (Kinney, 1997), the process remains labour-intensive and time consuming. The extension of transformation technology to other oil crops such as sunflower will undoubtedly occur in the near future but will be more difficult than in rapeseed, mainly due to problems of tissue culture and regeneration in these species.

The manipulation of oil quality traits in rapeseed and the application of this technology to other oil crops has also benefited from a general perception that many of these

Table 2. Fatty acid composition in some diverse oilseed species (%w/w)

		Cuphea species				Palm	Shea	Olive	Coriander	Castor
Fatty acid		*Cuphea cyanea*	*Cuphea Koehn-Eana*	*Cuphea poly-morpha*	*Cuphea epilobiifolia*	*Elaris guineensis*	*Butyro spermum parkii*	*Olea europea*	*Coriandrumsativum*	*Ricinusc ommunis*
Octanoic	8:0	**67.8**	0.1	—	—	—	—	—	—	—
Capric	10:0	24.4	**91.0**	7.4	0.3	—	—	—	—	—
Lauric	12:0	—	1.5	**80.1**	31.8	0.1	—	—	—	—
Myristic	14:0	—	0.6	3.6	**55.3**	0.9	0.1	—	0.2	—
Palmitic	16:0	1.4	1.3	2.0	5.2	**45.0**	3.3	10.3	3.0	1.0
Stearic	18:0	—	0.3	0.1	0.8	4.5	**44.3**	2.3	0.5	1.0
Oleic	$18{:}1_{9}$	2.4	1.1	2.5	1.2	39.5	45.6	**78.1**	5.9	3.0
Petroselinic	$18{:}1_{6}$	—	—	—	—	—	—	—	**75.3**	—
Ricinoleic	18:1-OH	—	—	—	—	—	—	—	—	**90.0**
Linoleic	$18{:}2_{9,12}$	3.1	3.1	4.3	4.9	9.6	5.5	7.3	13.3	4.0
Vernolic	$18{:}1_{9}$ epoxy$_{12,13}$	—	—	—	—	—	—	—	—	—
α-linolenic	$18{:}3_{9,12,15}$	—	0.2	—	0.1	0.1	—	0.6	—	tr
ϒ- linolenic	$18{:}3_{6,19,12}$	—	—	—	—	—	—	—	—	—
α-eleostearic	$18{:}3_{9,11,13}$	—	—	—	—	—	—	—	—	—
Δ_5-eicosenoic	$20{:}1_{5}$	—	—	—	—	—	—	—	—	—
Erucic	$22{:}1_{13}$	—	—	—	—	—	—	—	—	—
Nervonic	$24{:}1_{15}$	—	—	—	—	—	—	—	—	—

Table 2. (*continued*)

Fatty acid		Safflower Carthamus tinctorius	Vernonia *Vernonia anthelmintica*	Linseed *Linum usitatissimum*	Borage *Borago officianalis*	Tung *Aleurites fordii*	Meadow Foam *Limnanthes alba*	Crambe *Crambe abyssinica*	Honesty *Lunaria annua*
Octanoic	8:0	—	—	—	—	—	—	—	—
Capric	10:0	—	—	—	—	—	—	—	—
Lauric	12:0	—	—	—	—	—	—	—	—
Myristic	14:0	—	tr	—	—	—	—	0.1	—
Palmitic	16:0	6.5	2.0	6.1	11.9	3.1	0.3	1.7	4.0
Stearic	18:0	2.9	1.0	3.2	4.0	2.1	0.5	tr	2.0
Oleic	$18{:}1_9$	13.8	2.0	16.6	16.2	11.2	1.4	16.7	18.0
Petroselinic	$18{:}1_6$	—		—	—	—	—	—	—
Ricinoleic	18:1-OH	—	—	—	—	—	—	—	—
Linoleic	$18{:}2_{9,12}$	**75.3**	9.0	14.2	56.2	14.6	2.9	7.8	2.0
Vernolic	$18{:}1_9$ epoxy$_{12,13}$	—	**79.5**	—	—	—	—	—	—
α-linolenic	$18{:}3_{9,12,15}$	—	0.5	**59.8**	0.3	—	—	6.9	—
ϒ- linolenic	$18{:}3_{6,19,12}$	—	—	—	**22.8**	—	—	—	—
α-eleostearic	$18{:}3_{9,11,13}$	—	—	—	—	**69.0**	—	—	—
Δ_5-eicosenoic	$20{:}1_5$	—	—	—	—	—	**66.6**	—	—
Erucic	$22{:}1_{13}$	—	—	—	—	—	10.2	**55.7**	48.0
Nervonic	$24{:}1_{15}$	—	—	—	—	—	—	2.9	**24.0**

traits would involve a very small number of genes. This would make it technically and commercially feasible to effect desirable changes in such traits by the transfer of only one or two genes to create a novel transgenic rapeseed variety. Rapeseed biotechnology has also benefited specifically from the use of *Arabidopsis* reverse genetics for the characterisation and isolation of numerous genes encoding enzymes involved in the metabolic pathways of oil biosynthesis. Individual genes have been identified by the screening of chemically-treated or DNA-tagged *Arabidopsis* mutants which have alterations in their seed lipid metabolism (Arondel et al, 1992; Yadav et al, 1993; Okuley et al, 1994). Many of these genes encode membrane-bound proteins such as fatty acid desaturases, which have proved recalcitrant to biochemical purification. Other candidate genes potentially involved in the regulation of seed oil quality have been isolated by a rapid throughput automated cDNA sequencing of expressed sequence tags (ESTs) (van de Loo et al, 1995). Approximately 17,000 non-redundant ESTs are now publically available from the *Arabidopsis* Resource Centres in the USA and Europe. Due to the close phylogenetic relationship between rapeseed and *Arabidopsis*, many of these *Arabidopsis* clones can be used directly in rapeseed, particularly for down-regulating gene expression, without the need for re-cloning the rapeseed homologue.

SOCIETAL DRIVERS

While oilseed biotechnology has benefited from technical developments of the kind listed above, there have also been a number of societal developments which have had an impact which is likely to grow in magnitude over the next decade. Firstly, in relation to the edible uses of the plant oils, there is an increasing awareness and sophistication amongst consumers about the relative merits of different fatty acids and different chemical processing techniques used in the manufacture of food products. For example, during the past few years there has been increasing concern about the use of catalytic hydrogenation for the removal of double bonds from vegetable oils. This process is used for two reasons. Firstly, to alter oxidatively unstable fatty acids such as linolenic acid and secondly to produce solid fats, ie margarines and other spreads, from liquid oils. A by-product of the hydrogenation process is the formation of a significant number of *trans*-double bonds in the fatty acids of the vegetable oil. Although this has been known for many years, it is only relatively recently that medical evidence has begun to accumulate about the possible adverse consequences of high levels of *trans*-fatty acids in the human diet. We now find that *trans*-fatty acids are joining the catalogue of "bad foods" in much the same way as saturated fatty acids and high cholesterol products. This provides a strong incentive for the food industry to demand vegetable oils which do not require hydrogenation. In a related development the supposed benefits of "Mediterranean diet" have been much publicised. This diet contains olive oil as its main lipidic component and is associated with the increased longevity and lower incidence of heart disease in Mediterranean populations compared with those in Northern Europe and North America. This has led to a demand for vegetable oils similar in fatty acids composition to olive oil, ie high in monounsaturates.

In addition to vegetable oils, another traditional source of highly nutritious fats has been fish. Most fish accumulate relatively high levels of long-chain n-3 polyunsaturated fatty acids such as EPA and DHA, which play important roles in human nutrition. These fatty acids act as precursors for the formation of a variety of hormones including prostaglandins and leukotrienes. Over the past decade the annual catch of all the major marine fish has fallen appreciably due to population pressures and improved harvesting technologies, which have resulted in over-fishing. The outcome is a vastly reduced availability of fish products and an increase in their price, although culturing fish is now commonplace. An alternative strategy is to engineer oilseeds to produce long-chain n-3 polyunsaturated fatty acids similar to those accumulated in fish. Ironically, although they are known as fish oils, these products are actually produced by marine algae (Shiran et al, 1996; Cohen et al, 1997) which are then ingested by fish with a resulting build-up of the oils within their tissues. Using biotechnological methods it is now possible to consider the isolation of genes responsible for the biosynthesis of long-chain n-3 polyunsaturated fatty acids from marine algae and to transfer them to oilseed crops such as soybean and rapeseed so that they can be produced on a large scale and incorporated into edible oils for human nutrition.

Although only about 20% of globally traded vegetable oils are currently used to supply oleochemicals for the non-food sector, such oils are potentially able to serve as raw materials for a huge additional variety of products, many of which are currently obtained from petroleum-derived chemicals. Petrochemicals are still relatively cheap and abundant but are ultimately a non-renewable resource, which will probably begin to run out during the next century. Many petroleum-derived products will not biodegrade readily, some may contain toxins, and all of them create problems for disposal. The alternative use of plant-derived oleochemicals has a number of environmental benefits including better biodegradability, CO_2 neutrality, and an ability to be synthesised directly within the seed or fruit of the plant rather than via expensive and often polluting industrial processes. A good example is the use of oils for lubrication. Lubricating oils represent a partial-to-total loss system, depending on the rate of their leakage into the environment. Chain-saw oils used in forestry management are virtually totally lost to the system and the use of petroleum-derived lubricants has now been banned for forestry work in several European countries, such as Germany, mainly due to their persistence in and deleterious effect on the environment (Carruthers et al, 1995) . An oleochemical lubricant based on soybean or rapeseed oil has recently been developed which provides a CO_2-neutral and readily biodegradable alternative to the petroleum product. At present, most oleochemical products cannot compete directly with petrochemicals, particularly if they are treated as large-scale commodities. Nevertheless, the environmental lobby often supported by consumer choice of more environmentally-friendly products, is likely to lead to increasing financial and legislative constraints, eg carbon taxes, on petroleum products. Such developments will eventually favor the partial or total replacement of petrochemicals with oleochemical products, particularly in environmentally sensitive areas.

PROGRESS IN BIOTECHNOLOGY

The past decade has witnessed some striking examples of the manipulation of quality traits in transgenic oil crops using single-gene insertion events. A list of some of the rapeseed and soybean varieties with modified quality traits that are currently under development is shown in Table 3. Two of the most notable achievements in oil modification to date are the 40% stearic and 40% lauric rapeseed varieties first trialed by Calgene in 1993–1994 (Murphy, 1996). Rapeseed normally contains only 1–2% stearic acid and <0.1% lauric acid. The development of these transgenic varieties can be used to illustrate both the achievements and some of the challenges for oilseed biotechnology. The 40% lauric rapeseed variety was approved for unrestricted commercial release in the USA in 1995. These plants contain a lauroyl ACP thioesterase gene from the California bay plant.

Table 3. Transgenic rapeseed and soybean varieties currently in development around the world

Seed product	Industrial product	First field trials	Development status[a]
40% Stearic (18:0)	Margarine, cocoa butter	1994	***
40% Lauric (12:0)	Detergents	1994	***
60% Lauric (12:0)	Detergents	1996	***
85% Oleic (18:1 Δ_9)[b]	Food, lubricants, inks	1995	***
Petroselinic (18:1 Δ_6)	Polymers, detergents	1998(?)	**
'Jojoba' wax (C_{20}, C_{22})	Cosmetics, lubricants	1996(?)	**
40% Myristate (14:0)	Detergents, soaps, personal care	1996	**
90% Erucic (22:1)	Polymers, cosmetics, inks, pharmaceuticals	1998(?)	**
Ricinoleic (18:1-OH)	Lubricants, plasticisers, cosmetics, pharmaceuticals	1998(?)	**
Polyhydroxybutyrate	Biodegradable plastics	1998(?)	**
Phytase	Animal feed	1994	***
Industrial enzymes	Fermentation, paper manufacture, food processing	1998(?)	*
Novel peptides	Pharmaceuticals	1995	***

[a]Status: * = key genes already cloned, ** = transgenic plants produced, *** = field trials underway.
[b] This product is present in both transgenic rapeseed and soybean. The other products are present only in transgenic rapeseed

This causes premature chain termination of the growing fatty acid resulting in the accumulation of C12 (lauric) rather than the normal C18 oils. Lauric oils are mainly used in soaps and detergents, although their use in confectionery fats and infant milk formulas is also being developed. The major challenge now is to increase the level of lauric acid from 40% to something more like 60–90% in order to compete with cheaper tropical lauric oils, such as coconut and palm kernel oils (Voelker et al, 1996). The 40% stearic rapeseed contains an antisense copy of a *Brassica* stearate desaturase gene. This interferes with the function of the normal rapeseed stearate desaturase gene leading to an accumulation of stearic acid rather than allowing its desaturation to oleic acid (Knutzon et al, 1992). The resulting high stearic oil may have advantages over normal rapeseed oil for the production of certain solid fats such as margarines. The major challenge with the 40% stearic rapeseed variety is its relatively poor agronomic performance as exemplified by low germination rates. This is probably due to the "leakage" of some of the stearic acid into membrane lipids and the consequential decrease in membrane fluidity, particularly at low temperatures (Thompson & Li, 1997). In order to produce a stable and high yielding commercial variety of stearic rapeseed, it will be necessary to ensure that the stearic acid is channelled virtually exclusively to storage triacylglycerols rather than to membrane lipids.

Although many of the genes involved in the regulation of fatty acid biosynthesis have now been isolated, there are probably numerous additional genes which are still awaiting discovery. This is particularly the case in relatively exotic plants producing unusual and potentially highly valuable fatty acids (see Table 2). The use of EST libraries prepared from oilseeds which produce interesting and/or unusual fatty acids has already met with considerable success in the search for the key genes of interest. An EST library from *Borago officianalis* (borage) was used to isolate the Δ_6 linoleate desaturase gene responsible for the synthesis of γ-linolenic acid, which is an important therapeutic agent for treatment of conditions such as PMT and skin disorders (Beremand et al, 1997). An EST library from *Ricinus communis* (castor bean) was used to isolate the Δ_{12} oleate hydroxylase gene involved in the formation of ricinoleic acid, which has numerous industrial and pharmaceutical uses (van de Loo et al, 1995). Finally, a similar approach has been used in a *Vernonia* sp to isolate the epoxidase gene responsible for synthesis of vernolic acid, which is an epoxy acid with uses in the manufacture of coatings and polymers (Kinney, 1997).

PROTEIN ENGINEERING

Genes encoding many of the enzymes responsible for the amazing diversity of fatty acid composition in the natural world are now available and many more will be isolated in the near future. This will allow us to consider the medium-to-large-scale production of fatty acids containing hydroxyl groups, conjugated and acetylinic double bonds, branched-chains, very long or very short chains, epoxide groups and numerous other interesting functionalities. However, with the use of protein engineering we need not be limited even by this huge diversity of natural fatty acids in the natural world. By redesigning the enzymes responsible for fatty acid biosynthesis and modification, it is possible to envisage a virtually unlimited potential for inserting any given functionality into any given position of fatty acid of any given chain-length.

This prospect has come closer to reality recently with the resolution of the crystal structures of two important enzymes involved in fatty acid biosynthesis, ie enoyl-ACP reductase (Rafferty et al, 1995) and steroyl-ACP desaturase (Lindquist et al, 1996). There are also preliminary reports of crystallisation of the *E.coli* keto acyl-ACP synthase II (Dehesh, 1997) which is very similar to the plant homologue. The availability of these crystal structures in combination with relatively straightforward site-directed mutagenesis techniques and good over-expression systems for the production of large quantities of pure recombinant proteins will enable us to understand and hence manipulate the relationship between the tertiary structure and enzyme activity in such proteins.

The first steps along this road have already been taken in the case of the soluble fatty acid desaturases in plants. For example, domain swapping experiments between a Δ_9 stearoyl desaturase from castor bean and a Δ_6 palmitoyl desaturase from *Thunbergia alata* have shown that it is possible to manipulate desaturases in order to put double bonds in different locations (Cahoon & Shanklin, 1997; Shanklin, 1997). This is potentially of great interest to industrial chemists since naturally occurring fatty acids in plants tend to have double bonds only in a relatively restrictive number of locations along the fatty acid chain, namely Δ_6, Δ_9, Δ_{12} and Δ_{15}. The ability to insert double bonds into novel positions such as Δ_7 or Δ_8 will allow us to produce a whole range of new raw materials for the manufacture of products such as polymers, paints, varnishes and fine chemicals. The soluble fatty acid desaturases of plants as exemplified by the castor bean stearoyl-ACP desaturase have striking structural similarities with microbial ribonucleotide reductases and methane monooxygenases. The three dimensional structures of both of these proteins can be almost perfectly superimposed on that of the plant Δ_9 desaturase (Lindquist, 1996). All three proteins are examples of diiron enzymes which have similarities in their oxygen activation chemistry, although they use entirely different substrates. Such insights allow us to consider the manipulation of desaturases to carry out similar chemical modifications but using entirely different types of compounds from their normal fatty acid substrates.

While the characterisation of soluble desaturases has made great progress over the past few years, the analysis of the structure and regulation of membrane-bound desaturases presents many more challenges. However, even here there have been encouraging developments with a preliminary report of the desolubilisation purification of a membrane-bound alkane hydroxylase that shares the conserved 8-histidine motif with the membrane-bound desaturases (Shanklin, 1997). This protein contains 3 iron atoms per molecule of which 2 may comprise the diiron centre. If this enzyme is still catalytically active in the micellar state in which it is purified, it may be amenable to the kind of spectroscopic techniques which have been so informative in studying the reaction mechanism of the soluble desaturases. With recent progress in the crystallisation of a number of membrane-bound proteins it is even possible to envisage the atomic resolution of the structure of membrane-bound fatty acid desaturases within the next few years (Murphy and Piffanelli, 1998).

In addition to fatty acids, plants produce many other lipophilic components including sterols, flavonoids and glucosinolates, which may have important uses either in nutrition or for medicinal purposes. For example, one of the most characteristic attributes of the *Brassica* crops is the production of glucosinolates. These compounds are responsible for the astringent flavor of the popular condiment, mustard, and also for the highly characteristic odor and flavor of *Brassica* vegetables such as broccoli, cabbage and cauliflower. Glucosinolate breakdown products have also been found to have important anticarcinogenic roles both in epidemiological studies (Block et al, 1992) and in laboratory-based feeding studies (Tawfig et al, 1995; Zhang et al, 1994).

IMPACT OF GENOMICS

The application of genomics research is now having a particularly strong impact on our ability to manipulate both simple and complex genetic traits responsible for factors such as oil quality in crops. This is particularly relevant as the appreciation has dawned that single-gene insertion events will not necessarily produce the desired change in oil quality (Voelker et al, 1996; Murphy, 1996). Oil quality and oil yield are both complex quantitative traits, the analysis and elucidation of which will benefit greatly from recent advances in genomics (Murphy, 1998).

During the past ten years a great deal of plant research has been concentrated on the single model species, *Arabidopis thaliana*. *Arabidopsis* has the virtue of containing a relatively small genome of only 100 Mb. This genome is arranged on 5 chromosomes and contains relatively little repetitive DNA. In contrast, the major crop species maize and barley, have genome sizes of 2,500 and 5,000 Mb, respectively, while the hexaploid species, wheat, is 16,000 Mb. The relatively small genome size of *Arabidopsis* has made it the first plant target for a multi-national genome-sequencing project which should cover the entire genome by the year 2000. Already, physical maps of the *Arabidopsis* chromosomes are becoming available and the majority of the estimated 25,000 *Arabidopsis*-expressed genes have already been identified as ESTs (Somerville, 1996). In parallel with these genomic studies, a number of tagged mutant libraries are being created in addition to large numbers of chemically mutagenised lines of *Arabidopsis*. Ordered contiguous segments (contigs) of yeast artificial chromosomes (YACs) and bacterial artificial chromosomes (BACs) corresponding to much of the *Arabidopsis* genome have been created. This, in conjunction with detailed molecular maps, based on RFLP, cDNA and microsatellite markers, now allows for positional cloning of genes which have been mapped to within an accuracy of about 1cM.

Although the current *Brassica* lineage probably diverged from the *Arabidopsis* lineage about 10 million years ago, their genomes exhibit striking similarities. The diploid *Brassicas* appear to contain 3 more-or-less complete, although somewhat rearranged, copies of an *Arabidopsis*-sized genome in which the gene order and intergenic distances are substantially conserved (Parkin et al, 1997). In addition to this macro-synteny, homologous genes in *Brassica* and *Arabidopsis* encode proteins which are about 85% identical in sequence. Even regulatory elements such as gene promoters are often highly conserved between *Arabidopsis* and *Brassica* species (Sadanandom et al, 1996). In our own laboratories and elsewhere, *Arabidopsis* promoters have been used to drive to tissue specific expression in transgenic *Brassica* plants and *vice versa* (Parmenter et al, 1995). The close similarity between these genomes mean that molecular probes can be used interchangeably between *Brassica* and *Arabidopsis,* in order to identify, isolate and manipulate agronomically relevant genes.

QTLS FOR OIL QUALITY AND YIELD

As discussed previously, many characters involved in the regulation of oil quality and oil yield in oilseed crops are regulated by large numbers of genes and do not, therefore, segregate into simple Mendelian ratios as expected if only one or two genes were involved. During the past five years, quantitative trait locus (QTL) mapping has proved to be a powerful tool for studying the inheritance of such complex characters. This method relies on genetic linkage analysis and the availability of marker loci at a density of at least 20cM throughout the genome. The mapping of QTLs allows for the estimation of the number of loci controlling a trait, their chromosome location and the proportion of the phenotypic variation for which each locus is responsible.

An example of a complex trait relating to oil quality in rapeseed is the accumulation of lauric acid. Although up to 50% levels of lauric acid were achieved by the insertion of a single thioesterase gene from the California bay, it was not possible to achieve higher levels of lauric acid with a single-gene insertion. An important constraint is the selectivity of the acyltransferases responsible for incorporating fatty acids into triacylglycerols (Voelker et al, 1996). This problem has been partially overcome by the transfer of a second gene, this time encoding a coconut lysophosphatidic acid acyltransferase, which readily accepts lauric acid as a substrate as opposed to the endogenous rapeseed enzyme which is highly selective against lauric acid. Transgenic rapeseed plants expressing these two additional genes are able to accumulate up to 60% lauric acid in their seed oil (Dehesh, 1997). Higher levels of lauric acid are undoubtedly possible for example, *Cuphea polymorphia* can accumulate as much as 80% lauric acid in its seed oil (Table 2). It is possible that this *Cuphea* species contains additional genes responsible for the high lauric acid phenotype. Possible candidates for such genes include those encoding other lauric-specific acyltransferases, a lauric-specific acyl-CoA synthetase and enzymes such as phospholipases which may be involved in preventing the "leakage" of lauric acid from storage triacylglycerol biosynthesis into the membrane lipid pool. Any significant accumulation of lauric acid in membrane lipids is likely to have highly deleterious consequences for membrane function due to the detergent properties of this short chain fatty acid.

Therefore, while the production of < 40% of the single fatty acid in a seed oil is possible with a single-gene insertion, and the achievement of levels of up to 60% may be possible with the insertion of two genes, the production of near homogeneous seed oils containing up to 90% of a single fatty acid may well involve the transfer of numerous additional genes. While this is by no mean scientific and technical challenge, the fact that numerous seeds are able to accumulate almost homogeneous oils (see Table 2), should encourage us to persevere in attempting to reach this important biotechnological goal. In order to achieve this, it is probably best to take a combined biochemical and genetical approach, particularly since some of the genes regulating such characters may not necessarily encode readily identifiable enzymes involved in lipid metabolism, but rather regulatory proteins such as transcription factors.

The quality, ie the fatty acid composition, of a seed oil is obviously very important for the down-stream processors of the crop, eg the seed crushers and oil refiners. However, an equally important attribute is the oil yield of the crop concerned. Different crop types tend to accumulate different levels of seed oils in a way that is largely determined genetically and can, therefore, be manipulated beyond the natural variation of the particular species using transgenic methods. For example, most of the major oilseed crops contain between 20–45% oil in their seeds. Despite decades of effort by breeders, it has not been possible to obtain significantly higher levels of seed oil. However, there are many other oil rich plants which contain as much as 60–75% oil in their seeds. This implies that there is no biological barrier to increasing the yields of oil in a particular seed to as much as double their present levels. Efforts are now underway to identify and manipulate QTLs regulating oil yield in crop plants of agronomic interest. A higher yielding oilseed is more valuable both for the farmer and the processor since there are fewer by-products to be disposed of.

COMMERCIAL/MANAGEMENT CONSTRAINTS

A serious commercial barrier to the large scale production of novel transgenic oilseed varieties relates to the management and effective segregation of such crops. The vast majority of the major oilseed crops are grown for the production of bulk commodities. Following harvesting, seed batches from different farms are stored together in large mills and are then bulk-crushed in a continuous process, often involving millions of tonnes of oil. This large-scale continuous processing and the treatment of vegetable oils as a commodity to be traded by the tanker load achieves considerable economies of scale. As can be seen in Table 3, there are currently at least 10 transgenic varieties of rapeseed and soybean which have been or are nearing commercial release. While not all of these varieties will necessarily progress to large scale cultivation, there will still be several additional rapeseed cultivars in the next few years, which will be essentially identical in structure and growth habit to the conventional crops, but which will contain completely different seed products. These modified seed oils or proteins will often have been designed for very different and often incompatible downstream uses, ranging from pharmaceuticals to lubricants, detergents and margarines. A prerequisite for managing such crops is the establishment of rigorous systems of seed segregation and identity preservation, particularly during cultivation, harvesting and crushing, in order to prevent cross-contamination of different seed varieties.

The management of many transgenic oil crops destined for diverse end-uses will probably require the use of contract growers and some geographical segregation of the crops. It will also require the monitoring of the seeds following harvesting and the development of the rapid and inexpensive but accurate analytical techniques which can distinguish between, for example, a 60% oleic oil and a 40% lauric oil. It will also be important to be able to process the different transgenic seed varieties in batches at crushing mills and to eliminate or minimise contamination from previous crushes. All of these requirements will add to the overhead costs involved in oil production and processing and will make it more likely that commercially successful transgenic varieties will be used for relatively high-value niche markets rather than for lower cost, high- volume products.

OIL MANIPULATION IS GENERIC TECHNOLOGY

The first generation of transgenic crops with modified seed oils has been achieved with rapeseed, thanks largely to the kind of scientific and technical advances as discussed above. More recently, this technology has been extended successfully to soybeans (Kinney, 1997). Indeed, it has been shown that the insertion of the same lipid-related transgenes into either soybean or rapeseed gives rise to essentially the same modification in seed oil profiles in both species. This implies that the technology already developed for rapeseed will be generic for all oil crops. Transformation protocols for sunflower and linseed are being developed steadily and within the next 5 years it should be possible to transform all the major oilseed crops. It is also likely that gene constructs and binary vectors that have been developed for seed oil modification in rapeseed will be able to be used directly for transforming sunflower, linseed and other species that are amenable to *Agrobacterium*-mediated transformation.

The transformation of perennial oil crops such as oil palm, coconut and olive is more challenging but is under active development in several laboratories around the world (Cheah, 1997). Given that oil palm is predicted to account for almost 30% of global vegetable oil production by 2010, it will be important to extend transgenic technology to this major oil crop. Already, techniques such as projectile-mediated transformation have been used to transfer marker genes into oil palm explants. Regeneration of clonal copies of oil palm plants remains a challenge in view of the high incidence of some clonal abnormalities, particularly in the flowering structures, which result in sterility and a failure to produce fruit (Corley et al, 1986). Nevertheless, given the current pace of progress in plant biotechnology, it is possible to be cautiously optimistic that transgenic oil palms with altered oil quality will be a reality within the not-too-distant future.

FUTURE PROSPECTS

During the next two decades we can envisage the use of oil crops as environmentally friendly green factories for the production of a huge range of products for the edible, industrial and pharmaceutical sectors. This will involve the isolation of genes from a wide variety of oilseeds which make useful products and their transfer to major oil crops such as rapeseed, soybean, oil palm and sunflower. We will also be able to engineer proteins in order to produce an even greater diversity of fatty acid and related lipophilic products, many of which do not occur naturally at present. The challenge now is for biotechnologists to communicate more effectively with industrial chemists and others involved in the processing and utilisation of the products of oil crops. At present there is a significant knowledge gap whereby many synthetic chemists and others in relevant disciplines do not fully appreciate the possibilities now available from biotechnology. For example, in addition to producing large quantities of useful fatty acids, oil crops can also produce fairly elaborate lipid molecules with specific chiral properties which are difficult, if not impossible, to make by a synthetic route. Until we are able to overcome these communication problems, a great deal of the potential of oil crop biotechnology will remain unrealized.

REFERENCES

Arondel, V.; Lemieux, B.; Hwang, I.; Gibson, S.; Goodman, H. M.; Somerville, C. R. Map-based cloning of a gene controlling Omega-3 fatty acid desaturation in *Arabidopsis*. *Science* **1992**, *258*, 1353–1355.

Beremand, P. D.; Nunberg, A. N.; Reddy, A. S.; Thomas, T. L. Production of m-linolenic acid by transgenic plants expressing cyanobacterial or plant $Æ^6$-desaturase genes. In *Physiology, Biochemistry and Molecular Biology of Plant Lipids*; Williams J. P., Khan M. U, Lem N. W., Eds.; Kluwer: Dordrecht, **1997**; pp 351–353.

Block, G.; Patterson, B.; Suber, A. Fruit, vegetables, and cancer prevention - a review of the epidemiologic evidence. *Nutrition and Cancer* **1992**, *18*, 1–19.

Cahoon, E. B.; Shanklin, J. Approaches to the design of acyl-ACP desaturases with altered fatty acid chain-length and double bond positional specificities. In *Physiology, Biochemistry and Molecular Biology of Plant Lipids*; Williams J. P., Khan M. U, Lem N. W., Eds.; Kluwer: Dordrecht, **1997**; pp 374–376.

Carruthers, S. P.; Marsh, J. S.; Turner, P. W.; Ellis, F.; Murphy, D. J.; Slabas, T.; Chapman, B. A. *Industrial markers for UK produced oilseeds. Centre for Agricultural Strategy. Research Review No 0S9. University of Reading*; Centre for Agricultural Strategy: University of Reading, 1995.

Cheah, S. C. *22nd World Congress & Exhibition of the International Society for Fat Research, 8–12 Sept, 1997, Kuala Lumpur*; ISF, Champaign, Illinois, 1997.

Cohen, Z.; Shiran, D.; Khozine, I.; Heimer, Y. M. Fatty acid unsaturation in the red alga *Porphyridium cruentum*. Is the methylene interrupted nature of polyunsaturated fatty acids an intrinsic property of the desaturases? *Biochim Biophys Acta* **1997**, *1344*, 59–64.

Corley, R. H. V.; Lee, C. H.; Law, I. W.; Wong, C. Y. Abnormal flower development in oil palm clones. *Planter* **1986**, *62*, 233–240.

Dehesh, K. Molecular and biochemical analysis of medium-chain lipid biosynthesis. Abstract 100. *5th Int. Congr. of Plant Mol. Biol., 21–27 September 1997*; Kluwer, Dordrecht, 1997.

Kinney, A. J. Genetic engineering of fatty acid metabolism in soybeans. Abstract 99. *5th Int. Congr. of Plant Mol. Biol., 21–27 September 1997*; Kluwer, Dordrecht, 1997.

Knutzon, D. S.; Thompson, G. A.; Radke, S. E.; Johnson, W. B.; Knauf, V. C.; Kridl, J. C. Modification of *Brassica* seed oil by antisense expression of a stearoyl-acyl carrier protein desaturase gene. *Proc Natl Acad Sci USA* **1992**, *89*, 2624–2628.

Lindqvist, Y.; Huang, W.; Schneider, G.; Shanklin, J. Crystal structure of a delta-9 stearoyl-acyl carrier protein desaturase from castor seed and its relationship to other di-iron proteins. *EMBO Journal* **1996**, *15*, 4081–4092.

Moloney, M. M.; Walker, J.; Sharma, K. High efficiency transformation of *Brassica napus* using *Agrobacterium* vectors. *Plant Cell Rep.* **1989**, *8*, 238–242.

Murphy, D. J.; Piffanelli, P. Fatty acid desaturases: structure mechanism and regulation. In *Plant Lipid Biosynthesis: Recent Advances of Agricultural Importance*; Harwood, J. L., Ed.; Cambridge University Press, 1998, pp. 95–130.

Murphy, D. J. *Genomics: Commercial Opportunities from a Scientific Revolution, University of Cambridge*; Society of Chemical Industry, 1998, pp. 199–210.

Murphy, D. J. Biotechnological improvement of oil crops. *TIBTECH* **1996**, *14*, 206–213.

Okuley, J.; Lightner, J.; Feldmann, K.; Yadav, N.; Lark, E.; Browse, J. Arabidopsis *FAD2* gene encodes the enzyme that is essential for polyunsaturated lipid synthesis. *Plant Cell* **1994,** *6*, 147–158.

Parkin, I. A.; Cavell, A.; Oldknow, J.; Lydiate, D. J.; Trick, M. AGenome synteny between *Arabidopsis thaliana* and *Brassica napus.,@* John Innes Centre Annual Report, 1997.

Parmenter, D. L.; Boothe, J. G.; van Rooijen, G. J. H.; Yeung, E. C.; Moloney, M. M. Production of biologically active hirudin in plant seeds using oleosin partitioning. *Plant Mol.Biol.* **1995**, *29*, 1167–1180.

Rafferty, J. B.; Simon, J. W.; Baldock, C.; Artymiuk, P. J.; Baker, P. J.; Stuitje, A. R.; Slabas, A. R.; Rice, D. W. Common themes in redox chemistry emerge from the x-ray structure of oilseed rape (*Brassica napus*) enoyl acyl carrier protein reductase. *Struct.* **1995**, *3*, 927–938.

Sadanandom, A.; Piffanelli, P.; Knott, T.; Sharpe, A.; Lydiate, D.; Murphy, D. J.; Fairbairn, D. Identification of a peptide methionine sulfoxide reductase gene in an oleosin promotor from *Brassica napus*. *Plant Journal* **1996**, *10*, 235–242.

Shanklin, J.; Cahoon, E. B.; Whittle, E.; Lindqvist, Y.; Huang, W.; Schneider, G.; Schmidt, H. Structure-function studies on desaturases and related hydrocarbon hydroxylases. In *Physiology, Biochemistry and Molecular Biology of Plant Lipids*; Williams J. P., Khan M. U, Lem N. W., Eds.; Kluwer: Dordrecht, 1997; pp 6–10.

Shiran, D.; Khozin, I.; Heimer, Y. M. Biosynthesis of eicosapentaenoic acid in the microalga *Porphyridium cruentum* . I: The use of externally supplied fatty acids. *Lipids* **1996**, *31*, 1277–1282.

Somerville, C. The physical map of an *Arabidopsis* chromosome. *Trends in Plant Sci* **1996**, *1*, 2.

Tawfiq, N.; Heaney, R.; Plumb, J.; Fenwick, G.; Musk, S.; Williamson, G. Dietary glucosinolates as blocking agents against carcinogenesis: glucosinolate breakdown products assessed by induction of quinone reductase activity in murine hepa1c1c7 cells. *Carcinogenesis.* **1995**, *16*, 1191–1194.

Thompson, G. A.; Li, C. Altered fatty acid composition of membrane lipid in seeds and seedling tissues of high-saturate canolas. In *Physiology, Biochemistry and Molecular Biology of Plant Lipids*; Williams J. P., Khan M. U, Lem N. W., Eds.; Kluwer: Dordrecht, 1997; pp 313–315.

van de Loo, F. J.; Broun, P.; Turner, S.; Somerville, C. An oleate 12-hydroxylase from *Ricinus communis* L. is a fatty acyl desaturase homolog. *Proc Natl Acad Sci USA* **1995**, *92*, 6743–6747.

Voelker, T. A.; Hayes, T. R.; Cranmer, A. M.; Turner, J. C.; Davies, H. M. Genetic engineering of a quantitative trait: metabolic and genetic parameters influencing the accumulation of laurate in rapeseed. *Plant J.* **1996**, *9*, 229–241.

Yadav, N. S.; Wierzbicki, A.; Aegerter, M.; Caster, C. S.; Perez-Grau, L.; Kinney, A. J.; Hitz, W. D.; Booth, J. R. J.; Schweiger, B.; Stecca, K. L.; Allen, S. M.; Blackwell, M.; Reiter, R. S.; Carlson, T. J.; Russell, S. H.; Feldmann, K. A.; Pierce, J.; Browse, J. Cloning of higher plant Ω-3 fatty acid desaturases. *Plant Physiol* **1993**, *103*, 467–476.

Zhang, Y.; Kensler, T. W.; Cho, C. G.; Posner, G. H.; Talalay, P. *Proc Natl Acad Sci USA* **1994** 91, 3147–3150.

4

BIOSYNTHESIS OF RICINOLEATE IN CASTOR OIL

Thomas A. McKcon, Jiann-Tsyh Lin, and Allan E. Stafford

Western Regional Research Center
Agricultural Research Service
U.S. Dept. of Agriculture, Albany, California 94710

ABSTRACT

Castor oil is 90% ricinoleate (12-hydroxyoleate) and has numerous industrial uses. Components of castor bean (*Ricinus communis* L.) pose serious problems to processors. Other researchers have cloned the gene for the oleoyl hydroxylase, but transgenic plants produce only about 20% hydroxy fatty acid. To improve such transgenic substitutes for castor, we are using HPLC analysis of castor bean microsomal suspensions to follow the hydroxylase reaction and the movement of ^{14}C-ricinoleate through phospholipid into triacylglycerol. Most labeled ricinoleate is rapidly removed from the phospholipid fraction as free fatty acid and incorporated into triacylglycerol, with triricinolein predominating. Elucidation of the basis for high incorporation of ricinoleate and exclusion of oleate from triacylglycerols will identify genes that can be used to engineer high ricinoleate production in transgenic plants.

INTRODUCTION

Vegetable oils are major agricultural products that are used in food and as sources of industrial chemicals (McKeon et al., 1996). Many vegetable oils play an important role in nutrition, providing calories and essential fatty acids for the diet. The oils are used directly in food products such as salad dressings, fried or baked goods, or are partially hydrogenated and used as vegetable shortening or margarine. Currently, oils with high levels of oleate are considered most beneficial for health, with those containing α-linolenate serving as important sources of ω-3 fatty acids. Minor components of oils include tocopherols, tocotrienols or conjugated linoleate, which are beneficial as antioxidants. These same food oils are used for industrial purposes as well, for animal feed, soaps and detergents, drying

Chemicals via Higher Plant Bioengineering, edited by Shahidi *et al.*
Kluwer Academic / Plenum Publishers, New York, 1999.

oils, plastics, cosmetics and coatings. On the other hand, there are some vegetable oils that are used solely for industrial purposes; these include castor, rapeseed and tung oils. These oils contain fatty acids that impart useful properties for industrial applications but may have unpleasant or toxic effects when ingested.

Castor oil is composed of 85 to 90% ricinoleate (12-hydroxyoleate) and up to 70% of its triacylglycerol content is triricinolein. The hydroxyl group imparts high viscosity to the oil due to the interchain hydrogen bonding and provides a reactive mid-chain, functional group that makes ricinoleate a very useful chemical precursor (Caupin, 1997) . In addition to its continued use as a laxative, castor oil provides the basis for a number of other products. Its major strategic uses are in lithium grease and lipstick. Other important products derived from castor oil include engineering plastics (Nylon 11), anti-fungal compounds, plasticizers, drying oils, cosmetics and other coatings (Caupin, 1997). Castor bean produced by the plant *Ricinus communis* L. is unusual in producing an oil that is a technical grade of purity in ricinoleate, and also in containing up to 60% oil and, thus, among the richest oil producers. Unfortunately, the bean also contains the potent toxin ricin as 2% of the storage protein content as well as hyperallergenic glycoproteins (Thorpe et al., 1988). These constituents make castor an undesirable crop, as processing to meet U.S. needs for castor oil leads to the accumulation of 50–60 million pounds of highly toxic, biohazardous waste. Thus, development of an alternative crop that can produce high levels of ricinoleate would be of great benefit.

The most logical means to develop an alternative to castor is by way of a transgenic plant modified to produce ricinoleate. We will describe in this chapter the challenges presented by this goal and the progress in achieving it.

FATTY ACID BIOSYNTHESIS

Fatty acid biosynthesis in plants has been recently reviewed (Ohlrogge and Browse, 1995) and is summarized in Figure 1 . There have been several commercial successes in developing transgenic plants with altered fatty acid composition (Bornscheuer, 1998). These include laurate canola and high oleic soybean. A number of other transgenic oilseed crops are in commercial development. In most cases, the alteration in fatty acid composition is caused by a single gene change. It has become apparent that there is a limit to the amount of compositional change that can be brought about by a single gene change. This limitation is a function of the fatty acid to be produced and its interaction with the lipid biosynthetic system of the transgenic plant. One example is laurate canola, which reaches a peak of 40% laurate by weight in canola expressing a lauroyl-ACP thioesterase. However, levels of laurate can be increased to 55% by inclusion of a lysophosphatidic acid acyl transferase (LPAAT) from coconut that can put laurate in the *sn*-2 position of tricylglycerols (Knutzon et al., 1995). Hydroxy- and epoxy- fatty acids pose the additional problem of having a mid-chain polar functional group that could interfere with membrane function. Even plants that do not produce these fatty acids have a phospholipase A_2 that preferentially removes the oxygenated fatty acids (Stahl et al., 1995). This activity may serve in a membrane repair system to remove hydroperoxy fatty acids that could lead to accelerated free radical damage of the membrane. The released fatty acid is then replaced by one of the common fatty acids, principally oleate, using an acyl transferase that adds a fatty acid to the lyso-phospholipid to complete the membrane repair. The released fatty acid is incorporated into coenzyme A and inserted into triacylglycerol or undergoes β-oxidation. When a cDNA for the oleoyl-12-hydroxylase from castor was introduced into Ni-

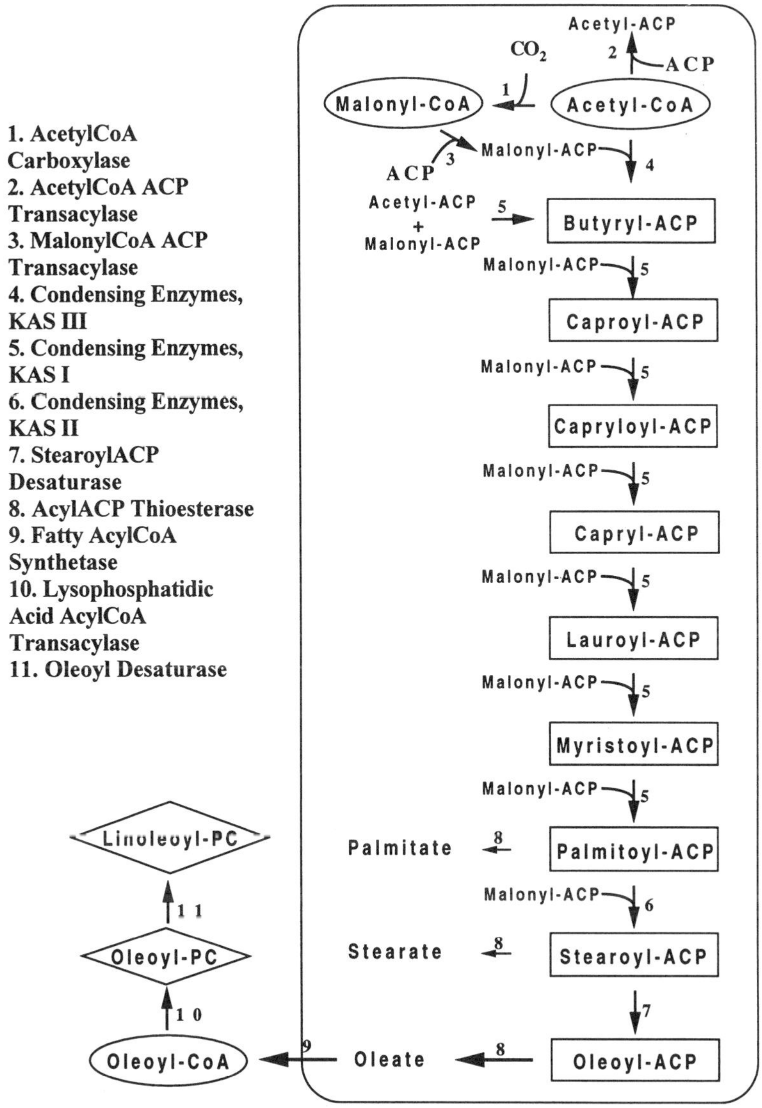

Figure 1. Fatty acid biosynthesis in plants.

cotiana, only 0.1% of the fatty acid produced was ricinoleate (Van de Loo et al., 1995), suggesting that the ricinoleate produced may have been rapidly metabolized. The highest levels reported for expression of this cDNA in canola is 20% (Broun et al., 1996). Specific reasons for the low levels produced in these cases are not known. However, it appears that plants that do produce significant amounts of oxygenated fatty acids (ricinoleate from castor; 12,13-epoxy oleate from *Vernonia*) have specific enzymatic systems not just for production of the fatty acid but also for incorporating it into triacylglycerols. We describe here our research and review related research to report the specific factors that lead to high levels of ricinoleate production and incorporation into triacylglycerols by castor.

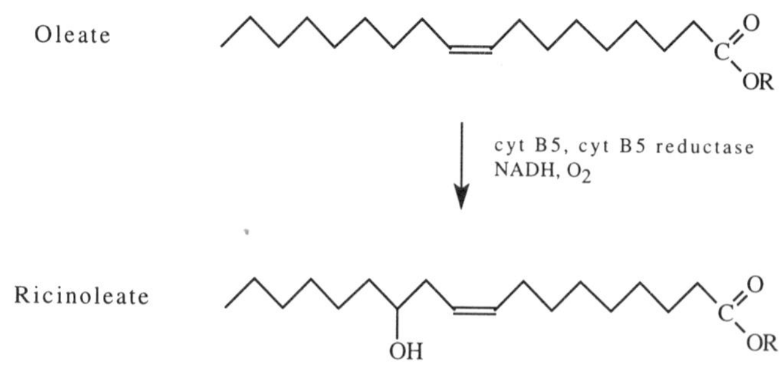

Figure 2. Oleoyl-12-hydroxylase reaction.

RICINOLEATE BIOSYNTHESIS

The oleoyl-12-hydroxylase catalyzes the hydroxylation of oleate to ricinoleate (Figure 2). This enzyme has been under study for over 30 years (Yamada and Stumpf, 1964; James et al., 1965). Although it has not yet been purified, its cDNA has been cloned and a great deal is known about hydroxylase enzymology based on *in-vitro* studies.

Initial efforts in elucidating hydroxylase demonstrated that it was membrane associated, with activity remaining in the microsomal fraction (Galliard and Stumpf, 1966). Full activity required a substrate, oleoyl-CoA, and the reducing agents NADH or NADPH (Galliard and Stumpf, 1966). Later research indicated that the oleate from oleoyl-CoA is rapidly incorporated into phosphatidylcholine, leading to the conclusion that the substrate for the enzyme is actually oleoyl-PC (Moreau and Stumpf, 1981). This conclusion was further elaborated to show that oleate in the *sn*-2 position of PC was the substrate. Our work definitively established this as the substrate for hydroxylation by directly using *sn*-2 oleoyl PC as the substrate and isolating *sn*-2 ricinoleoyl as product from the incubation with castor microsomes (Lin et al., 1998B). The requirement for NADH arises from the need for a membrane-based reducing system consisting of cytochrome b_5 and NADH-dependent cytochrome b_5 reductase; NADPH is not as effective, but allows partial activity via the NADPH-dependent cytochrome P_{450} reductase (Smith et al., 1992).

In initial studies on ricinoleate biosynthesis, it was found that significant amounts of hydroxylase activity remained associated with microsomal fractions (Galliard and Stumpf, 1966; Moreau and Stumpf, 1981). Since the microsomal fraction is a heterogeneous collection of plant membrane types, several studies have been undertaken to determine the membrane fraction associated with hydroxylase activity. This has proven to be difficult, due to the instability of the enzyme. However, one study was able to achieve separation of hydroxylase activity from activity in other membrane markers (mitochondria and peroxisome). The hydroxylase activity was found to be associated in the gradient fractions with endoplasmic reticulum marker enzymes (Richards et al., 1993)

The association of the mammalian stearoyl-CoA desaturase with membrane fractions and its requirement for the membrane-based cytochrome b_5-reducing system led Moreau and Stumpf to suggest a similar system in plants operating in hydroxylation (1981). The later demonstration that cytochrome b_5 was involved in oleate desaturation (Kearns et al., 1991) by inhibition of desaturation with antibodies to cytochrome b_5 provided a model for assessing the role of cytochrome b_5 in oleate hydroxylation. Smith et al. (1992) demonstrated that antibodies to cytochrome b_5 blocked hydroxylation, coinciding with its ability

Table 1. Assay conditions for optimal hydroxylation using oleoyl-PC as substrate

Sodium phosphate buffer (0.1 M, pH 6.3), 1 ml volume
NADH (0.5 mmole)
ATP (0.5 mmole)
$MgCl_2$ (0.5 mmole)
Catalase (1000 units)*
Castor microsomes (90 mg)
1-Palmitoyl-2-[^{14}C]oleoyl-sn-glycero-phosphocholine (2.2 nmol) in 20 ml ethanol, added last with rapid mixing.
Incubate mixture with shaking at 22°C

to block other enzyme reactions dependent on reduced cytochrome b_5. They also demonstrated that both NADH-dependent cytochrome b_5 reductase and NADPH-dependent cytochrome P_{450} reductase can reduce cytochrome b_5 for use in oleate hydroxylation, although the P_{450} reductase is less effective.

Efforts to purify the hydroxylase have not yet suceeded. Membrane- bound enzymes often prove difficult to purify and the hydroxylase has been reported to be unstable to storage (Bafor et al., 1991). Since the activity is dependent on a membrane-bound reducing system and probably one or more other membrane-associated enzyme activities, these pose additional complicating factors in hydroxylase isolation. We have made some progress in stabilizing the hydroxylase. Catalase, as previously reported (Moreau and Stumpf, 1981), and the protease inhibitor antipain (McKeon et al., 1997) which blocks papain-like activities allow long term storage of the microsome at -80°C. Use of lower amounts of microsomal protein in the assay (10 to 20 µg versus 300µg to >1 mg) and shorter incubation times enable us to determine optimal levels of activity and evaluate potential inhibitors of the reaction. When added to microsomal preparations at a level of 0.5%, the detergent Triton X-100 solubilizes the hydroxylase, based on complete loss of hydroxylase activity from the microsomes after recentrifugation. As a result of the low levels of microsomal protein needed for the assay, we are able to reconstitute hydroxylase activity by dilution of the detergent (McKeon et al., 1997). Thus, we have achieved at least the first step leading to solubilization and purification of active hydroxylase.

The hydroxylase assays used over the years measure the conversion of ^{14}C-labeled oleate to ^{14}C-labeled ricinoleate. Previous workers have used GC or TLC to separate the oleate and ricinoleate, and reactions used ^{14}C-oleoyl-CoA as the substrate. We have introduced a quick HPLC separation, using a 5 cm C18 reverse-phase column eluted with aqueous methanol (Lin et al., 1996), coupled with radioflow detection of the ^{14}C-labeled fatty acid methyl esters. This system provides resolution in 10 min and simultaneous determination of hydroxylation and desaturation. Oleoyl-CoA added as the source of labeled oleate provides the highest measurable levels of activity (McKeon et al., 1997), but we have developed assay conditions that maximize activity on the actual substrate *sn*-2 oleoyl-PC (Table 1). Using these conditions, we have shown that added substrate is converted to a PC product that we have identified as *sn*-2 ricinoleoyl-PC (Figure 3; Lin et al., 1998B). We have also demonstrated that the glycol analogue of the substrate, oleoyloxy ethylphosphocholine (Figure 4) is an effective inhibitor of hydroxylation in assays using added ^{14}C-oleoyl-CoA or *sn*-2 ^{14}C-oleoyl-PC as the source of labeled oleate for hydroxylation (Lin et al., 1996; McKeon et al., 1997).

Because of the limited progress in purification of the hydroxylase, an alternative approach to cloning the hydroxylase was taken (Van de Loo et al., 1995). Based on presumed analogy of the reaction to the oleoyl-12-desaturase reaction, sequences of cDNA

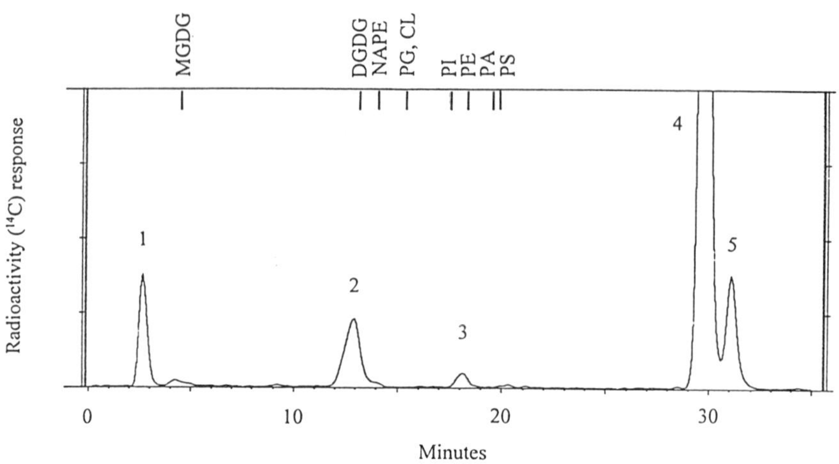

Figure 3. Isolation of *sn*-2 Ricinoleoyl-PC Peak 1: ^{14}C-free fatty acid and ^{14}C-triacylglycerol Peak 4: ^{14}C-oleoyl-PC and ^{14}C-linoleoyl-PC Peak 5: ^{14}C-ricinoleoyl-PC.

from developing castor bean were screened for the presence of histidine clusters similar to those in the oleoyl and linoleoyl desaturases. The prospective hydroxylase cDNA's were then introduced into tobacco and the seeds were screened for the production of ricinoleate. Although very low levels were produced by one of the cDNA's, it did correspond to the hydroxylase, based on homology to a cDNA from another hydroxy fatty acid producing plant, *Lesquerella*, and in later work produced up to 20% hydroxy fatty acid in transgenic rapeseed (Van de Loo et al., 1995; Broun et al., 1997) The contention that the hydroxylase and desaturase reactions are analogous (Van de Loo et al., 1995) has also been proven (Shanklin et al., 1997). Both reactions are brought about by a hypervalent diiron-oxo species differing only in whether two protons are removed by the activated oxo-group or if the hydroxyl formed by the first proton abstraction is inserted in the radical intermediate. The inter-relationship of these reactions has been proven by converting desaturases to hydroxylases through site-specific mutagenesis (Shanklin et al., 1997).

LIPID BIOSYNTHESIS

The general pathway for triacylglycerol biosynthesis in plants is presented in Figure 5. Pathways for lipid biosynthesis in plants have been recently reviewed (Ohlrogge and

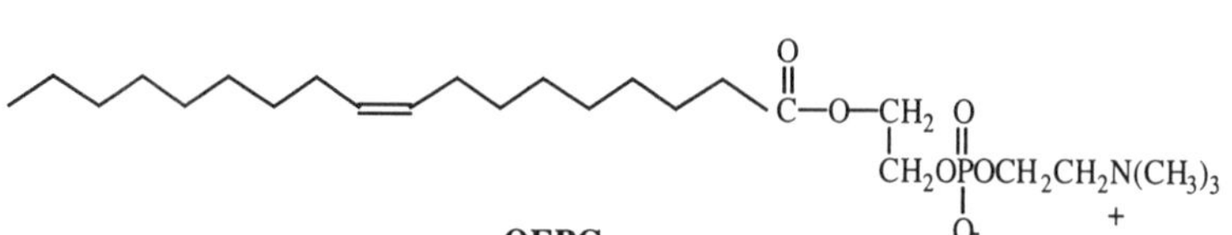

OEPC

Figure 4. Oleoylethylphosphocholine.

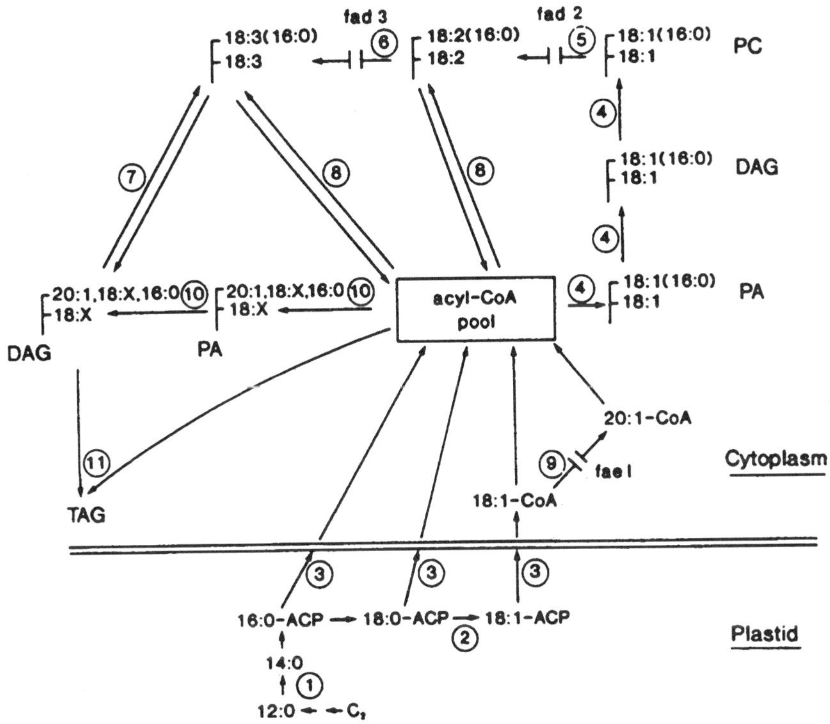

Figure 5. Triacylglycerol biosynthesis in plants (Browse and Somerville, 1991).

Browse, 1995) and will be briefly summarized here. After synthesis of the fatty acid as an acyl ACP in the plastid, the fatty acid is cleaved by an acyl-ACP thioesterase and exported to the cytoplasm and converted to an acyl CoA. The fatty acid can then be added to diacylglycerol to form the triacylglycerol by the action of diacylglycerol acyltransferase. The fatty acid can also be incorporated into phospholipid by way of several possible enzymatic steps, undergo modifications such as further desaturation, and eventually wind up as part of the diacylglycerol to be converted to triacylglycerol. Because of interrelationships among the numerous substrates and products in the biosynthetic pathway to triacylglycerol, there is opportunity for competing pathways to make a product and for a product to be used in alternate pathways. Moreover, the need for phospholipids as membrane components concomitant with the production of triacylglycerol during seed development indicates interdependence of the two processes. The result is an interconversion of acylated intermediates that remarkably leads to production of triacylglycerols with a composition that is characteristic of the crop.

Because castor oil contains up to 90% ricinoleate and approximately 4% oleate and 4% linoleate plus linolenate, the conversion of oleate to ricinoleate in castor must be extremely efficient and competition for oleate as a substrate for desaturation or inclusion in triacylglycerol must be very limited. In research approaches to develop transgenic crops which duplicate the fatty acid composition of castor oil, the obvious target was the oleoyl-12-hydroxylase. The cDNA for this enzyme was eventually cloned as described above. Yet, when the transgenic rapeseed expressed this cDNA, the level of hydroxy fatty acid produced was limited to 20%, composed of a mixture of 18, 20 and 22 carbon chain fatty

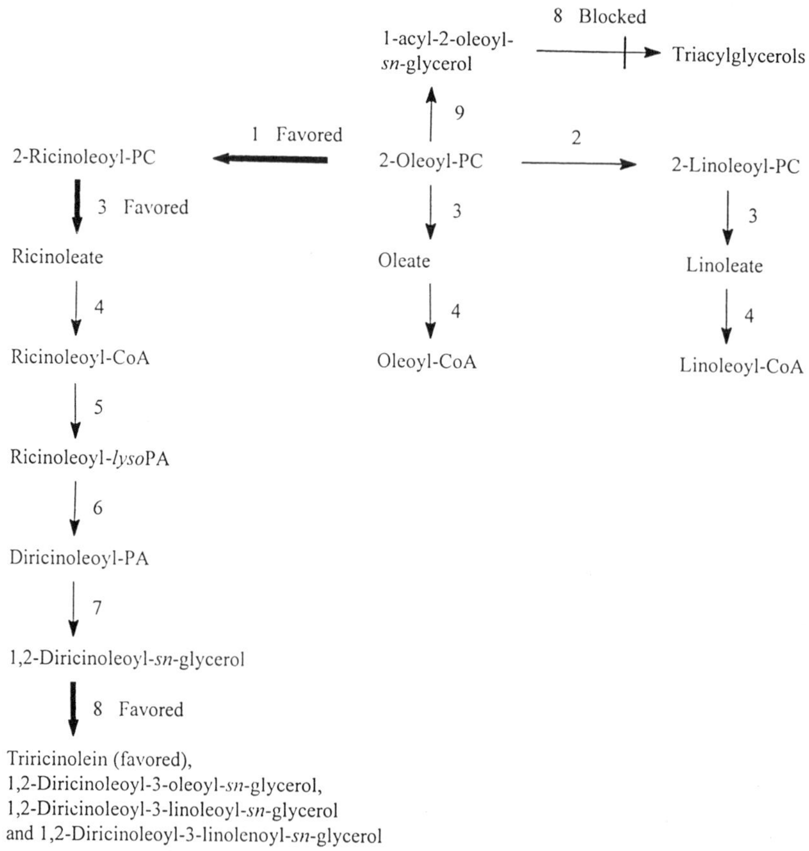

Figure 6. Triricinolein biosynthesis.

acids (Broun et al., 1997). Clearly, the castor bean is equipped to produce ricinoleate, so, to achieve further progress in developing a castor oil replacement, we have identified several enzymatic steps in castor bean that lead to predominant production of triricinolein (Lin et al., 1998B).

Figure 6 summarizes our findings. The steps leading to high production of ricinoleate and incorporation into triacylglycerol include:

a. the lyso-phosphatidylcholine acyltransferase (LPCAT) which transfers oleate from oleoyl CoA into the *sn*-2 position of PC for action by:
b. the oleoyl-12-hydroxylase hydroxylates the oleate and the *sn*-2 ricinoleoyl-PC is hydrolyzed by:
c. phospholipase A2 which preferentially removes ricinoleate from the *sn*-2 position and releases lyso-PC for reincorporation of oleate by LPCAT and the free fatty acid which is preferentially incorporated into ricinoleoyl-containing diacylglycerol by:
d. the diacylglycerol acyltransferase (DAGAT) which preferentially incorporates ricinoleate to form diricinoleins and triricinolein.

Since most investigations of ricinoleate biosynthesis have used oleoyl-CoA as the substrate donor, the LPCAT has, as a consequence, been relied on to generate the PC substrate for the reaction. The first recognition of oleoyl-PC as the likely substrate occurred with the demonstration by Moreau and Stumpf (1981) that most of the labeled oleate from ^{14}C-oleoyl-CoA is found in the PC fraction shortly after addition to the microsomal reaction mix. Other research groups added increasing levels of evidence supporting the same basic conclusion (Bafor et al., 1991). By isolating the labeled product *sn*-2-ricinoleoyl-PC and demonstrating that labeled ricinoleate is poorly incorporated into PC, we have identified the LPCAT as the first essential step in maintaining oleate as part of the substrate pool for conversion to ricinoleate (Lin et al., 1998B). The enzyme was solubilized by several different detergent treatments (Moreau and Stumpf, 1982). A potentially related enzyme has been purified from coconut (Knutzon et al., 1995), and is responsible for the insertion of medium-chain fatty acids in the *sn*-2 position of PA and, ultimately, triacylglycerol.

The next key step in castor oil biosynthesis is the hydroxylation of the *sn*-2-oleate on the PC. This step has been described in an earlier section of this chapter.

Release of the ricinoleate from PC is the next step leading to triricinolein production. The phospholipase A_2 from castor has been shown to have a strong preference for release of ricinoleate as opposed to oleate (Bafor et al., 1991). This feature is common among oilseed phospholipases A_2, even those from plants that do not make oxygenated fatty acids in their oils (Stahl et al. 1995). The activity is thought to be part of a membrane "proofreading" system that removes such fatty acids from PC to prevent their incorporation into membrane. The presence of such fatty acids with mid-chain polar groups would disrupt the otherwise non-polar fatty acid chains present in the membrane between the polar layers of the outer membrane. Our results have elaborated on the role of phospholipase A_2 in castor by demonstrating that free ricinoleate is a major labeled product detected in microsomal incubations of *sn*-2–^{14}C-oleoyl-PC. Additionally, when *sn*-2–^{14}C-ricinoleoyl-PC, the putative endogenous substrate of phospholipase A_2, is added to microsomal incubations, ^{14}C-ricinoleate is also a major product (Lin et al., 1998B).

The final step that we have identified as being a key to high ricinoleate content is also the final step in triacylglycerol biosynthesis. The diacylglycerol:acyl-CoA transferase (DAGAT) transfers the acyl group from acyl-CoA to the *sn*-3 position of *sn*-1,2-diacylglycerol. In microsomal incubations with ^{14}C-ricinoleate we observed labeled ricinoleate incorporation in triacylglycerol peaks corresponding to di- and triricinolein, with minor levels in monoricinolein. In contrast, in incubations with *sn*-2–^{14}C-oleoyl-PC, the principal labeled triacyglycerols contained labeled ricinoleate. Diricinolein was the only peak significantly labeled with ^{14}C-oleate (Lin et al., 1998B). This distribution pattern of label indicates a strong preference for ricinoleate incorporation into triacylglycerol already acylated with one or two ricinoleates, especially given the high ratio of ^{14}C-oleate to ^{14}C-ricinoleate in such microsomal incubations. Previous work by Vogel and Browse (1996) indicated at least a three-fold preference for transferring oleate from oleoyl-CoA into diricinoleoylglycerol over other diacylglycerols as measured by the DAGAT assay.

It was formerly thought that triacylglycerols were the end product of lipid biosynthesis and unchangeable. It is now apparent that there is still an opportunity to adjust their composition. Acyl exchange reactions allow transfer of acyl groups among triacylglycerols as well as from them to mono- and diacylglycerols. A recent study (Mancha and Stymne, 1997) using castor microsomes labelled with ^{14}C-glycerol lipids or ^{14}C-ricinoleoyl-lipids demonstrated that acyl groups are exchanged, with ricinoleoyl- tending to transfer from di- or triricinolein to mono- or diricinolein. Perhaps more importantly, they demonstrated that the castor endosperm has two pools of triacylglycerol, the microsomal, where acyl-exchange is active, and the oil bodies, which are rich in triricinolein. This finding indicates that the triacylglycerol can be actively tailored in the microsomes before being deposited in final form to the oil bodies.

The ultimate test of this approach lies in developing transgenic plants that express the castor enzymes that lead to high ricinoleate production. Such plants should be able to produce the high levels of ricinoleate needed to develop a transgenic crop that replaces castor. There are a number of products that can be produced from ricinoleate but, due to limited or unreliable supplies of castor oil, the products have not been produced commercially (Caupin, 1997). A transgenic crop could expand the industry dependent on ricinoleate, as the "castor-substitute" oil could be produced in greater quantity without the hazards involved in castor cultivation.

REFERENCES

Bafor, M., Smith, M.A., Jonsson, L., Stobart, K.; Stymne, S. Ricinoleic acid biosynthesis and triacylglycerol assembly in microsomal preparations from developing castor-bean (*Ricinus communis*) endosperm. *Biochem. J.*, **1991**, 280, 507–514.

Bornscheuer, U. Toward new fats, oils through biotechnology. *INFORM*, **1998** 485–488.

Broun, P., Hawker, N., Somerville, C.R. Expression of castor and *L. fendlerii* oleate 12-hydroxylase in transgenic plants: effects on lipid metabolism and inferences on structure-function relationships in fatty acid hydroxylases. *in*: John P. Williams, Mobashsher U. Khan and Nora W. Lem (eds.) Physiology, Biochemistry and Molecular Biology of Plant Lipids. Kluwer Academic Publishers, Dordrecht. **1997** pp. 342–344.

Browse, J., Somerville, C. Glycerolipid Synthesis: Biodensity and Regulation. *Ann. Rev. Pl. Physiol. Pl. Mol. Biol.* **1991**, 42, 467–501.

Caupin, H.J. Products from castor oil: past, present, and future. *in* Frank D. Gunstone and Fred B. Padley (eds.) Lipid Technologies and Applications. Marcle Dekker, Inc. New York **1997** pp. 787–795.

Galliard, T.; Stumpf, P.K. Fat metabolism in plants. XXX. Enzymatic synthesis of ricinoleic acid by a microsomal preparation from developing *Ricinus communis* seeds. J. Biol. Chem., **1966**, 241, 5806–5812.

Goodrich-Tanrikulu, M., Stafford, A.E., Lin, J.T.; McKeon, T.A. Metabolism of ricinoleate by *Neurospora crassa. Appl. Microbiol. and Biotechnol.*, **1996**, 46, 382–387.

James, A.T., Hadaway, H.C. and Webb, J.P.W. The biosynthesis of ricinoleic acid. Biochem. J. **1965**, 95, 448–452.

Kearns, E.V., Hugly, S., Somerville, C.R. The role of cytochrome b_5 in D12 desaturation of oleic acid by microsomes of safflower (*Carthamus tinctorius* L.). *Arch. Biochem. Biophys.* **1991**, 284, 431–436.

Knutzon, D.S., Lardizabal, K.D., Nelsen, J.S., Bleibaum, J.L., Davies, H.M. and Metz, J.G. Cloning of a coconut endosperm cDNA encoding a 1-acyl-*sn*-glycerol-3-phosphate acyltransferase that accepts medium-chain-length substrates. *Pl. Physiol.* **1995**, 109, 999–1006.

Lin, J.T., McKeon, T.A.; Stafford, A.E. Gradient reversed-phase high performance liquid chromatography of saturated, unsaturated and oxygenated free fatty acids and their methyl esters. *J. Chromotog.*, **1995**, 699, 85–91.

Lin, J.T., Snyder, L.R.; McKeon, T.A. Prediction of relative retention times of triacylglycerols in non-aqueous reversed-phase high-performance liquid chromatography. Journal of Chromatography A, **1998A**, 808, 43–49.

Lin, J.T., McKeon, T.A., Goodrich-Tanrikulu, M.; Stafford, A.E., Characterization of oleoyl-12-hydroxylase in castor microsomes using the putative substrate, 1-acyl-2-oleoyl-*sn*-glycero-3-phosphocholine. *Lipids*, **1996**, 31: 571–577.

Lin, J.T., Woodruff, C.L., Lagouche, O.J., McKeon, T.A., Stafford, A.E. Goodrich-Tanrikulu, M., Singleton, J.A.; Haney, C.A. Biosynthesis of triacylglycerols containing ricinoleate in castor microsomes using 1-acyl-2-oleoyl-*sn*-glycerol-3-phosphocholine as the substrate of oleoyl-12-hydroxylase. *Lipids*, **1998B**, 33, 59–69.

Lin, J.T., Woodruff, C.L.; McKeon, T.A. Non-aqueous reversed-phase high-performance liquid chromatography of synthetic triacylglycerols and diacylglycerols. *Journal of Chromatography A* 782, **1997**, 41–48.

Mancha, M.; Stymne, S. Remodelling of triacylglycerols in microsomal preparations from developing castor bean (*Ricinus communis* L.) endosperm. *Planta*, **1997**, 203, 51–57.

McKeon, T. A., Lin, J.T., Goodrich-Tanrikulu, M.; Stafford, A. E. Ricinoleate biosynthesis in castor microsomes. *Industrial Crops and Products*, **1997**, 6, 383–389.

McKeon, T.A., Lin, J.T., Goodrich-Tanrikulu, M.; Stafford, A.E. Genetic modification of oilseed crops to produce industrial chemicals. *in*: Fuller, G., McKeon, T.A. and Bills, D. (eds.) Agricultural Materials as Renewable Resources: Non-food and Industrial Applications. **1996**, ACS Symposium Series 647, American Chemical Society, Washington, DC. pp. 158–178

Moreau, R.A.; Stumpf, P.K. Recent studies of the enzymic synthesis of ricinoleic acid by developing castor beans. *Plant Physiol.*, **1981**, 67, 672–676.

Moreau, R.A.; Stumpf, P.K. Solubilization and characterization of an acyl-CoA A: O-lysophospholipid acyltransferase from the microsomes of developing safflower seeds. *Plant Physiol.*, **1982**, 69, 1293–1297.

Murphy, D.J., Richards, D., Taylor, R., Capdevielle, J., Guillemot, J., Grison, R., Fairbairn, D., Bowra, S. Manipulation of seed oil content to produce industrial crops. *Industrial Crops and Products*, **1994**, 3, 17–27.

Ohlrogge, J.; Browse, J. Lipid biosynthesis. *Plant Cell*, **1995**, 7, 957–970.

Richards, D.E., Taylor, R.D.; Murphy, D.J. Localization and possible substrate requirement of the oleate-12-hydroxylase of developing *Ricinus communis* seeds. *Plant Physiol. Biochem.*, **1993**, 31: 89–94.

Shanklin, J., Cahoon, E.B., Whittle, E., Lindqvist, Y., Huang, W., Schneider, G.; Schmidt, N. Structure-Function studies on desaturases and related hydroxylases. in John P. Williams, Mobasher U. Khan and Nora (eds.) Physiology, Biochemistry, and Molecular Biology of Plant Lipids, Kluwer Academic Publishers, Dordrecht, **1997** pp. 6–10.

Smith, M.A., Jonsson, L., Stymne, S.; Stobart, K. Evidence for cytochrome b_5 as an electron donor in ricinoleic biosynthesis in microsomal preparations from developing castor bean (*Ricinus communis* L). *Biochem. J.*, **1992**, 287, 141–144.

Stahl, U., Banas, A.; Stymne, S. Plant microsomal phospholipid acyl hydrolases have selectivities for uncommon fatty acids. *Plant Physiol.*, **1995**, 107, 953–962.

Thorpe, S.C., Kemeny, D.M., Panzani, R.C., McGurl, B., Lord, M. Allergy to castor bean. II. Identification of the major allergens in castor bean seeds. *J. Allergy Clin. Immunol.* **1988**, 82: 67–72.

Van de Loo, F.J., Broun, P., Turner, S.; Somerville, C. An oleate 12-hydroxylase from *Ricinus communis* L. is a fatty acyl desaturase homolog. *Proc. Natl. Acad. Sci. USA*, **1995**, 92, 6743–6747.

Vogel, G.; Browse, J. Cholinephosphotransferase and diacylglycerol acyltransferase. *Plant Physiol.*, **1996**, 110, 923–931.

Yamada, M.; Stumpf, P.K. Enzymic synthesis of ricinoleic acid by extracts of developing *Ricinus communis* L. seeds. *Biochem. Biophys. Res. Comm.* **1964**, 14, 165–171.

5

MONOTERPENES IN ESSENTIAL OILS

Biosynthesis and Properties

Herminia Loza-Tavera*

Departamento de Bioquímica
Facultad de Química
Universidad Nacional Autónoma de México
Cd. Universitaria, México, D.F. 04510 México

ABSTRACT

Monoterpenes are compounds found in the essential oils extracted from many plants, including fruits, vegetables, spices and herbs. These compounds contribute to the flavor and aroma of plant from which they are extracted. Monoterpenes are acyclic, monocyclic, or bicyclic C_{10} compounds synthesized by monoterpene synthases using geranyl pyrophosphate (GPP) as substrate. GPP is also the precursor in the synthesis of farnesyl pyrophosphate (FPP) and geranyl-geranyl pyrophosphate (GGPP), two important compounds in cell metabolism of animals, plants and yeast. Monoterpene cyclases produce cyclic monoterpenes through a multistep mechanism involving a universal intermediate, a terpinyl cation which can be transformed to several compounds. Experimental studies, using animal cancer models, have demonstrated that some monoterpenes possess anticarcinogenic properties, acting at different cellular and molecular levels. From these discoveries it seems clear that monoterpenes could be considered as effective, nontoxic dietary antitumorigenic agents that hold promise as a novel class of anticancer drugs.

INTRODUCTION

Essential oils (EO) are mixtures of compounds characterized by their capacity to generate flavor or aroma. Generally EO are obtained from spices, aromatic herbs, fruits or flowers. Many of them are used for seasoning and flavoring food, as basis of perfumes and

* telephone (5)622 5280; Fax (5)622-5329; e-mail hlozat@servidor.unam.mx

Chemicals via Higher Plant Bioengineering, edited by Shahidi *et al.*
Kluwer Academic / Plenum Publishers, New York, 1999.

other household items, and in aromatherapy for the treatment of several affections. EO can be extracted from the whole plant or a part of it, and although the method of extraction is very important to yield an essential oil capable of producing "almost as the raw plant" aroma (Moyler, *et al.*, 1994), the plant biological component is the decisive factor to obtain an essential oil with specific organoleptic characteristics (Irving and Adams, 1973). However, the biochemical and genetic aspects involved in the synthesis of different compounds present in EO have been barely studied. It is also important to note that other factors such as climate, soil and growing conditions influence the quality and concentration of components of an essential oil (Müller-Riebau *et al.*, 1997). Thus, to ensure consistency in a formulated product containing natural extracts, it is important to specify the country of origin of a plant material ingredient as well as its botanical name (Moyler, 1994).

As previously mentioned, EO can be extracted using different processes which must result in a minimal chemical change of the compounds present in the oil, in order to maintain its natural aroma. The economy of the process and the yield and recovery of active components are important. Some techniques of interest for extraction of aroma compounds are steam and water destillation, hydrodiffusion, and water infussion as well as use water or steam as solvents. Other methods use alcoholic tinctures, extraction of oleoresins, using organic solvents such as methanol, ethanol, isopropanol, ethyl acetate or acetone (Moyler, 1994). However, the use of non-toxic solvents during the extraction process is an important issue because of regulatory restrictions (Marion *et al.*, 1994). New techniques using supercritical or subcritial CO_2 have been developed with significant advantages compared to the use of organic solvents (Moyler *et al.*, 1994; Meireles and Nikolov, 1994). Methods such as gas- and high-performance liquid chromatography, mass spectrometry and nuclear magnetic resonance have been used to determine the composition of oils, the quantities present in the extracts, as well as the nature of the oils components (Tateo *et al.*, 1994).

Analysis of EO show that, of the different compounds forming them, terpenoids are most abundant and are present either as hemiterpenes, monoterpenes or sesquiterpenes, and as their derivatives like alcohols, esters, acetates, and others. Terpenoids form a unique group in the sense of the range and diversity of compounds they represent. Structural types and their derivatives comprise thousands of compounds, being the vastest group in nature (Connolly and Hill, 1992). In many spices, monoterpenes and their derivatives are by far the most abundant compounds in the EO, being responsible for the characteristic flavor and aroma the plant possesses.

Monoterpenes in their natural environment give different properties to the plant in which they are present. Citral, for example, is an important constituent of the smell of lemon. Thymol is involved in the flavor of mandarin oranges. Other monoterpenes, such as limonene and geranyl are constituents of flower scents and attract plant pollinators. High concentrations of monoterpenes in plants will repel many potential predators, but may attract other animals. Animals use citronellal, citral, and α- and β-pinenes as feeding deterrents and geraniol, geranyl esters, myrcene, and terpinolene as pheromones. Some others, such as 1,8-cineole and camphor, are involved in plant-plant interactions. Some commercially important monoterpenes are menthol, camphor, carvone, thymol, fenchone and α-pinene (Luckner, 1984). Recently application of monoterpenes as anticancer drugs based on the studies showing this effect in animal models, has been demonstrated (Gould, 1995).

In this chapter a short review is presented about some of the main monoterpenes found in essential oils, the biosynthetic reactions involved in their production, and their properties as antimicrobial agents and anticancer compounds.

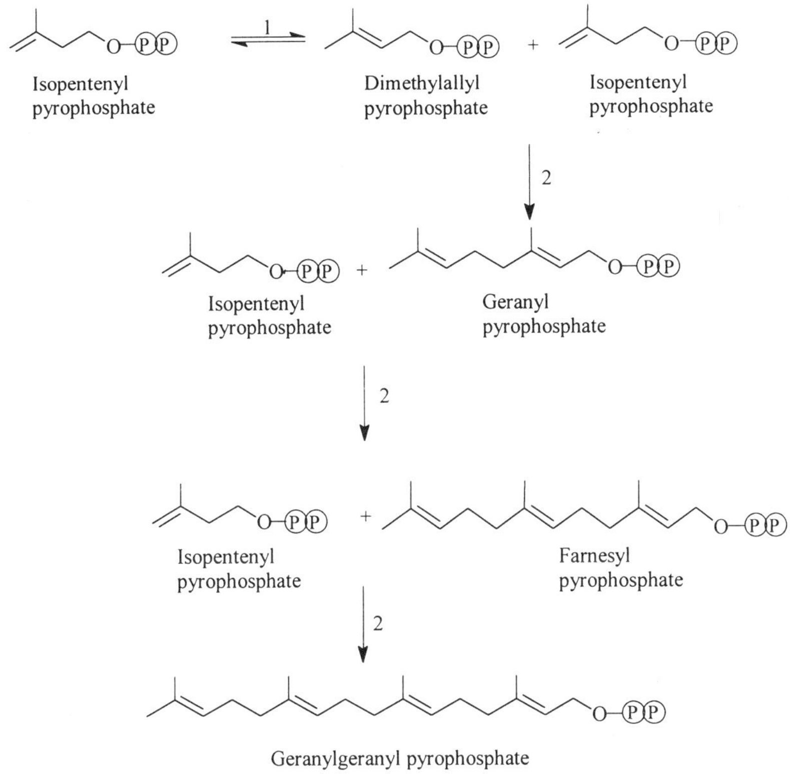

Figure 1. Synthesis of prenyl phosphates. Prenyl phosphates are synthesized by polymerization of isopentenyl pyrophosphate units. 1. Isopentenyl pyrophosphate isomerase; 2. Prenyl transferase.

SYNTHESIS OF MONOTERPENES IN PLANTS

Monoterpenes are terpenoid compounds that are produced from isoprene (2-methylbutadiene) units. Isoprene is synthesized as an activated molecule, isopentenyl pyrophosphate (IPPP), from Acetyl-CoA through a series of steps which includes an important intermediate molecule, mevalonic acid, synthesized by the widely studied enzyme hydroxymetyl glutaryl-Coenzyme A (HMG-CoA) Reductase. By the action of prenyltransferases, IPPP polymerizes to form prenyl pyrophosphates (Figure 1). The synthesis of prenyl pyrophosphates begins with the formation of the IPPP isomer, dimethylallyl pyrophosphate (DAPP), and continues with successive additions of IPPP units. The head to tail condensation of DAPP and IPPP produces geranyl pyrophosphate (GPP). Successive additions to GPP of IPPP units produce farnesyl pyrophosphate (FPP) and geranyl, geranyl pyrophosphate (GGPP) (Luckner, 1984). The metabolic pathway for the synthesis of prenyl phosphates is very important in cell metabolism, since it provides the substrates for the synthesis of a wide variety of compounds (Figure 2). Among these are the heme A group, cholesterol, and other steroids in animal cells and terpenoids, carotenoids, and abscicic acid in plants. Moreover, FPP and GGPP are molecules involved in post-translational modifications of key proteins functioning in cell growth, such as the cell cycle controlling protein, $p21^{ras}$, G proteins, and nuclear lamins. The farnesylation or geranyl-geranylation

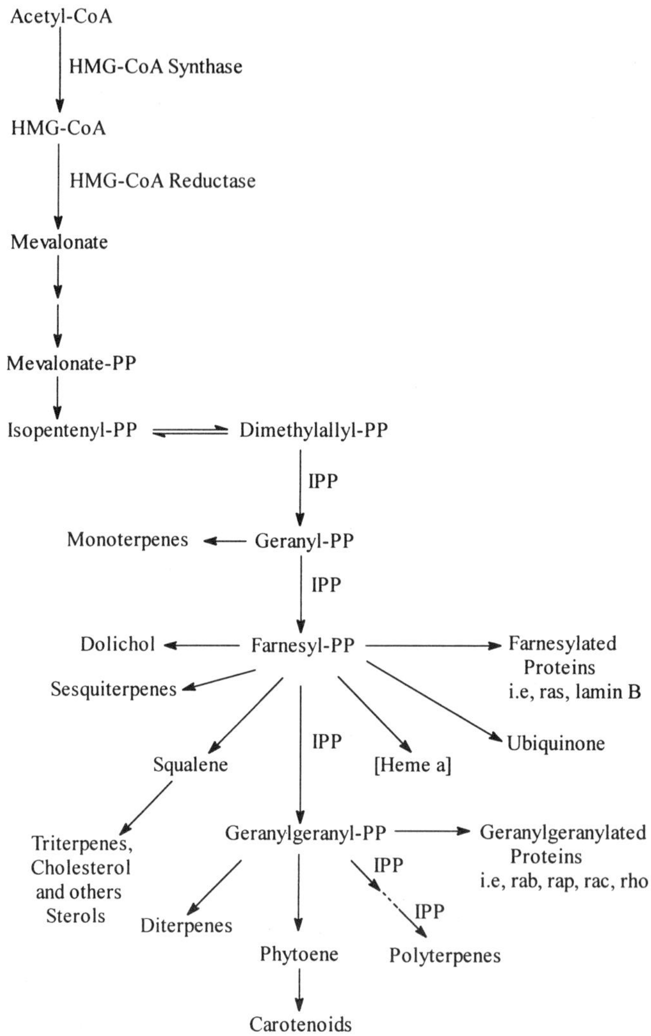

Figure 2. Mevalonate pathway for the synthesis of isoprenoids. HMG-CoA, Hydroxymethyl glutaryl Coenzyme A; IPP, isopentenyl pyrophosphate.

work as signals that facilitate the anchoring of the modified proteins to the membrane of several cellular compartments (Casey, 1995).

Recently, it has been demonstrated that a mevalonate-independent pathway for isoprene biosynthesis operates in several eubacteria and green algae (Rohmer *et al.*, 1993). In this pathway, pyruvate and glyceraldehyde 3-phosphate (GAP), instead of Acetyl CoA, are the substrates for the synthesis of isoprene. In plastids, some evidence has arisen showing that several isoprenoids of plants, including carotenoids and the prenyl side chains of chlorophyll and plastoquinone (Lichtenthaler *et al.*, 1997) as well as isoprene, monoterpenes (Eisenreich *et al.*, 1997) and diterpenes (Eisenreich *et al.*, 1996) are synthesized via the mevalonate independent pathway. In this pathway, decarboxylated pyruvate is bound to GAP by means of an intermediate thiamin pyrophosphate compound, producing 1-de-

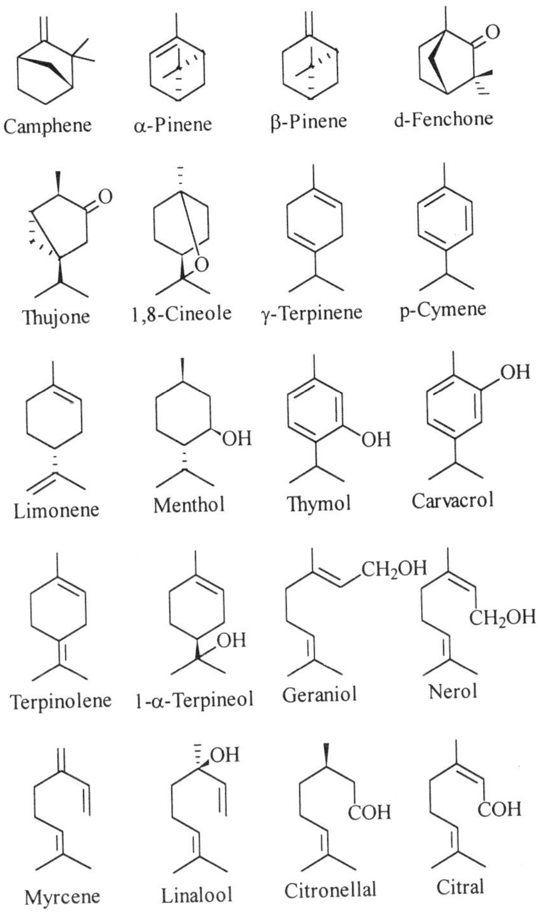

Figure 3. Chemical structures of some important monoterpenes.

oxyxylulose-5 phosphate (DXP). This compound, after several transformations, produces isoprene. The enzyme producing DXP, DXP synthase, is a transketolase. The gene encoding DXP synthase has been isolated from peppermint (Lange *et al.*, 1998) and *E. coli* (Lois *et al.*, 1998). Interestingly this gene has previously been isolated in an albino mutant of *Arabidopsis thaliana* with chloroplast morphology altered (*Cla* 1). In this mutant, chloroplast development is arrested at an early stage, suggesting the importance of the mutated gene product in chloroplast development (Mandel *et al.*, 1996).

Monoterpenes are typical products of plant metabolism, but are also synthesized in animal cells and in few microorganisms. They are aliphatic or cyclic isoprenoid C_{10} compounds, having GPP as precursor. Hydrolytic cleavage of the pyrophosphate bond of GPP produces geraniol or its cis-isomer nerol, which is synthesized via the transient formation of the corresponding aldehyde. Derivatives of these two compounds constitute the acyclic monoterpenes with different substitutent groups (e.g. nerolic acid, myrcene, linalool, citronellal, citral, ocimene and geranic acid). Cyclization of GPP or nerol pyrophosphate by polyfunctional enzymes (carbocyclases) produce mono- (e.g. limonene, γ-terpinene, terpinolene, 1-α-terpineol, 1,8-cineole, p-cymene, menthol, thymol, carvacrol) or bicyclic (e.g. pinene, camphene, sabinene, d-fenchone, thujone, borneol) monoterpenes having different substitutents (Figure 3).

ENZYMATIC MECHANISM FOR THE SYNTHESIS OF MONOTERPENES

It is well established that GPP, the ubiquitous C_{10} intermediate of the general isoprenoid pathway, functions as the acyclic precursor of most monoterpenes (Croteau, 1987). The enzyme responsible for the synthesis of this compound, the geranyl pyrophosphate synthase has been characterized from sage oil glands (Croteau and Purkett, 1989). Enzymes called monoterpene synthases (also termed cyclases when they catalyze the formation of cyclic products) catalyze the transformation of GPP to monoterpenes. A monoterpene synthase involved in the synthesis of linalool, an acyclic monoterpene, has been isolated and characterized from *Clarkia breweri* flowers (Pichersky *et al.*, 1995).

An unusual feature of many monoterpene synthases is their ability to synthesize multiple products from the GPP substrate. For example, pinene synthases purified from leaves of common sage (*Salvia officinalis*) generate multiple olefins at the same active site (Croteau *et al.*, 1987; 1988a; Gambliel and Croteau, 1984). (+)-Pinene synthase (cyclase I) catalyses the conversion of GPP to (+)-α-pinene and (+)-camphene and to lesser amounts of the stereochemically related monocyclic olefin (+)-limonene and several acyclic olefins. (−)-Pinene synthase (cyclase II) catalyzes the transformation of GPP to (−)-α-pinene and (−)-β-pinene and to the configurationally related products (−)-camphene and (−)-limonene. Pinene synthase III, which has been isolated from seeds, produces (+)-α-pinene (36%), (+)-β-pinene (45%) and about 20% monocyclic and acyclic olefins (Wagschal *et al.*, 1994).

From studies with different cyclases responsible for the formation of bicyclic monoterpenes of the pinane, bornane, fenchane, camphane, and thujane families, a general stereochemical model for this reaction type was proposed. This mechanism has been shown to work also in some monocyclic monoterpene synthases as limonene synthase (Rajaonarivony *et al.*, 1992) and 1,8-cineole synthase (Croteau *et al.*, 1994). In this proposed mechanism, monoterpene synthases catalyze the conversion of the acyclic precursor GPP to the cyclic parents of the various monoterpene skeletal types (Figure 4). This multistep transformation of GPP is considered to involve the initial ionization of the primary allylic ester to generate an ion pair that, by suprafacial rearrangement, yields the enzyme-bound tertiary allylic isomer, linalyl pyrophosphate (LPP). Rotation about the newly generated C2-C3 single bond to the cis conformer overcomes the original stereochemical restriction against cyclization of GPP, while subsequent ionization of this tertiary allylic isomer allows C1-C6 cyclization of the *anti,endo*-conformer generating a monocyclic α-terpinyl cation. Further reactions from this universal intermediate may involve additional electrophilic cyclizations and/or rearrangements before termination of the cationic reaction sequence. Variations of this mechanism are considered to account for the formation of some 30–40 individual cyclic products from GPP under the catalysis of a nearly equal number of distinct cyclases (Croteau 1987; Croteau *et al.*, 1986; 1988a; 1988b; 1989; 1990).

Acyclic monoterpenes synthases have been studied to a lesser extent than cyclases. From studies on linalool synthase, it was initially proposed that the synthesis of linalool, having as substrate GPP, proceeded through the intermediate linalyl pyrophosphate (LPP), as occurs in monoterpene cyclases. However, it was demonstrated that LPP is not the intermediate, although there is a common intermediate in both reactions. This intermediate is the linalyl cation which, by the capture of water, is transformed to linalool directly, without the involvement of LPP. This reaction is considerably simpler in the acyclic monoterpene synthases than the coupled isomerization-cyclization sequence catalyzed by monoterpene cyclases (Pichersky *et al.*, 1995).

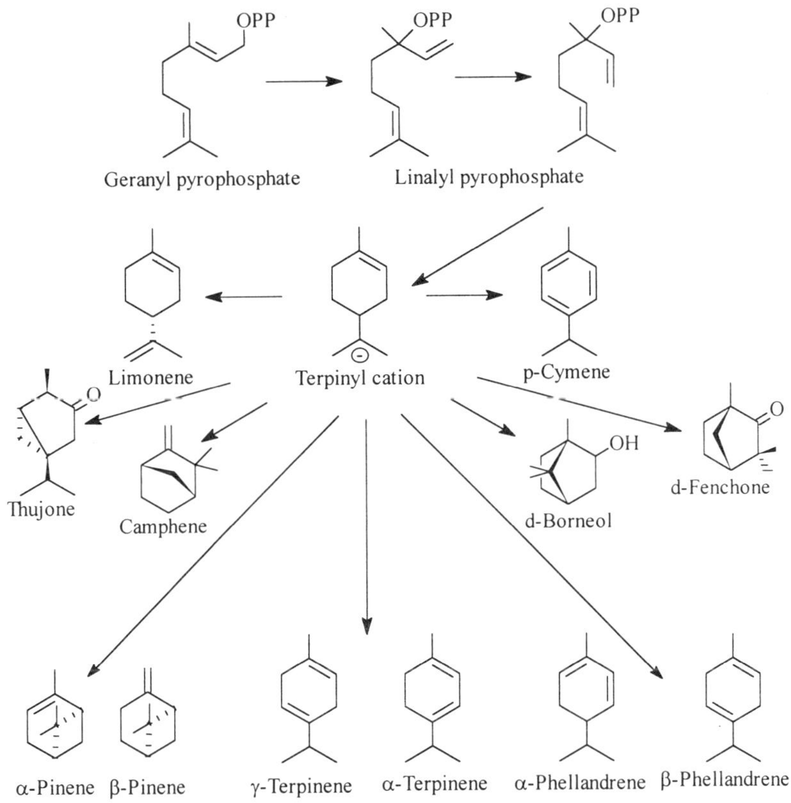

Figure 4. Schematic representation of reactions involved in the synthesis of mono and bicyclic monoterpenes. Geranyl pyrophosphate is the precursor that through an intermediate terpinyl cation is transformed to different monoterpenes by the action of several monoterpenes cyclases.

PLANT MONOTERPENE SYNTHASES

For a long time it was thought that isoprene was not synthesized in plants. However, it has been demonstrated that isoprene is expelled to the atmosphere by plant leaves. This is a natural process that contributes more hydrocarbon to the atmosphere than all of the anthropogenic sources (Sharkey, 1996). Monoterpenes and some of their biosynthetic enzymes have been found in the nonparticulate fraction of plastids (Luckner, 1984). Although there are several open reading frames whose products have not been identified, the analyses of the chloroplast genomes sequenced until now, which include maize, rice, tobacco, *Pinus thunbergii*, *Epifagus virginiana*, and some algi, show that any enzyme involved in monoterpene or isoprene biosynthesis is encoded in the plastid genome. Monoterpene synthases and other enzymes involved in the isoprene pathway must be nuclear encoded, synthesized in the cytoplasm and targeted to plastids (Luckner, 1984). This has been confirmed by the isolation of limonene synthase and linalool synthase genes from genomic libraries. Other specific enzymes for the synthesis of essential oils have been found free in vacuoles, cytoplasm and extra-cytoplasmic space (Luckner, 1984).

Several monoterpene synthases have been isolated from *Mentha spicata* and *Salvia officinalis*. The higher plant monoterpene cyclases appear to be of two general types: homodimers of M_r ~100,000 and monomers in the range M_r ~50,000–70,000 (Alonso and Croteau, 1993). The molecular weight for linalool synthase isolated from *Clarkia breweri* has been determined to be between 73 and 90 kDa (Pichersky *et al.*, 1995).

The genes encoding 4S-limonene synthases from spearmint (*Mentha spicata*) (Colby *et al.*, 1993) and *Perilla frutescens* (Yuba *et al.*, 1996) have been isolated. (–)-Limonene is the cyclic olefin intermediate in the biosynthesis of *p*-menthane monoterpenes in *Mentha* species, including the characteristic components of the essential oil of peppermint (i.e. (–)-menthol) and the essential oil of spearmint (i.e. (–)-carvone). A linalool synthase gene has also been isolated from *Clarkia breweri* flower (Dudareva *et al.*, 1996). The cDNAs encoding the previously mentioned genes have homology to the epi-aristolochene synthase, a sesquiterpene synthase from tobacco (Facchini and Chapell, 1995) and the casbene synthase, a diterpene cyclase from castor bean (Mau and West, 1994). All of these sequences have a DDXXD domain, which has been proposed to be a metal ion-substrate complex binding domain. Similarities in amino acid sequence with limonene synthase, aristolochene synthase, and casbene synthase were also observed when a peptide obtained after cianogen bromide treatment of a purified protein fraction containing monoterpene cyclase activities was compared. This fraction purified from culinary sage contained pinene synthase and bornyl pyrophosphate synthase (McGeady and Croteau, 1995). These similarities are consistent with the idea that an active site peptide is conserved in enzymes involved in terpenoid biosynthesis.

ANTIMICROBIAL PROPERTIES OF ESSENTIAL OILS AND THEIR MONOTERPENES

It has been observed that EO from many spices and aromatic herbs have antimicrobial properties. Detailed analyses using the components of the oils revealed that these properties reside in part on the monoterpene constituents; several studies have confirmed these observations (for an extensive review see Nakatani, 1994). However, the mode of action of monoterpenes as antimicrobial agents has not yet been elucidated.

Thirteen EO were tested against the mycelial growth of, and aflatoxin synthesis by *Aspergillus parasiticus*. Concentrations as low as 0.1% of cinnamon, thyme, oregano, and cumin EO were able to stop mycelial growth, while curcumin, ginger, lemon, and orange EO were unable to totally inhibit the growth even at a 1% concentration. Coriander, black pepper, mugwort, bay, and rosemary EO stop the mycelial growth at concentrations between 0.2 and 1%. The EO, most active upon mycelial growth, were also most active against aflatoxinogenesis. However, aflatoxin synthesis was inhibited by all the EO to a higher extent than mycelial growth (Tantaoui-Elaraki and Beraoud, 1994).

It has been postulated that the use of spices in food processing was adopted because of taste and also because they prevented the spoilage of food. This aspect has been analyzed in a work where the antibacterial activity of some EO was examined against bacteria involved in meat spoilage. EO of clove, cinnamon, pimento, and rosemary were found to be the most active inhibitors. A relationship was found between the inhibitory effect of EO and the presence of eugenol and cinnamaldehyde (Ouattara *et al.*, 1997).

Monoterpens from cyclic and acyclic configuration (cineole, citral, geraniol, linalool and menthol) were tested against different bacterial strains including Gram-positive cocci and rods, Gram-negative rods, and yeast-like and filamentous fungi. The analysis showed that linalool exerted the most antimicrobial effect, followed by cineole, geraniol and the aromatic compounds menthol and citral. However, the citral and geraniol oils were the most effective against fungi, followed by linalool, cineole and menthol (Pattnaik *et al.*, 1997). EO of thyme or its constituent thymol has also been shown to be effective in decreasing the viable counts of *Salmonella typhimurium* and *Staphylococcus aureus*, especially under anaerobic conditions. In contrast to the phenolic constituents of the oil, thymol and carvacrol, the chemically related terpenes p-cymene and γ-terpinene had no antagonistic effects against *S. typhimurium* (Juven *et al.*, 1994).

ANTICARCINOGENIC PROPERTIES OF SOME MONOTERPENES

Several monoterpenes have been shown to have preventive and anticarcinogenic effects on mammary, liver, lung and other types of cancers. The preventive action in cancer development occurs either at the initiation or progression stages. The anticarcinogenic effect has been observed to cause complete regression of the cancer (Crowell *et al.*, 1996). Several cellular, metabolic and molecular activities are associated with anticarcinogenic effect of monoterpenes. Based on the therapeutic efficacy of monoterpenes in experimental animal systems, clinical evaluation of this class of compounds has begun in advanced cancer patients (Gould, 1995).

Monoterpene Structure-Antitumorigenic Activity Relationships

The unsubstituted monocyclic monoterpene S-limonene and some of its hydroxylated derivatives such as perillyl alcohol, carveol, menthol, uroterpenol and sobrerol have been shown to have chemotherapeutic activity against liver, pancreatic, mammary, and prostatic tumors, and preventive activity against a 7,12-dimethylbenz[α]anthracene (DMBA)-induced mammary cancer (Crowell *et al.*, 1992; 1996). Studies on DMBA-induced mammary cancer model to elucidate the antitumorigenic activity of other monoterpene types, have shown that acyclic (linalool and β-myrcene) and bicyclic monoterpenes (1,8-cineole and (+/–)-alpha-pinene) did not have significant chemopreventive activities, as the monocyclic compounds shown (Russin *et al.*, 1989). It has also been observed that hydroxylation of monoterpenes affects chemopreventive potential, thus presence of two hydroxyl groups provides for a more effective anticarcinogen than one hydroxyl group which is still more effective than when no hydroxyl group in present (Crowell *et al.*, 1992). More recent studies have shown that other isoprenoid related compounds such as farnesol and some tocotrienols are also inhibitors of cell proliferation in a melanoma model (He *et al.*, 1997).

Effect of Monoterpenes in Apoptosis and Differentiation

Recent evidence supports the concept that tumor growth depends on evasion of normal homeostatic control mechanisms that operates through induction of cell death by apoptosis (Hall *et al.*, 1994). By this reason, several studies on monoterpene effect on tumor regression have been focused to analyze apoptosis in tumor cells. The effect of feeding perillyl alcohol (PA) to rats with azoxymethane-induced colon carcinogenesis was studied, using light microscopy, in order to analyze colon tumors to detect apoptosis. It was observed that diets of 1g of PA/kg diet inhibit the incidence and multiplicity of invasive colon adenocarcinomas, whereas 2g/kg diet totally inhibited them. These studies also showed that the colon tumors of animals fed PA, exhibited increased apoptosis as compared to those fed the control diet (Reddy *et al.*, 1997). In another study, liver tumors from animals treated with PA showed a 10-fold decrease in tumor weight compared to those of untreated animals. As in the previously described research, PA treatment did not influence tumor cell proliferation, instead it increased the apoptotic index approximately 10-fold (Mills *et al.*, 1995). It has also been observed that, in a neuroblastoma line, treatment with limonene and PA induced differentiation of neuroblastoma cells to the Neuro-2A cell line (Shi and Gould, 1995).

The differentiation and growth of normal and tumor tissues have been shown to be regulated by numerous growth factors from which transforming growth factor beta (TGF-β) has been suggested to inhibit the growth of breast cancer (Jirtle *et al.*, 1993) and liver tumors (Mills *et al.*, 1995). TGF-β is secreted from cells as a latent complex containing phosphomannosyl residues and it has to be activated by proteolysis, releasing the factor from this complex. This activation can be provided by the mannose 6-phosphate/insulin-like growth factor II (M6P/IGF-II) receptor. This receptor has a dual function, to activate TGF-β a potent mitoinhibitor, which has been suggested as a signal for induction of cell death by apoptosis (Bursch *et al.*, 1993), and to degrade a potent mitogen, the IGF-II. After TGF-β is activated, it binds to TGF-β receptors where the signal transduction pathway begins (Wrana *et al.*, 1994). Analysis of the effect of PA on mammary tumors showed that regressing tumors have an increased cellular concentration of TGF-β and M6P/IGF-II receptor (Jirtle, *et al.*, 1993). In a more detailed analysis on liver tumors, it was observed that mRNA levels for M6P/IGF-II and TGF-β receptors increased significantly in the tumors of PA-treated animals, compared to the levels in the normal tissue surrounding the tumors and in the tumors of untreated animals (Mills *et al.*, 1995).

Effect of Monoterpenes in Farnesylation

The post-translational modification of cellular proteins by isoprenylation, either by farnesylation or geranylgeranylation, is an important mechanism by which their subcellular localization is specified. Interference of this post-translational modification has shown to perturb cellular physiology. Avoiding isoprenylation of the small G protein-activated *ras*, by mutagenesis on the farnesylation site, this oncogenic protein is unable to cause cellular transformation (Jackson *et al.*, 1990). Based on this observation it has been postulated that inhibition of small G proteins prenylation could be the basis of the anti-tumor effects of limonene and PA. In a study to test the ability of limonene and its oxygenated analogs to inhibit protein prenylation enzymes *in vitro*, limonene and PA and their major *in vivo* metabolite, perillic acid, were weak inhibitors of both mammalian and yeast farnesyl transferase (FT) and geranylgeranyl transferase. In contrast, a minor metabolite of limonene and PA, perillic acid methyl ester, was a potent inhibitor of both enzymes. Perillic acid methyl ester is a competitive inhibitor of yeast FT with respect to farnesyl pyrophosphate. These studies suggest that if the inhibition of protein prenylation is a mechanism for limonene and PA anti-cancer activities, these monoterpenes may be prodrugs that are converted into pharmacologically-active substances by metabolic modification (Crowell *et al.*, 1992; Gelb *et al.*, 1995).

When a more detailed study was performed analyzing not only farnesylated/ unfarnesylated *ras*, but also the levels of newly synthesized *ras*, it was determined that PA did not affect the levels of farnesylated *ras* by inhibiting FT, but by increasing *ras* turnover either decreasing *ras* transcription/translation or increasing its degradation (Hohl and Lewis, 1995).

However, in another study, the growth inhibitory effect of limonene and other related monoterpenes on PANC-1 pancreas carcinoma cells (carrying a K-*ras* mutation) and on 12V-H-*ras*-transformed rat fibroblasts was examined. PA, 7-methyl-perillyl alcohol, limonene oxide and perillic acid methyl ester were all found to efficiently inhibit cell growth at 1 mM, whereas limonene caused an approximately 50% growth reduction at 5 mM. However, BZA-5B, an inhibitor of *ras* FT, was found to induce morphological reversion of 12V-H-*ras*-transformed cells, PA and limonene did not induce reversion. Furthermore, monoterpenes did not decrease MAP kinase enzyme activity or collagenase promoter activity in PANC-1 cells; two functions known to be down-stream from *ras*. The conclusion of this work was that, although effective in inhibiting the growth of tumor cells harboring activated *ras* oncogenes, limonene and PA are unlikely to act by inhibiting *ras* function (Karlson *et al.*, 1996).

Effect of Monoterpenes in Hepatic Detoxifying Enzymes

Other effects of monoterpenes have been observed in hepatic detoxifying enzymes. β-myrcene (MYR), an acyclic monoterpene found in the essential oils of several plants such as lemongrass (*Cymbopogon citratus*), hop, bay, verbena and others, as well as lemongrass oil, blocked the metabolic activation of some promutagens (e.g., cyclophosphamide and aflatoxin B1) in *in vitro* genotoxicity assays. The potent inhibitory effects of several monoterpenes on CYP4502B1 suggest that they interfere with the metabolism of xenobiotics which are substrates for this isoenzyme (De-Oliveira *et al.*, 1997a). The evaluation of the inhibitory effects of MYR and some other monoterpenoid compounds on microsomal enzymes involved in the activation of genotoxic substances showed that this compound is an inducer of isoenzymes belonging to CYP2B monoxygenase subfamily, specifically pentoxy-resorufin O-dealkylase and benzyloxy-resorufin O-dealkylase (De-Oliveira *et al.*, 1997b).

CONCLUDING REMARKS

Essential oils and their monoterpenes are "old compounds" with "new" interesting effects on human physiology. Many of them have been used for a long time in different cultures for seasoning meals and because of their antimicrobial and antioxidant effects. The discovery of a new pathway for the synthesis of isoprene in bacteria and plastids has opened a new insight in the study of these compounds. However, very little is known about the regulatory mechanisms controlling the activities of the several enzymes involved in the biosynthetic pathways, but less is known about the many genes encoding the enzymes and how these genes are regulated. Also, studies on monoterpene degradation mechanisms have just begun. On the other hand, the function of many monoterpene compounds in plant metabolism and the way they are involved in the interactions between organisms and the environment are still to be elucidated. The fact that monoterpenes have been found to have anticancer properties in animal models, make these compounds interesting subjects for more detailed studies on their mode of action as possible chemopreventive and anticancer compounds.

ACKNOWLEDGMENTS

I am grateful to Sandra Robles Hernández for drawing of the figures and to Gustavo Fuentes for critical review of the manuscript.

REFERENCES

Alonso, W. R. and Croteau, R. Prenyl transferases and cyclases In: *Methods in Plant Biochemistry*. Lea, P. J., Ed.; Academic Press Ltd., London, England **1993**, 239–260.

Bursch, W.; Oberhammer, F.; Jirtle, R. L.; Askari, M.; Sedovu, R.; Grasl-Kraupp, B.; Purchio, A. F.; Schulte-Hermann, R. Transforming growth factor-β1 as signal for induction of cell death by apoptosis. *Br. J. Can.* **1993**, *67*, 531–536

Casey, P. J. Protein lipidation in cell signaling. *Science*. **1995,** *268*, 221–224.

Colby, S. M.; Alonso, W. R.; Katahira, E. J.; McGarvey, D. J.; Croteau, R. 4S-limonene synthase from the oil glands of spearmint (*Mentha spicata*). cDNA isolation, characterization, and bacterial expression of the catalytically active monoterpene cyclase. *J. Biol. Chem.* **1993**, *268*, 23016–23024.

Connolly, J. D.; Hill, R. A. *Dictionary of Terpenoids*. Chapman & Hall, New York, **1992**.

Croteau, R. Biosynthesis and catabolism of monoterpenoids. *Chem. Rev.* **1987**, *87*, 929–954.

Croteau, R.; Purkett, P. T. Geranyl pyrophosphate synthase: Characterization of the enzyme and evidence that this chain-length specific prenyltransfarase is associated with monoterpene biosynthesis in sage (*Salvia officinalis*). *Arch. Biochem. Biophys.* **1989**, *271*, 524–535.

Croteau, R.; Alonso, W. R.; Koepp A. E.; Johnson, M. A. Biosynthesis of Monoterpenes. Partial purification, characterization, and mechanism of action of 1,8-cineole synthase. *Arch. Biochem. Biophys.* **1994**, *309*, 184–192.

Croteau, R.; Gershenzon, J.; Wheeler, C. J.; Satterwhite, D. M. Biosynthesis of Monoterpenes. Stereochemistry of the coupled isomerization and cyclization of geranyl pyrophosphate to camphene and isocamphene monoterpenes. *Arch. Biochem. Biophys.* **1990**, *277*, 374–381.

Croteau, R; Satterwhite, D. M.; Cane, D. E, Chang C. C. Biosynthesis of Monoterpenes. Enantioselectivity in the enzymatic cyclizations of (+)- and (–)-linalyl pyrophosphate to (+)- and (–)-bornyl pyrophosphate. *J. Biol. Chem.* **1986**, *261*, 13438–13445.

Croteau, R.; Satterwhite, D. M.; Cane, D. E.; Chang, C. C. Biosynthesis of Monoterpenes. Enantioselectivity in the enzymatic cyclizations of (+)-linalyl pyrophosphate to (+)- and (–)-pinene and (–)-camphene. *J. Biol. Chem.* **1988a**, *263*, 10063–10071.

Croteau, R.; Satterwhite, D. M.; Wheeler, C. J.; Felton, N. M. Biosynthesis of Monoterpenes. Stereochemistry of the enzymatic cyclizations of geranyl pyrophosphate to (–)-*endo*-fenchol. *J. Biol. Chem.* **1988b**, *263*, 15449–15453.

Croteau, R.; Satterwhite, D. M.; Wheeler, C. J.; Felton, N. M. Biosynthesis of Monoterpenes. Stereochemistry of the enzymatic cyclizations of geranyl pyrophosphate to (+)-α-pinene and (–)-β-pinene. *J. Biol. Chem.* **1989**, *264*, 2075–2080.

Croteau, R. B.; Wheeler, C. J.; Cane, D. E.; Ebert, R.; Ha, H.-J. Isotopically sensitive branching in the formation of cyclic monoterpenes: Proof that (–)-α-pinene and (–)-β- pinene are synthesized by the same monotepene cyclase via deprotonation of a common intermediate. *Biochemistry* **1987**, *26*, 5383–5389.

Crowell, P. L.; Kennan, W. S.; Haag, J. D.; Ahmad, S.; Vedejs, E.; Gould, M. N. Chemoprevention of mammary carcinogenesis by hydroxylated derivatives of d-limonene. *Carcinogenesis* **1992**, *13*, 1261–1264.

Crowell, P. L.; Siar Ayoubi, A.; Burke, Y. D. Antitumorigenic effects of limonene and perillyl alcohol against pancreatic and breast cancer. *Adv. Exp. Med. Biol.* **1996**, *401*, 131–136.

De-Oliveira, A. C.; Ribeiro-Pinto, L. F.; Otto, S. S.; Goncalves, A.; Paumgartten, F. J. Induction of liver monooxygenases by beta-myrcene. *Toxicology* **1997a**, *124*, 135–140.

De-Oliveira, A. C.; Ribeiro-Pinto, L. F.; Paumgartten, J. R. *In vitro* inhibition of CYP2B1 monooxygenase by beta-myrcene and other monoterpenoid compounds. *Toxicol. Lett.* **1997b**, *92*, 39–44.

Dudareva, N.; Cseke, L.; Blanc, V. M.; Pichersky, E. Evolution of floral scent in *Clarkia*: Novel patterns of S- linalool synthase gene expression in the *C. breweri* flower. *The Plant Cell.* **1996**, *8*, 1137–1148.

Eisenreich, W.; Menhard, B.; Hylands, P. J.; Zenk, J. H.; Bacher, A. Studies on the biosynthesis of taxol: The taxane carbon skeleton is not of mevalonoid origin. *Proc. Natl. Acad. Sci. USA*. **1996**, *93*, 6431–6436.

Eisenreich, W.; Sagner, S.; Zenk, M. H.; Bacher, A. Monoterpenoid essential oils are not of mevalonoid origin. *Tetrahedron Lett.* **1997**, *38*, 3889–3892.

Facchini, P. J.; Chapell, J. Gene family for an elicitor induced sesquiterpene cyclase in tobacco. *Proc. Natl. Acad. Sci. USA*. **1992**, *89*, 11088–11092.

Gambliel, H.; Croteau, R. Pinene Cyclases I and II. Two enzymes form sage (*Salvia officinalis*) which catalyze stereospecific cyclizations of geranyl pyrophosphate to monoterpene olefins of opposite configuration. *J. Biol. Chem.* **1984**, *259*, 740–748.

Gelb, M. H.; Tamanoi, F.; Yokoyama, K.; Ghomashchi, F.; Esson, K.; Gould, M. N. The inhibition of protein prenyltransferases by oxygenated metabolites of limonene and perillyl alcohol. *Cancer Lett.* **1995,** *91,* 169–175.

Gould, M. N. Prevention and therapy of mammary cancer by monoterpenes. *J. Cell Biochem.* **1995,** Suppl *22,* 139–144.

Hall, P.; Coates, P. J.; Ansari, B.; Hopwood, D. Regulation of cell number in the mammalian gastrointestinal tract: the importance of apoptosis. *J. Cell Sci.* **1994,** *107,* 3569–3577.

He, L.; Mo, H.; Hadisusilo, S.; Qureshi, A. A.; Elson, C. E. Isoprenoids suppress the growth of murine B16 melanomas *in vitro* and *in vivo*. *J. Nutr.* **1997,** *127,* 668–674.

Hohl, R. J.; Lewis, K. Differential effects of monoterpenes and lovastatin on ras processing. *J. Biol. Chem.* **1995,** *270,* 17508–17512.

Irving, R. S.; Adams, R. P. Genetic and biosynthetic relationships of monoterpenes. In: *Terpenoids: Structure, biogenesis, and distribution.* Runeckles, V. C.; Mabry, T. J. Eds. Academic Press. New York and London. **1973,** 187–214.

Jackson, J. H.; Cochrane, C.G.; Bourne, J. R.; Solski, P. A.; Buss, J. E.; Der, C. J. Farnesol modification of Kirsten-ras exon 4B protein is essential for transformation. *Proc. Natl. Acad. Sci. USA.* **1990,** *87,* 3042–3046.

Jirtle, R.L.; Haag, J. D.; Ariazi, E. A.; Gould, M. N. Increased mannose 6-phosphate/insulin -like growth factor II receptor and transforming growth factor β1 levels during monoterpene-induced regression of mammary tumors. *Can. Res.* **1993,** *53,* 3849–3852.

Juven, B. J.; Kanner, J.; Schved, F.; Weisslowicz, H. Factors that interact with the antibacterial action of thyme essential oil and its active constituents. *J. Appl. Bacteriol.* **1994,** *76,* 626–631.

Karlson, J.; Borg-Karlson, A. K.; Unelius, R.; Shoshan, M. C.; Wilking, N.; Ringborg, U.; Linder, S. Inhibition of tumor cell growth by monoterpenes *in vitro*: Evidence of a Ras-independent mechanism of action. *Anticancer Drugs.* **1996,** *7,* 422–429.

Lange, B. M.; Wildung, M. R.; McCaskill, D.; Croteau, R. A family of tansketolases that directs isoprenoid biosynthesis via a mevalonate independent pathway. *Proc. Natl. Acad. Sci.* **1998,** *95,* 2100–2104.

Lichtenthaler, R.; Schwender, J.; Disch, A.; Rohmer, M. Biosynthesis of isoprenoids in higher plant chloroplasts proceeds via a mevalonate-independent pathway. *FEBS Lett.* **1997,** *400,* 271–274.

Lois, L. M.; Campos, N.; Putra, S. R.; Danielsen, K.; Rohmer, M.; Boronat, A. Cloning and characterization of a gene from *Escherichia coli* encoding a transketolase-like enzyme that catalyzes the synthesis of d-1-deoxyxylulose 5-phosphate, a common precursor for isoprenoid, thiamin, and pyridoxol biosynthesis. *Proc. Natl. Acad. Sci. USA.* **1998,** *95,* 2105–2110.

Luckner, M. *Secondary metabolism in microorganisms, plants, and animals.* Second Edition. Springer-Verlag, Berlin, Germany **1984,** 35–37, 41–42, 199–210.

Mandel, A.; Feldmann, K.; Herrera-Estrella, L.; Rocha-Sosa, M.; León, P. Cla1, a novel gene required for chloroplast development, is highly conserved in evolution. *Plant J.* **1996,** *9,* 649–658.

Mau, C. J. D.; West, C. A. Cloning of casbene synthase cDNA: Evidence for conserved structural features among terpenoid cyclases in plants. *Proc. Natl. Acad. Sci. USA.* **1994,** *91,* 8497–8501.

Marion, M. J. P.; Audrin, A.; Miagnial, L.; Brevard, H. Spices and their extracts: Utilization, selection, quality control and new developments. In: *Spices, Herbs and Edible Fungi*; Charalambous, G., Ed.; Elsevier. Science B.V. Amsterdam, The Netherlands **1994,** 71–95.

McGeady, P.; Croteau, R. Isolation and characterization of an active-site peptide from a monoterpene cyclase labeled with a mechanism-based inhibitor. *Arch. Biochem. Biophys.* **1995,** *317,* 149–155.

Meireles, M. A. A.; Nikolov, Z. L. Extraction and fractionation of essential oils with liquid carbon dioxide. In: *Spices, Herbs and Edible Fungi*; Charalambous, G., Ed.; Elsevier Science B.V. Amsterdam, The Netherlands **1994,** 1–70.

Mills, J. J.; Chari, R. S.; Boyer, I. J.; Gould, M. N.; Jirtle, R. L. Induction of apoptosis in liver tumors by the monoterpene perillyl alcohol. *Cancer Res.* **1995,** *55,* 979–983.

Moyler, D. A. Spices-Recent Advances. In: *Spices, Herbs and Edible Fungi*; Charalambous, G., Ed.; Elsevier Science BV. Amsterdam, The Netherlands. **1994,** 1–70.

Moyler, D. A.; Browing, R. M.; Stephens, M. A. CO_2 extraction of esential oils: Part V, nutmeg and mace oils. In: *Spices, Herbs and Edible Fungi*; Charalambous, G., Ed.; Elsevier Science BV, Amsterdam, The Netherlands **1994,** 145–170.

Müller-Riebau, F. J.; Berger, B. M.; Yegen, O; Cakir, C. Seasonal variations in the chemical compositions of essential oils of selected aromatic plants growing wild in Turkey. *J. Agric. Food Chem.* **1997,** *45,* 4821–4825.

Nakatani, N. Antioxidative and antimicrobial constituents of herbs and spices. In: *Spices, Herbs and Edible Fungi*; Charalambous, G., Ed.; Elsevier Science B.V. Amsterdam, The Netherlands **1994,** 251–271.

Ouattara, B; Simard, R. E.; Holley, R. A.; Piette, G. J.; Begin, A. Antibacterial activity of selected fatty acids and essential oils against six meat spoilage organisms. *Int. J. Food Microbiol.* **1997,** *37,* 155–162.

Pattnaik, S.; Subramanyam, V. R.; Bapaji, M.; Kole, C. R. Antibacterial and antifungal activity of aromatic constituents of essential oils. *Microbios.* **1997**, *89*, 39–46.

Pichersky, E.; Lewinsohn, E.; Croteau, R. Purification and characterization of S-linalool synthase, an enzyme involved in the production of floral scent in *Clarkia breweri*. *Arch. Biochem. Biophys.* **1995**, *316*, 803–807.

Rajaonarivony, J. I. M.; Gershenzon, J.; Croteau, R. Characterization and mechanism of (4S)- limonene synthase, a monoterpene cyclase from the glandular trichomes of peppermint (*Mentha* x *piperita*). *Arch. Biochem. Biophys.* **1992**, *296*, 49–57.

Reddy, B. S.; Wang, C. X.; Samaha, H.; Lubet, R.; Steele, V. E.; Kelloff, G. J.; Rao, C. V. Chemoprevention of colon carcinogenesis by dietary perillyl alcohol. *Cancer Res.* **1997**, *57*, 420–425.

Rohmer, M.; Knani, M.; Simonin, P.; Sutter, B.; Sahm, H. Isoprenoid biosynthesis in bacteria: a novel pathway for the early steps leading to isopentenyl diphosphate. *Biochem. J.* **1993**. *295*, 517–524.

Russin, W. A.; Hoesly, J. D.; Elson, C. E.; Tanner, M. A.; Gould, M. N. Inhibition of rat mammary carcinogenesis by monoterpenoids. *Carcinogenesis.* **1989**, *10*, 2161–2164.

Sharkey, T. D. Isoprene synthesis by plants and animals. *Endeavour.* **1996**, *20*, 74–78.

Shi, W.; Gould, M. N. Induction of differentiation in neuro-2A cells by the monoterpene perillyl alcohol. *Cancer Lett.* **1995**, *95*, 1–6.

Tantaoui-Elaraki; A.; Beraoud L. Inhibition of growth and aflatoxin production in *Aspergillus parasiticus* by essential oils of selected plant materials. *J. Environ. Pathol. Toxicol. Oncol.* **1994**, *13*, 67–72.

Tateo, R.; Ferrillo, A.; Orlandi, A. 1994. Analysis of natural flavourings: Problems and actual methods. In: *Spices, Herbs and Edible Fungi*; Charalambous, G., Ed.; Elsevier Science, B.V. Amsterdam, The Netherlands **1994**, 97–129.

Wagschal, K. C.; Pyun, H.-J.; Coates, R. M.; Croteau, R. Monoterpene biosynthesis: Isotope effects associated with bicyclic olefin formation catalyzed by pinene synthases from sage (*Salvia officinalis*). *Arch. Biochem. Biophys.* **1994,** *308*, 477–487.

Wrana, J. L.; Attisano, L.; Wieser, R.; Ventura, F.; Massagué, J. Mechanism of activation of the TGF-β receptor. *Nature (Lond).* **1994**, *370*, 341–347.

Yuba, A.; Yazaki, K.; Tabata, M.; Honda, G.; Croteau, R. cDNA cloning, characterization, and functional expression of 4S-(−)-limonene synthase from *Perilla frutescens*. *Arch. Biochem. Biophys.* **1996**, *332*, 280–287.

6

BIOCHEMICAL AND MOLECULAR TOOLS FOR THE PRODUCTION OF USEFUL TERPENE PRODUCTS FROM PEPPER (*CAPSICUM ANNUUM*)

Edmundo Lozoya-Gloria

Genetic Engineering Department
CINVESTAV-IPN / Irapuato Unit
P.O. Box 629, C.P.
36500 Irapuato, Gto., México

ABSTRACT

Among other natural products such as colorants and flavorants, natural fungicides like the pepper phytoalexin capsidiol, and the related biochemical pathways, may be used for practical approaches. Key enzymes such as 3-hydroxy-3-methylglutaryl Coenzyme A: reductase, the farnesyl pyrophosphate synthase and and farnesyl pyrophosphate cyclases are known and some related genes have been isolated. However, specific enzymes for important and final modifications as methylation and others, are still to be studied. Construction of chimeric enzymes allowed already the synthesis of different products and the possibilities of designing new enzymes by gene manipulation to produce unknown and useful chemicals are open.

INTRODUCTION

Pepper (*Capsicum annuum*) fruits have been traditionally used for the production of natural color (carotenoids) and pungent flavor (capsaicinoids), either directly as the fruit itself or after processing (Govindarajan, 1986). However, there are many other pepper products (Table 1) with interesting biological effects and most of them have been identified in fruits. Other pepper tissues may also produce chemicals according to the role these products play in the plant physiology (Rhodes, 1994).

Chemicals via Higher Plant Bioengineering, edited by Shahidi *et al.*
Kluwer Academic / Plenum Publishers, New York, 1999.

Table 1. Chemicals and their biological activities in Capsicum annum L. (Solanaceae "Bell Pepper")

Chemical	Tissue	Biological activity
24-methylene-cycloartan	seed	antirheumatic, cancer-preventive, hypocholesterolemic
acetylcholine	pericarp seed	allergenic, antiamblyopic, antidiuretic, antiileus, antitachycardic, cardiodepressant, cerebrostimulant, histaminic, hypotensive, miotic, myocontractant, myopic, myotonic, neurotransmitter, parasympathomimetic, peristaltic, sialogogue, spasmogenic, vasodilator
α-carotene	fruit	colorant, previtamin-A
α-linolenic-acid	fruit	antiaggregant, antiinflammatory, antileukotriene, antimenorrhagic, antimetastatic, antiprostatitic, cancer-preventive, hypotensive, immunostimulant, lymphocytogenic
α-phellandrene	fruit	dermal, flavor, insectiphile, irritant, perfumery
α-pinene	fruit	allelochemic, allergenic, antiflu, antiinflammatory, antiviral, bactericide, cancer-preventive, expectorant, flavor, herbicide, insectifuge, insectiphile, irritant, perfumery, sedative, tranquilizer.
α-terpineol	fruit	anticariogenic, antiseptic, bactericide, flavor, motor-depressant, nematicide, perfumery, sedative, termiticide.
antheraxanthin,	fruit	colorant.
apiin	fruit	aldose-reductase-inhibitor, antiarrhythmic, antibradykinic,
asparagine	fruit	antisickling, diuretic
β-pinene	fruit	allergenic, flavor, herbicide, insectifuge, perfumery.
betaine	fruit	abortifacient, antigastritic, antihomocystinuric, antimyoatrophic, bruchiphobe, emmenagogue, hepatoprotective, lipotropic.
caffeic acid	fruit	allergenic, analgesic, antiHIV, antiadenovirus, antiaggregant, anticarcinogenic, antiedemic, antiflu, antigonadotropic, antihemolytic, antihepatotoxic, antiherpetic, antihypercholesterolemic, antiinflammatory, antimutagenic, antinitrosaminic, antioxidant, antioxidant, antiperoxidant, antiradicular, antiseptic, antispasmodic, antistomatitic, antithiamin, antithyroid, antitumor, antitumor-promoter, antiulcerogenic, antivaccinia, antiviral, bactericide, CNS-active, cancer-preventive, carcinogenic, cholagogue, choleretic, clastogenic, co-carcinogenic, DNA-active, diuretic, fungicide, hepatoprotective, hepatotropic, histamine-inhibitor, insectifuge, leukotriene-inhibitor, lipoxygenase-inhibitor, metal-chelator, ornithine-inhibitor, prostaglandigenic, sedative, spasmolytic, sunscreen, tumorigenic, viricide, vulnerary.
camphene	fruit	antilithic?, antioxidant, flavor, hypocholesterolemic?, insectifuge, spasmogenic.
capsaicin	fruit	5-lipoxygenase-inhibitor, analgesic, anaphylactic, anesthetic, antiaggregant, anticolonospasmic, antiinflammatory, antimastalgic, antineuralgic, antineuritic, antinitrosaminic, antinociceptive, antioxidant, antipsoriatic, antiseptic, antiulcer, cancer-preventive, carcinogenic, cardiotonic, catabolic, cyclooxygenase-inhibitor, diaphoretic, digestive, endocrinactive, flavor, hypothermic, irritant, neurotoxic, repellent, respirasensitizer, rubefacient, sialogogue, spasmolytic, vasodilator.
capsanthin	fruit	colorant
capsidiol	fruit	fungicide, phytoalexin.
capsorubin	fruit	colorant

caryophyllene	fruit	antiasthmatic, anticariogenic, antiedemic, antifeedant, antiinflammatory, antitumor, bactericide, flavor, insectifuge, perfumery, spasmolytic, termitifuge.
chlorogenic acid	fruit	allergenic, analgesic antiebv, antihiv, antifeedant, antigonadotropic, antihemolytic, antihepatotoxic, antiherpetic, antihypercholesterolemic, antiinflammatory, antimutagenic, antinitrosaminic, antioxidant, antioxidant, antiperoxidant, antipolio, antiradicular, antiseptic, antithyroid, antitumor, antitumor-promoter, antiulcer, antiviral, bactericide, CNS-active, CNS-stimulant, cancer-preventive, cholagogue, choleretic, clastogenic, diuretic, fungicide, hepatoprotective, histamine-inhibitor, immunostimulant, insectifuge, interferon-inducer, juvabional, larvistat, leukotriene-inhibitor, lipoxygenase-inhibitor, metal-chelator, ornithine-decarboxylase-inhibitor, oviposition-stimulant, sweetener.
cinnamic acid	tissue culture	allergenic, anesthetic, antiinflammatory, antimutagenic, bactericide, cancer-preventive, choleretic, dermatitigenic, flavor, fungicide, herbicide, laxative, lipoxygenase-inhibitor, spasmolytic, vermifuge.
cryptoxanthin	fruit	antimutagenic.
cycloartenol	seed	antiinflammatory, antirheumatic, antistaphylococcic, bactericide, hypocholesterolemic.
δ-3-carene	fruit	allergenic?, antiinflammatory, bactericide, dermatitigenic, flavor, insectifuge.
dihydrocapsaicin	fruit	hypocholesterolemic
eugenol	fruit	allergenic, analgesic, anesthetic, antiaggregant, anticonvulsant, antiedemic, antifeedant, antiinflammatory, antimitotic, antimutagenic, antinitrosating, antioxidant, antiprostaglandin, antiradicular, antiseptic, antithromboxane, antitumor, antiulcer, apifuge, bactericide, CNS-depressant, cancer-preventive, candidicide, carcinogenic?, carminative, choleretic, cytotoxic, dermatitigenic, enterorelaxant, febrifuge, flavor, fungicide, herbicide, insecticide, insectifuge, irritant, juvabional, larvicide, motor-depressant, nematicide, neurotoxic, perfumery, sedative, spasmolytic, trichomonistat, trypsin-enhancer, ulcerogenic, vermifuge.
γ-terpinene	fruit	antioxidant, flavor, insectifuge, perfumery.
glucosamine	fruit	antiarthritic.
hesperidin	fruit	anti-DNA, anti-RNA, antiallergenic, antibradykinic, antiflu, antiinflammatory, antiinflammatory, antioxidant, antistomatitic, antivaccinia, antiviral, capillariprotective, catabolic, choleretic, vasopressor.
hexanal	fruit	flavor.
hexanoic acid	fruit	acidulant, flavor.
lanosterol	seed	cancer-preventive.
limonene	fruit	ache-inhibitor, allergenic, antialzheimeran?, anticancer, antiflu, antilithic, antimutagenic, antitumor, antiviral, bactericide, cancer-preventive, candidistat, enterocontractant, expectorant, fungiphilic, fungistat, herbicide, insecticide, insectifuge, irritant, nematicide, sedative, spasmolytic, viricide.
linalol	fruit	allergenic, antiallergic, anticariogenic, antihistamine, antiseptic, antishock, antiviral, bactericide, bronchorelaxant, cancer-preventive, candidistat, expectorant, flavor, fungicide, insecticide, insectifuge, motor-depressant, nematicide, perfumery, sedative, spasmolytic, termitifuge, tumor-promoter, viricide.
lupeol	seed	antiedemic, antiflu, antihyperglycemic, antiinflammatory, antilithic, antimalarial, antioxalate, antioxidant, antiperoxidant, antirheumatic, antitumor, antiurethrotic, antiviral, cytotoxic, fptase-inhibitor, hypotensive.

(continued)

Table 1. *(continued)*

myrcene	fruit	allergenic, analgesic, antimutagenic, antinociceptive, antinociceptive, antioxidant, bactericide, flavor, fungicide, insectifuge, perfumery, spasmolytic.
octanoic acid	fruit	candidicide, flavor, fungicide, perfumery.
oxalic acid	fruit	antiseptic, CNS-paralytic, fatal, hemostatic, irritant, renotoxic.
p-coumaric acid	fruit	allelopathic, antifertility, antihepatotoxic, antioxidant, antiperoxidant, antitumor, bactericide, cancer-preventive, choleretic, cytotoxic, diaphoretic?, fungicide, lipoxygenase-inhibitor, prostaglandigenic, prostaglandin-synthesis-inhibitor.
p-cymene	fruit	analgesic, antiflu, antirheumatalgic, bactericide, flavor, fungicide, herbicide, insectifuge, viricide.
piperidine	fruit	CNS-depressant, insectifuge, spinoconvulsant, urate-solvent.
pulegone	fruit	ache-inhibitor, antialzheimeran?, antihistaminic, antipyretic, avifuge, bactericide, cancer-preventive, candidicide, cerebrotoxic, encephalopathic, flavor, fungicide, glutathionalytic, hallucinogenic, hepatotoxic, herbicide, insecticide, insectifuge, nephrotoxic, neurotoxic, perfumery, pulifuge, pulmonotoxic, sedative.
sabinene	fruit	perfumery
scopoletin	fruit	allelochemic, analgesic, antiasthmatic, antiedemic, antifeedant, antiinflammatory, antileukotrienogenic, antiseptic, antitumor, CNS-stimulant, cancer-preventive, hypoglycemic, hypotensive, myorelaxant, phytoalexin, spasmolytic, uterosedative.
silver	fruit	astringent, bactericide.
solanidine	fruit	antifeedant
solanine	fruit	antiasthmatic, antibronchitic, antiepileptic, antiparkinsonian, cardiodepressant, cardiotonic, hemolytic, memranolytic, myostimulant, narcotic, sedative, teratogenic.
solasodine	fruit	antiandrogenic, antiinflammatory, antispermatogenic, antitumor, contraceptive, cytotoxic, fungicide, hepatoprotective, teratogenic.
terpinen-4-ol	fruit	antiallergic, antiasthmatic, antiseptic, antitussive, bactericide, bacteriostatic, diuretic, fungicide, herbicide, insectifuge, nematicide, spermicide, vulnerary.
terpinolene	fruit	deodorant, flavor, fungicide, perfumery.
tetramethylpyrazine	fruit	antiaggregant, antianginal, anticholinergic, antidysmenorrheic, antiembolic, antihistaminic, antihypertensive, arteriodilator, bactericide, cardiac, cardiotonic, diuretic, hypotensive, schistosomicide, spasmolytic, uterosedative.
toluene	fruit	encephalopathic.
trigonelline	seed	anticancer, antihyperglycemic, antimigraine, antiseptic, hypocholesterolemic, hypoglycemic.
xanthophyll	plant	bruchifuge, insectifuge.

Adapted from Agricultural Research Service. Phytochemical and Ethnobotanical Databases
Phytochemeco Database - USDA - ARS - NGRL
Stephen M. Beckstrom-Sternberg and James A. Duke
Web site: http://sun.ars-grin.gov/~ngrlsb/.
Specific references are available at the Web site.

Among physiological processes, one of the most important shared by all plants, is the defense response against potentially damaging biotic and abiotic agents. A plant defense response is recognized by the presence of dark or black spots on the surface of infected leaves or stems. These spots are formed by dead plant cells and several plant products to avoid the spreading of infecting microorganisms. This is the hypersensitive response (HR) triggered by avirulent (non-pathogenic) organisms. If microorganisms overcome the plant defense mechanisms and escape HR, the infection process spreads causing disease and death of the whole plant (Dixon and Harrison, 1990).

Different signals recognized somehow by *Capsicum annuum* after pathogen's attack or similar stimuli, elicit the corresponding HR responses. For example, some lipopolysaccharides from enteric bacteria and *Xanthomonas campestris* prevent the pepper HR response elicited by avirulent *X campestris* strains (Newman et al., 1997). Also, the coat protein but not the RNA from tobamovirus is required for the elicitation of the *Capsicum* L2 gene-mediated resistance (de la Cruz et al., 1997). Other proteins related with defense responses have been identified in pepper as a 34 kDa beta-1,3-glucanase with antifungal activity against *Phytophthora capsici*. This enzyme acted synergistically with a basic pepper 32 kDa chitinase to inhibit hyphal growth of *F. oxysporum* and *P. capsici* (Kim and Hwang, 1997). The gene encoding a pepper endo-beta-1,4,glucanase was isolated although it was related to fruit ripening and enhanced transcription rate was achieved by ethylene-treatment of immature green fruits (Harpster et al., 1997). A fruit specific defensin-type protein J1 with antifungal properties and the respective gene, were isolated and characterized. The mature J1 protein contained eight cysteines forming four intramolecular disulfide bridges. This protein was not detected in healthy green fruits but it was strongly accumulated in wounded green fruits and during ripening (Meyer et al., 1996).

However, although these proteins are directly related with plant defense responses, there is also the elicited production of chemicals with antibiotic or deterrent effect on plant pathogens which plays a main role in this response. These chemicals called phytoalexins, are absent or present at only very low amounts in healthy tissues (Darvill and Albersheim, 1984).

The most well known pepper phytoalexin is the bicyclic sesquiterpene capsidiol. It has a recognized fungicide effect but it was also reported an inhibitory effect of capsidiol, on the contraction induced by histamine, acetylcholine and other agonists in isolated guinea-pig ileum and trachea and, on prostaglandin synthesis *in vitro* (Stoessl et al., 1972; Nasiri et al., 1993). The use of capsidiol as a natural fungicide instead of polluting agrochemicals is attractive. Also, the molecular mechanism involved in the elicitation of capsidiol production after precise stimuli recognition may be manipulated to control the expression of other genes. Finally, the specificity of the enzymes involved in the biochemical reactions related to the capsidiol pathway may be modified to synthesize other useful products.

SESQUITERPENE BIOSYNTHETIC PATHWAY AND KEY ENZYMES

Terpenes are part of quite different plant natural substances and constitute the volatiles and essential oils as well as the natural rubber. They are among the most useful and widespread products in nature. The biological effects of some terpenes are known not only in plants but also for insects and animals. Terpenes are classified according to the number of isoprene units (one isoprene = five carbons). Monoterpenes contain two isoprene units, sesquiterpenes contain three isoprene units. Diterpenes, triterpene and tetraterpene contains four, six and eight isoprene units, respectively. Terpenoids are the corresponding derivatives including different groups like alcohols, aldehydes or ketones (Banthorpe and Charlwood, 1980; Chappell, 1995).

3-Hydroxy-3-methylglutaryl Coenzyme A : reductase (HMGR)

The synthesis of mevalonic acid (MVA) from acetyl Coenzyme A is the first key step of sesquiterpene biosynthetic pathway (Fig. 1). This step is catalyzed by the 3-hydroxy-3-methylglutaryl Coenzyme A : reductase (HMGR) enzyme encoded by plant gene families and differentially expressed according to the recognized stimuli (developmental stage, pathogen's attack or other stress conditions) (Weissenborn et al., 1995).

HMGR isoenzymes have been located in mitochondria, chloroplast and microsomes membranes (Brooker and Russell, 1975; Wong et al., 1982). Genomic and cDNA-HMGR clones from several plant species has been reported (Learned and Fink, 1989; Narita and Gruissem 1989; Ferrer et al., 1990; Bach et al., 1991; Chye et al., 1991; Stermer et al., 1991; Choi et al., 1992; Genschik et al., 1992; Maldonado-Mendoza et al., 1992; Park et al., 1992; Aoyagi et al., 1993; Enjuto et al., 1994; Joost et al., 1995). Plant HMGR enzymes are 60–65 kDa average size. In spite of their demonstrated membrane location, the amino acid sequences at the amino terminus is highly divergent between plant species and none of these shows any signal or transit peptide for targeting to the mitochondria or plastid. On the other hand, the carboxy terminus contains the active site and the transmembrane hydrophobic domain and comparison of amino acid sequences at this region shows a high similarity of more than 90% (Cane, 1981; Campos and Boronat, 1995; Denbow et al., 1996).

Genes encoding for specific HMGR isoforms are related to the sesquiterpenic phytoalexin production and accumulation in Solanaceae (tobacco, cotton, potato and tomato) (Stermer et al., 1991; Choi et al., 1992; Chappell et al., 1995; Joost et al., 1995; Weissenborn et al., 1995). HMGR isoenzymes are also regulated at the protein level as shown by transgenic tobacco plants holding foreign HMGR genes (Chappell et al., 1995; Weissenborn et al., 1995).

Farnesyl Pyrophosphate Synthase (FPS)

After mevalonate is produced, it is converted to isopentenyl pyrophosphate (IPP) by mevalonate kinases and mevalonate 5-diphosphate decarboxylase enzymes. The IPP isomerase enzyme catalyze the production of the IPP isomer, the dimethylallyl pyrophosphate (DMAPP). The condensation of DMAPP with IPP produces geranyl pyrophosphate (GPP), which is the precursor of all monoterpenes. Subsequent addition of IPP molecules gives rise to pyrophosphate intermediates such as farnesyl pyrophosphate (FPP), a common precursor of all sesquiterpenes and sterols as well as geranylgeranyl pyrophosphate (GGPP), the precursor of carotenoids and gibberellins. The farnesyl pyrophosphate synthase (FPS), a prenyltransferase enzyme, is responsible for the production of FPP (Chappell, 1995).

Prenyltransferases are homodimers of 70–80 kDa and contain a number of conserved regions. The pyrophosphate moiety of the IPP and one ion metal (Mg^{2+} or Mn^{2+}) are bound to two aspartate-rich domains involved in the catalytic reaction (Chen et al., 1994; Chappell, 1995). FPS-DNA sequences have been cloned from *A. thaliana* cDNA (Delourme et al., 1994; Cunillera et al., 1996), *Lupinus albus* (Attucci et al., 1995), *Capsicum annuum* (Hugueney et al., 1996), *Hevea brasiliensis* (Adiwilaga and Kush, 1996)

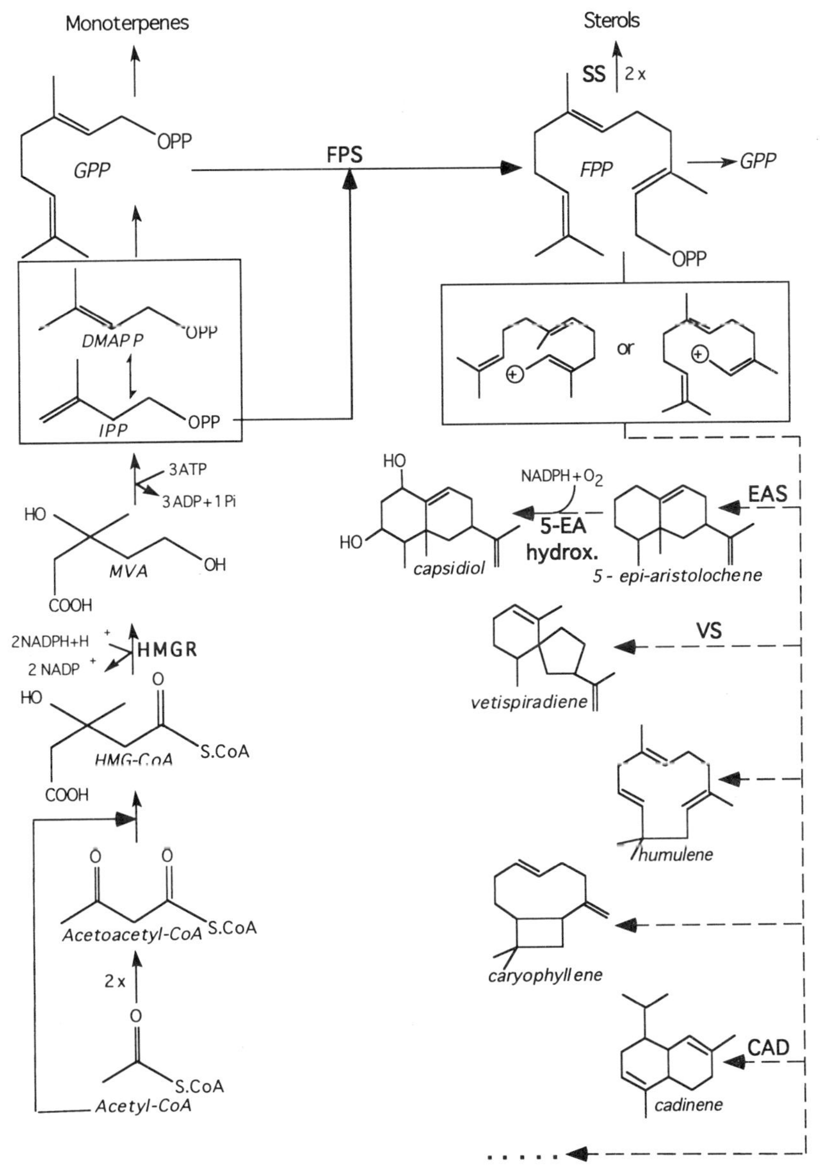

Figure 1. Sesquiterpene biosynthetic pathway. Biosynthesis initiates with acetylk-Coenzyme A (Acetyl-CoA), intermediates and final products are shown: acetoacetyl-Coenzyme A (Acetoacetyl-CoA), 3-hydroxy-3-methylglutaryl Coenzyme A (HMG-CoA), mevalonic acid (MVA), isopentenyl pyrophosphate (IPP), dimenthylallyl pyrophosphate (DMAPP), geranyl pyrophosphate (GPP), farnesyl pyrophosphate (FPP). Enzymes ate key steps are: 3-hydroxy-3-methyl glutaryl Coenzyme A: reductase (HMGR), farnesyl pyrophosphate synthase (FPS), squalene synthase (SS), 5-*epi*-aristolochene syntahse (EAS*)*, 5-*epi*-aristolochene hydroxylases (5-EA hydrox.), vetispiradiene synthase (VS), caninene synthase (CAD). Condensation of two molecules is indicated (2x) for acetyl-CoA and FPP.

and *Artemissia annua* (Matsushita et al., 1996). The *C. annuum*-FPS gene was expressed in *E. coli* (Hugueney et al., 1996).

This enzyme provides the substrate for sesquiterpenes and other isoprenoid pathways yielding main end products involved in growth and development of plants such as phytosterols, dolicols, ubiquinones and abscissic acid, among others. Channeling of FPS substrates and products may be directed toward specific pathways as sterols, cyclic sesquiterpenes or GGPP to produce longer terpenes. After cellulase treatment mimicking fungal attack, the pepper FPS was synthesized *de novo* to increase the FPP pool required for the capsidiol production by the farnesyl pyrophosphate cyclase (FPC) enzyme activity (Hugueney et al., 1996).

Farnesyl Pyrophosphate Cyclases (FPC) or Sesquiterpene Cyclases

After FPP production at least three enzymes are participating in the same branching step, the squalene synthase (SS), the farnesyl pyrophosphate cyclases (FPC) and the farnesyl protein transferases. The SS enzyme activity initiates the sterols biosynthesis, and the farnesyl protein transferases are related to primary metabolism. The FPCs or sesquiterpene cyclase enzymes initiates secondary metabolism where sesquiterpenic phytoalexins, essential oils and other final products are included (Chappell, 1995). A large variety of cyclic sesquiterpenes are synthezised by plants suggesting a broad stereospecificity of the enzymes involved in their biosynthesis (McCaskill and Croteau, 1995; Hugueney et al., 1996). These enzymes are often unstable, hard to isolate and produced as mixtures. Most of the necessary substrates are not commercially available and must be chemically synthesized, sometimes at high cost.

In spite of the broad stereospecificity, the catalytic activity of sesquiterpene cyclases is a common feature present in all terpene cyclases. The cyclic molecule results after joining a carbon atom at the diphosphate moiety of the FPP molecule to another carbon atom double bond-associated on the same molecule. The selection of the carbon atom double bond-associated is specific for each cyclase activity in order to yield particular sesquiterpenes (Croteau, 1992).

Sesquiterpene cyclase has been purified from *Salvia officinalis* synthesizing the sesquiterpenic essential oils, humulene and caryophyllene (Dehal and Croteau, 1988). The tobacco 5-*epi*-aristolochene synthase (tEAS) synthesizing the precursor of capsidiol was elicitor-inducible and purified from cellulase-treated suspension cell cultures (Vögeli and Chappell, 1990). The patchoulol synthase was purified from *Pogostemon cablin* (patchouli) and was responsible for the (-)-patchoulol production, another essential oil (Munck and Croteau, 1990). From these enzymes only the tobacco EAS is related to plant defense responses and although cell cultures have been proved as a quite convenient system, it is not very similar to what fungal pathogens do at the beginning of the infection process.

The plant roots are the first tissues directly affected by different factors like fertilizers, pesticides, a complex mixture of microorganisms and are also essentials for the water and nutrient uptake. This means that the adaptability and defense mechanisms of roots should be very efficient in order to protect the rest of the plant. However, phytoalexin production in roots has not been studied as deep as in other plant organs despite of the importance of these tissues.

We studied the capsidiol production of *in vitro* pepper root's cultures. After elicitation with cellulase treatment, capsidiol was the only chemical produced and excreted by roots to the culture media, in contrast to pepper fruits or tobacco cell cultures which release other substances as detected by thin layer chromatography (TLC). The carbon supply (e.g. sucrose) has a significant effect on the phytoalexin root's production. Pepper capsidiol is also synthesized from 5-*epi*-aristolochene and the presence of the inducible pEAS enzyme correlated with the capsidiol production as shown by immunodetection assays with monoclonal antibodies rose against the pure tEAS. The maximum pEAS enzyme activity was found between 6 to 8 hours after elicitation, this assay was carried out just after addition of 0.1 to 1.0 % of insoluble polyvinylpolypirrolidone (PVPP) to eliminate the phenolic compounds formed by grinding of the root tissue (Chávez-Moctezuma and Lozoya-Gloria, 1996).

The finding of capsidiol excretion from roots to the surrounding environment could be helpful to find out a practical procedure to stimulate the phytoalexin production in field or greenhouse conditions. This fact may avoid or decrease the use of polluting agrochemicals but ensuring the plant protection against pathogen's attack.

The inducible pEAS was partially purified from cellulase-treated pepper fruits by a procedure involving filtration of cell homogenates through a column of polyvinylpolypyrrolidone and a combination of ion exchange and hydrophobic interaction chromatographies. This method yielded a final enzyme recovery of 5% and a 91-fold enrichment over the starting material. The pEAS was a 60 kDa polypeptide and catalyzed the conversion of FPP to a single product which comigrated on argentation-TLC plates with that formed by tobacco EAS. The pepper enzyme required Mg^{2+} for activity and other cations such as Mn^{2+} and Ca^{2+} failed to activate it. A K_m value of 6.2 μM for FPP was calculated.

The effect of related isoprenoids on the cyclase activity was determined. While squalene showed no effect, farnesol and capsidiol were competitive (K_i, 20 μM) and noncompetitive inhibitors, respectively. The competitive inhibition of EAS by farnesol, suggested that early interaction of the substrate FPP with the enzyme catalytic pocket may not involve the pyrophosphate residue. These results were consistent with the observed failure of nucleotides triphosphates to affect EAS activity. The mechanism of EAS inhibition by capsidiol was analyzed and a non-competitive type of inhibition would be consistent with our observations. The physiological implications of EAS inhibition by capsidiol are difficult to predict but considering that capsidiol produced by pepper root cultures was excreted to the extracellular medium, the enzyme and inhibitor mixing is most probably an unlikely event. Furthermore, capsidiol is rapidly degraded when added to unelicited chili pepper cells (Stoessl et al., 1977). These findings suggest that EAS inhibition by capsidiol may merely reflect an *in vitro* effect with no implications for *in vivo* regulation (Cano-Camacho et al., 1997).

Two nearly identical elicitor-inducible tobacco genomic genes (gEAS3 and gEAS4) were isolated and a full size tEAS-cDNA was expressed in *E. coli* to produce a soluble and active tEAS enzyme (Facchini and Chappell, 1992; Back et al., 1994). One genomic gene (gVS1) from *Hyoscyamus muticus* coding for the inducible vetispiradieno synthase was isolated and one complete cVS1-cDNA was expressed in *E. coli* (Back and Chappell, 1995). The (+)-d-cadinene synthase (CAD) enzyme catalyzes the production of the phytoalexin gossypol. Three CAD-cDNAs (pXC1, pXC14 and *cad*1-A) from cell suspension cultures from *G. arboreum* were isolated. The pXC1 enzyme expressed in *E. coli*, catalyzed the synthesis of (+)-d-cadinene, an intermediate in the biosynthesis of gossypol and lacinilene C phytoalexins in cotton (Chen et al., 1995; Chen et al., 1996).

Comparison of cDNA sequences of sesquiterpene cyclases, shows an aspartate-rich consensus domain (DDXXD) as well as the cysteine and histidine residues as part of the catalytic site (Chappell, 1995). Domain swapping with exchange of specific DNA regions between cDNAs of tEAS and VS was attempted to verify the above hypothesis between the tEAS and VS cDNAs. The constructs were expressed in *E. coli* and the reaction products of the coded chimeric enzymes were identified. The exon 4 of the tEAS gene was specific for the reaction product of the tobacco enzyme (5-*epi*-aristolochene) and the exon 6 of the VS gene for the *Hyoscyamus* enzyme product (vetispiradiene). The combination of these two domains in the surrounding nucleotide sequence of tobacco resulted in a new enzyme able to synthesize the two different reaction products. A possible explanation is that many terpene cyclases yield multiple reaction products. The putative organization of two separated and different domains into the chimeric peptide, could produce a new activity and new products (Back and Chappell, 1996). Interestingly the multifunctional quimeric enzyme exhibited both a tighter binding of FPP and a faster conversion of FPP to products than either of its wild-type parents (Mathis et al., 1997). The crystal structure of tEAS alone and complexed with FPP analogs was analyzed. The enzymic mechanism for the bicyclic product, 5-*epi*-aristolochene, was proposed as well as the basis to understand the stereochemical selectivity displayed by other cyclases (Starks et al., 1997).

Specific Enzymes of Final Steps

After the specific cyclization of the particular sesquiterpenes, other modifications occurs as methylation, hydroxylation, demethylation, isomerization, hydration and reduction reactions. The enzymes catalyzing the 1- and 3-hydroxylation reaction of the 5-*epi*-aristolochene produced by pepper were studied. This reaction is NADPH dependent, requires molecular O_2 (Whitehead et al., 1989) and was suppressed by inhibitors of cytochrome P-450 (Hoshino et al., 1994). A 5-*epi*-aristolochene 3-hydroxylase was partially purified and located in microsomes of elicited green pepper fruits (Hoshino et al., 1995).

CONCLUSIONS

Pepper (*Capsicum annum*) is one of the major crops in the third world countries and used worldwide mainly because of the desirable flavor and color of its fruits. A large number of chemicals and corresponding biological activities have been identified in pepper fruits (Table 1). However, other tissues like roots are also producers of useful chemicals. Here we were focused only in some terpenes but other biosynthetical pathways producing phenylpropanoids, alkaloids and other secondary metabolites are functioning at the same time.

Many different terpenes need to be synthesized in quite different tissues, during several developmental stages of the plant in addition to those terpenes needed to respond to a variety of stimuli from the surrounding biosystem. This could be understood considering many specific isoprenoid pathways. Some of these pathways would be producing terpenes constitutively, some could be triggered for the fast synthesis of specific isoprenoids and some others will be producing tiny amounts of highly active terpenes.

Isoprenoid-related isoenzymes and the existence of gene families for key enzymes such as HMGR, FPS and FPC may explain the functioning of specific pathways. The right genes must be expressed from the respective promoters in fine concert to end with the right chemical for the required function and at the right time (Yin et al., 1997). These genes and others to be identified opens up many possibilities for manipulation of the production of specific and new useful terpenes for human beings. The gene manipulation directed to produce chimeric or new enzymes (Back and Chappell, 1996; Zook et al., 1996) would allow the production of new molecules for medicine, flavors, fragrances, food additives, essences, natural pesticides and others.

Also, there is much more work to be done in order to understand how the produced chemicals function by themselves. In animals, farnesylation of HMGR enzymes has been reported as being the responsible step for an increased enzyme degradation (Correll et al., 1994). In plants, farnesol and other isoprenoids are also involved in prenylation of some proteins (Shipton et al., 1995). Brassinosteroids are related to the signal transduction mechanism of plant development (Li and Chory, 1997; Li et al., 1997). Other more complex relations have been found where natural products from insects and plants interact, such as volicitin, a normal compound found in the oral secrertion of the beet armyworm caterpillar, which elicits plant volatiles attracting natural enemies of the caterpillar (Turlings et al., 1995; Alborn et al., 1997).

Finally, besides the genetic analysis, much remains to be done to understand the biochemistry, and spatial regulation of the enzymes and proteins related to isoprenes and other plant natural products (Heintze et al., 1990; Stachelin, 1997).

REFERENCES

Adiwilaga, K.; Kush, A. Cloning and characterization of cDNA encoding farnesyl diphosphate synthase from rubber tree (*Hevea brasiliensis*). *Plant Mol. Biol.* **1996,** *30,* 935–946.

Alborn, H.T.; Turlings, T.C.J.; Jones, T.H.; Stenhagen, G.; Loughrin, J.H.; Tumlinson, J.H. An elicitor of plant volatiles from beet armyworm oral secretion. *Science* **1997,** *276,* 945–949.

Aoyagi, K.; Beyou, A.; Moon, K.; Fang, L.; Ulrich, T. Isolation and characterization of cDNAs encoding wheat 3-hydroxy-3-methylglutaryl coenzyme A reductase. *Plant Physiol.* **1993,** *102,* 623–628.

Attucci, S.; Aitken, S.M.; Ibrahim, R.K.; Gulick, P.J. A cDNA encoding farnesyl pyrophosphate synthase in white lupin. *Plant Physiol.* **1995,** *108,* 835–836.

Bach, T. J.; Wettstein, A.; Boronat, A.; Ferrer, A.; Enjuto, M.; Gruissemm, W.; Narita, J. O. Properties and molecular cloning of plant HMG-CoA reductase. In *Physiology and Biochemistry of Sterols*; Patterson, G.W., Nes, W.D., Eds.; American Oil Chemists Society: Champaign, IL, USA, 1991; pp 29–49.

Back, K.; Chappell, J. Cloning and bacterial expression of a sesquiterpene cyclase from *Hyoscyamus muticus* and its molecular comparison to related terpene cyclases. *J. Biol. Chem.* **1995,** *270,* 7375–7381.

Back, K.; Chappell, J. Identifying functional domains within terpene cyclases using a domain-swapping strategy. *Proc. Natl. Acad. Sci. USA* **1996,** *93,* 6841–6845.

Back, K.; Yin, S.; Chappell, J. Expression of a plant sesquiterpene cyclase gene in *Escherichia coli*. *Arch. Biochem. Biophys.* **1994,** *315,* 527–532.

Banthorpe, D.V.; Charlwood, B.V. The isoprenoids. The terpenoids. In *Encyclopaedia of Plant Physiology. Secondary Plant Products*; Bell, E.A., Charlwood, B.V., Eds.; Springer-Verlag: Berlin, Heidelberg, Vol. 8, 1980; pp. 185–220.

Brooker, J.; Russell, D.W. Subcellular localization of 3-hydroxy-3-methylglutaryl coenzyme A reductase in *Pisum sativum* seedlings (peas). *Arch. Biochem. Biophys.* **1975,** *167,* 730–737.

Campos, N.; Boronat, A. Targeting and topology in the membrane of plant 3-hydroxy-3-methylglutaryl coenzyme A reductase. *Plant Cell* **1995,** *7,* 2163–2174.

Cane, D.E. Biosynthesis of sesquiterpenes. In *Biosynthesis of Isoprenoid Compounds*; Porter; J.W., Spurgeon, S. L. Eds.; John Wiley and Sons: New York, 1981; pp 283–374.

Cano-Camacho, H.; López-Romero, E.; Lozoya-Gloria, E. Partial purification and characterization of an elicitor stimulated sesquiterpene cyclase from chili pepper (*Capsicum annuum* L.) fruits. *Plant Sci.* **1997,** *124,* 23–31.

Chappell, J. Biochemistry and molecular biology of the isoprenoid biosynthetic pathway in plants. *Ann. Rev. Plant Physiol. Plant Mol. Biol.* **1995,** *46,* 521–547.

Chappell, J.; Wolf, F.; Proulx, J.; Saunders, C. Is the reaction catalyzed by 3-hydroxy-3-methylglutaryl Coenzyme A reductase a rate-limiting step for isoprenoid biosynthesis in plants?. *Plant Physiol.* **1995,** *109,* 1337–1343.

Chávez-Moctezuma, M.P.; Lozoya-Gloria, E. Biosynthesis of the sesquiterpene phytoalexin capsidiol in elicited root cultures of chili pepper (*Capsicum annuum*). *Plant Cell Rep.* **1996,** *15,* 360–366.

Chen, A.; Kroon, P.A.; Poulter, C.D. Isoprenyl diphosphate synthases: protein sequence comparisons, a phylogenetic tree, and predictions of secondary structure. *Protein Sci.* **1994,** *3,* 600–607.

Chen, X.Y.; Chen, Y.; Heinstein, P.; Davisson, V.J. Cloning, expression, and characterization of (+)-delta-cadinene synthase: a catalyst for cotton phytoalexin biosynthesis. *Arch. Biochem. Biophys.* **1995,** *324,* 255–266.

Chen, X.Y.; Wang, M.; Chen, Y.; Davisson, V.J.; Heinstein, P. Cloning and heterologous expression of a second (+)-delta-cadinene synthase from *Gossypium arboreum*. *J. Nat. Prod.* **1996,** *59,* 944–951.

Choi, D.; Ward, B.L.; Bostock, R.M. Differential induction and suppression of potato 3-hydroxy-3-methylglutaryl coenzyme A reductase genes in response to *Phytophthora infestans* and to its elicitor arachidonic acid. *Plant Cell* **1992,** *4,* 1333–1344.

Chye, M.L.; Kush, A.; Tan, C.T.; Chua, N.H. Characterization of cDNA and genomic clones encoding 3-hydroxy-3-methylglutaryl-coenzyme A reductase from *Hevea brasiliensis*. *Plant Mol. Biol.* **1991,** *16,* 567–577.

Correll, C.C.; Ng, L.; Edwards, P.A. Identification of farnesol as the non-sterol derivative of mevalonic acid required for the accelerated degradation of 3-hydroxy-3-methylglutaryl-coenzyme A reductase. *J. Biol. Chem.* **1994,** *269,* 17390–17393.

Croteau, R. Biochemistry of monoterpenes and sesquiterpenes of the essential oils. In *Herbs, Spices, and Medicinal Plants: Recent Advances in Botany, Horticulture, and Pharmacology*; Craker, L.E., Simon, J.E. Eds.; The Haworth Press, Inc.: New York, Vol. 1, 1992; pp 81–133.

Cunillera, N.; Arro, M.; Delourme, D.; Karst, F.; Boronat, A.; Ferrer, A. *Arabidopsis thaliana* contains two differentially expressed farnesyl-diphosphate synthase genes. *J. Biol. Chem.* **1996,** *271,* 7774–7780.

Darvill, A.G.; Albersheim, P. Phytoalexins and their elicitors. *Ann. Rev. Plant Physiol.* **1984,** *35,* 243–275.

de la Cruz, A.; Lopez, L.; Tenllado, F.;Diaz-Ruiz, J.R.; Sanz, A.I.;Vaquero, C.; Serra, M.T.; Garcia-Luque, I. The coat protein is required for the elicitation of the *Capsicum* L2 gene-mediated resistance against the tobamoviruses. *Mol. Plant Microbe. Interact.* **1997,** *10,* 107–113.

Dehal, S.S.; Croteau, R. Partial purification and characterization of two sesquiterpene cyclases from sage (*Salvia officinalis*) which catalyze the respective conversion of farnesyl pyrophosphate to humulene and caryophyllene. *Arch. Biochem. Biophys.* **1988,** *261,* 346–356.

Delourme, D.; Lacroute, F.; Karst, F. Cloning of an *Arabidopsis thaliana* cDNA coding for farnesyl diphosphate synthase by functional complementation in yeast. *Plant Mol. Biol.* **1994,** *26,* 1867–1873.

Denbow, C.J.; Lang, S.; Cramer, C.L. The N-terminal domain of tomato 3-hydroxy-3-methylglutaryl coenzyme A reductases. *J. Biol. Chem.* **1996,** *271,* 9710–9715.

Dixon, R.A.; Harrison, M.J. Activation structure and organization of genes involved in microbial defense in plants. *Adv. Genet.* **1990,** *28,* 165–234.

Enjuto, M.; Balcells, L.; Campos, N.; Caelles, C.; Arro, M.; Boronat, A. *Arabidopsis thaliana* contains two differentially expressed 3-hydroxy-3-methylglutaryl-CoA reductase genes, which encode microsomal forms of the enzyme. *Proc. Natl. Acad. Sci USA* **1994,** *91,* 927–931.

Facchini, P.J.; Chappell, J. Gene family for an elicitor-induced sesquiterpene cyclase in tobacco. *Proc. Natl. Acad. Sci. USA* **1992,** *89,* 11088–11092.

Ferrer, A.; Aparicio, C.; Nogues, N.; Wettstein, A.; Bach, T.J.; Boronat, A. Expression of catalytically active radish 3-hydroxy-3-methylglutaryl coenzyme A reductase in *Escherichia coli*. *FEBS Lett.* **1990,** *266,* 67–71.

Genschik, P.; Criqui, M.C.; Parmentier, Y.; Marbach, J.; Durr, A.; Fleck, J.; Jamet, E. Isolation and characterization of a cDNA encoding a 3-hydroxy-3-methylglutaryl coenzyme A reductase from *Nicotiana sylvestris*. *Plant Mol. Biol.* **1992,** *20,* 337–341.

Govindarajan, V.S. *Capsicum* - Production, technology, chemistry and quality. Part III - Chemistry of the color, aroma and pungency stimuli. *CRC Crit. Rev. Food Sci. and Nutr.* **1986,** *24,* 245–255.

Harpster, M.H.; Lee, K.Y.; Dunsmuir, P. Isolation and characterization of a gene encoding endo-beta-1,4-glucanase from pepper (*Capsicum annuum* L.). *Plant Mol. Biol.* **1997,** *33,* 47–59.

Heintze, A.J.; Görlach, C.; Leuschner, P.; Hoppe, P.; Hagelstein, D.; Schulze-Siebert, D.; Schulz, G. Plastidic isoprenoid synthesis during chloroplast development. Change from metabolic autonomy to a division-of-labor stage. *Plant Physiol.* **1990,** *93,* 1121–1127.

Hoshino, T.; Chida, M.; Yamaura, T.; Yoshizawa, Y.; Mizutani, J. Phytoalexin induction in green pepper cell cultures trated with arachidonic acid. *Phytochemistry* **1994,** *36,* 1417–1419.

Hoshino, T.; Yamaura, T.; Imaishi, H.; Chida, M.; Yoshizawa, Y.; Higashi, K.; Ohkawa, H.; Mizutani, J. 5-*epi*-aristolochene 3-hydroxylase from green pepper. *Phytochemistry* **1995,** *38,* 609–613.

Hugueney, P.; Bouvier, F.; Badillo, A.; Quennemet, J.; d'Harlingue, A.; Camara, B. Developmental and stress regulation of gene expression for plastid and cytosolic isoprenoid pathways in pepper fruits. *Plant Physiol.* **1996,** *111,* 619–626.

Joost, O.; Bianchini, G.; Bell, A.A.; Benedict, C.R.; Magill, C.W. Differential induction of 3-hydroxy-3-methylglutaryl CoA reductase in two cotton species following inoculation with *Verticillium*. *Mol. Plant Microbe Interact.* **1995,** *8,* 880–885.

Kim, Y.J.; Hwang, B.K. Isolation of a basic 34 kiloDalton beta-1,3-glucanase with inhibitory activity agains *Phytophthora capsici* from pepper stems. *Physiol. Mol. Plant Pathol.* **1997,** *50,* 103–115.

Learned, R.M.; Fink, G.R. 3-Hydroxy-3-methylglutaryl-coenzyme A reductase from *Arabidopsis thaliana* is structurally distinct from the yeast and animal enzymes. *Proc. Natl. Acad. Sci. USA* **1989,** *86,* 2779–2783.

Li, J.; Chory, J. A putative leucine-rich repeat receptor kinase involved in brassinosteroid signal transduction. *Cell* **1997,** *90,* 929–938.

Li, J.; Biswas, M.G.; Chao, A.; Russell, D.W.; Chory, J. Conservation of function between mammalian and plant steroid 5 alpha-reductases. *Proc. Natl. Acad. Sci. USA* **1997,** *94,* 3554–3559.

Maldonado-Mendoza, I.E.; Burnett, R.J.; Nessler, C.L. Nucleotide sequence of a cDNA encoding 3-hydroxy-3-methylglutaryl coenzyme A reductase from *Catharanthus roseus. Plant Physiol. 1992, 100,* 1613–1614.

Mathis, J.R.; Back, K.; Starks, C.; Noel, J.; Poulter, C.D.; Chappell, J. Pre-steady-state study of recombinant sesquiterpene cyclases. *Biochemistry* **1997,** *36,* 8340–8348.

Matsushita, Y.; Kang, W.; Charlwood, B.V. Cloning and analysis of a cDNA encoding farnesyl diphosphate synthase from *Artemisia annua. Gene* **1996,** *172,* 207–209.

McCaskill, D.; Croteau, R. Monoterpene and sesquiterpene biosynthesis in glandular trichomes of peppermint (*Mentha* X*piperita*). *Planta* **1995,** *197,* 49–56.

Meyer, B.; Houlne, G.; Pozueta-Romero, J.; Schantz, M.L.; Schantz, R. Fruit-specific expression of a defensin-type gene family in bell pepper. Upregulation during ripening and upon wounding. *Plant Physiol.* **1996,** *112,* 615–622.

Munck, S.L.; Croteau, R. Purification and characterization of the sesquiterpene cyclase patchoulol synthase from *Pogostemon cablin. Arch. Biochem. Biophys.* **1990,** *282,* 58–64.

Narita, J.O.; Gruissem, W. Tomato hydroxymethylglutaryl-CoA reductase is required early in fruit development but not during ripening. *Plant Cell* **1989,** *1,* 181–190.

Nasiri, A.; Holt, A; Bjork, L. Effects of the sesquiterpene capsidiol on isolated guinea-pig ileum and trachea, and on prostaglandin synthesis in vitro. *Planta Med.* **1993,** *59,* 203–206.

Newman, M.A.; Daniels, M.J.; Dow, J.M. The activity of lipid A and core components of bacterial lipopolysaccharides in the prevention of the hypersensitive response in pepper. *Mol. Plant Microb. Interact.* **1997,** *10,* 926–928.

Park, H.S.; Denbow, C.J.; Cramer, C.L. Structure and nucleotide sequence of tomato HMG2 encoding 3-hydroxy-3-methylglutaryl coenzyme A reductase. *Plant Mol. Biol.* **1992,** *20,* 327–331.

Rhodes, M.J. Physiological roles for secondary metabolites in plants: some progress, many outstanding problems. *Plant Mol. Biol.* **1994,** *24,* 1–20.

Shipton, C.A.; Parmryd, I.; Swiezewska, E.; Andersson, B., Dallner, G. Isoprenylation of plant proteins in vivo. Isoprenylated proteins are abundant in the mitochondria and nuclei of spinach. *J. Biol. Chem.* **1995,** *270,* 566–572.

Staehelin, L.A. The plant E.R.: a dynamic organelle composed of a large number of discrete functional domains. *Plant J.* **1997,** *11,* 1151–1165.

Starks, C.M.; Back, K.; Chappell, J.; Noel, J.P. Structural basis for cyclic terpene biosynthesis by tobacco 5-*epi*-aristolochene synthase. *Science* **1997,** *277,* 1815–1820.

Stermer, B.A.; Edwards, L.A.; Edington, B.V.; Dixon, R.A. Analysis of elicitor-inducible transcript encoding 3-hydroxy-3-methylglutaryl coenzyme A reductase in potato. *Physiol. Mol. Plant Pathol.* **1991,** *39,* 135–145.

Stoessl, A.; Robinson, J.R.; Rock, G.L.; Ward, E.W.B. Metabolism of capsidiol by sweet pepper tissue: some possible implications for phytoalexin studies. *Phytopathology* **1977,** *67,* 64–66.

Stoessl, A.; Unwin, C.H; Ward, E.W.B. Capsidiol, an antifungal compound from *Capsicum frutescens. Phytopath. Z.* **1972,** *74,* 141–152.

Turlings, T.C.; Loughrin, J.H.; McCall, P.J.; Rose, U.S.; Lewis, W.J.; Tumlinson, J.H. How caterpillar-damaged plants protect themselves by attracting parasitic wasps. *Proc. Natl. Acad. Sci. USA* **1995,** *92,* 4169–4174.

Vögeli, U.; Chappell, J. Regulation of a sesquiterpene cyclase in cellulase-treated tobacco cell suspension cultures. *Plant Physiol.* **1990,** *94,* 1860–1866.

Weissenborn, D.L.; Denbow, C.L.; Laine, M.; Lang, S.S.; Yang, Z.; Yu, X.; Cramer, C.L. HMG-CoA reductase and terpenoid phytoalexins: molecular specialization within a complex pathway. *Physiol. Plantarum* **1995,** *93,* 393–400.

Whitehead, I.M.; Threlfall, D.R.; Ewing, D.F. 5-*epi*-aristolochene is a common precursor of the sesquiterpenoid phytoalexins capsidiol and debneyol. *Phytochemistry* **1989,** *28,* 775–779.

Wong, R.J.; McCormack, D.K.; Russell, D.W. Plastid 3-hydroxy-3-methylglutaryl coenzyme A reductase has distinctive kinetic and regulatory features: properties of the enzyme and positive phytochrome control of activity in pea seedlings. *Arch. Biochem. Biophys.* **1982,** *216,* 631–638.

Yin, S.; Lei, L.; Newman, J.; Back, K.; Chappell, J. Regulation of sesquiterpene cyclase gene expression. *Plant Physiol.* **1997,** *115,* 437–451.

Zook, M.; Johnson, K.; Hohn, T.; Hammerschmidt, R. Structural characterization of 15-hydroxytrichodiene, a sesquiterpenoid produced by transformed tobacco cell suspension cultures expressing a trichodiene synthase gene from *Fusarium sporotrichioides*. *Phytochemistry* **1996,** *43,* 1235–1237.

7

TOWARDS THE PRODUCTION OF SALT-TOLERANT CROPS

B. J. Barkla, R. Vera-Estrella, and O. Pantoja*

Departamento de Biología Molecular de Plantas
Instituto de Biotecnología, UNAM,
Apartado Postal 510-3, Col. Miraval
Cuernavaca, Morelos, Mexico

ABSTRACT

Crop production is affected by numerous environmental factors, with soil salinity and drought having the most detrimental effects. Attempts to improve yield under stress conditions by plant breeding have been unsuccessful, primarily due to the multigenic origin of the adaptive responses. The transfer of genes through genetic engineering of crop plants appears more feasible. Important adaptive mechanisms targeted for potential gene transfer would be the tonoplast Na^{+}/H^{+} antiport, compatible solute synthesis and, regulation of water channel activity and expression, mechanisms involved in cellular osmoregulation. In this review we discuss recent advances in our understanding of these adaptive mechanisms.

INTRODUCTION

Salt-stress is one of the most serious environmental factors limiting the productivity of crop plants. Although plant species differ in their sensitivity to salt, crop plants are predominantly sensitive to the presence of high concentrations of salts in the soil. Salinity affects more than 40 percent of the world's irrigated lands, including the most productive agricultural areas of the Mediterranean basin, California and the South of Asia, where use of poor quality irrigation water has led to the progressive concentration of salts in the soil (Flowers and Yeo, 1995). Approximately 10 million hectares of irrigated land are thought

* Phone: (52-5) 622 7656, Fax:(52-73) 13 9988, e-mail: omar@ibt.unam.mx

Chemicals via Higher Plant Bioengineering, edited by Shahidi *et al.*
Kluwer Academic / Plenum Publishers, New York, 1999.

to be abandoned each year because of the adverse effects of secondary salinization (Szaboles, 1987). Although the engineering technology is readily available to combat this problem through drainage and supply of higher quality water, these measures are extremely costly and may not always be feasible (van Schilfgaarde, 1990). Thus, a more practical answer may lie with a biological solution to the problem.

One strategy may be the development of salt-resistant crops. Traditional plant breeding, however, presents several problems; as discussed in this review, salt-tolerance does not appear to depend on a single identifiable trait, and in fact, the transfer of single traits has been shown to improve salt-tolerance only marginally (Tarczynski et al., 1993; Delauney and Verma, 1993). Evidence supports the view that salt-tolerant plants employ diverse adaptive mechanisms. Therefore, genetic crossing would be extremely time consuming and would require the existence of wild related species which possess the genes of importance.

With the recent advances in plant genetic engineering it is now possible to consider a strategy of multiple gene transfers. The challenge lies in the identification of key processes involved in the adaptive responses to salinity. It is our view that knowledge of the mechanisms present in salt-tolerant plants (halophytes) could be used for the identification of genes to be engineered into crops.

MESEMBRYANTHEMUM CRYSTALLINUM: AN IDEAL MODEL SYSTEM?

The halophyte, *Mesembryanthemum crystallinum* (Aizoaceae), presents itself as an ideal model plant system for the identification of salt-adaptive traits. *M. crystallinum* is endemic of the Namibian desert in South Africa where it is adapted to high salt, low water potential soils. This plant has been introduced into several areas with similar climate conditions, including, coastal areas of the Mediterranean, California, Baja California and Australia, where it has rapidly displaced native plant species (El-Ghareeb, 1991).

M. crystallinum grown under greenhouse conditions shows a characteristic developmental pattern in which two clearly defined stages can be recognized, the juvenile and adult phases. The juvenile phase is characterized by the presence of obovate leaves appearing in pairs (primary leaves), alternating 90° from the main axis, with numbers varying between five to seven pairs in total (Cushman and Bohnert, 1996). The adult stage can be identified by the appearance of side shoots that emerge as early as five weeks after germination from axillary buds found between leaf pairs. Leaves that develop from the shoots (secondary leaves) are smaller and irregular in shape, appearing first in opposite position, later changing to an alternate arrangement. At the same time, older leaves start to senesce in an acropetal direction, as indicated by the loss of turgor and yellowing, and flowers appear from the shoots. One unique characteristic of *M. crystallinum* is the presence of specialized epidermal cells, the bladder cells. These cells are more conspicuous in the abaxial side of the leaves than in the adaxial surface and are also found on the stems and flower pods. Exposure of *M. crystallinum* to concentrations of 200 mM NaCl results in the stimulation of plant growth (Fig 1a). Plants in the juvenile phase respond to Na^+ with an enlargement of the primary leaves which show noticeable succulence. Adult plants respond by the accelerated development of side shoots. As an additional response to salt exposure, bladder cells swell, reaching dimensions in the millimeter range, with the largest bladder cells surrounding shoots and flower pods (Fig. 1b). These can be spherical, cylindrical or pear shaped, depending upon the location. The highest accumulation of NaCl occurs in

Figure 1. NaCl stimulates growth in *M. crystallinum*. a) Addition of 200 mM NaCl to the soil of 6 week old plants for two weeks stimulated plant growth (right) in comparison to plants not exposed to NaCl (left). b) Epidermal bladder cells from leaves of a 12-week old plant treated with 200 mM NaCl for 8 weeks, (magnification 30 X).

these cells, where concentrations of 1.2 M have been reported (Adams et al., 1992). Flower number seems to have a direct correlation with salinity, and this may be related to the stimulation of growth observed under high salinity conditions.

Mesembryanthemum crystallinum exhibits several mechanisms that allow its growth under saline conditions. These include, the age-dependent establishment of crassulacean acid metabolism (CAM) and, the ability to accumulate high Na^+ concentrations in the vacuole of young, growing, aerial parts of the plant, balanced by the synthesis of compatible solutes in the cytoplasm. These adaptive responses allow *M. crystallinum* to increase water use efficiency through the regulation of stomatal functioning and osmotic potential.

CRASSULACEAN ACID METABOLISM (CAM)

The salt- and drought-induced switch from C_3 metabolism to CAM in *M. crystallinum* has been extensively characterized at the biochemical and molecular levels (Cushman and Bohnert, 1997). During CAM water loss is prevented by the closing of stomata during the day and carbon fixation occurs through the Calvin cycle employing the CO_2 derived from the decarboxylation of malate by malic enzyme. At night, stomata open and allow for the influx of CO_2 which is fixed into malate by phospho-*enol*-pyruvate carboxylase (PEPC) and stored in the vacuole.

Increased nocturnal accumulation of malate in the vacuole is the result of the induction of enzymes directly related with malate metabolism. One example is the CAM-specific PEPC isogene (*Ppc1*) whose expression is transcriptionally induced by salt stress (Cushman et al., 1989), while that for the isogene *Ppc2* displays an expression pattern that is salt-insensitive (Cushman et al., 1989). Additional examples of transcriptional activation of CAM-specific genes include NAD-glyceraldehyde-3-phosphate dehydrogenase (*Gdp1*; Ostrem et al., 1990), a cytosolic NADP-malic enzyme (*Mod1*; Cushman, 1992), and a chloroplast-localized NADP-malate dehydrogenase (*Mdh1*; Cushman, 1993). Similar to the response observed for *Ppc1*, these latter genes showed an increase in mRNA accumulation that is partially due to an increase in transcriptional rate (Cushman and Bohnert, 1996).

The plant growth regulator, abscisic acid (ABA), has been shown to participate in the adaptive responses of plants to adverse environmental conditions including drought and salinity. Under water stress conditions, ABA has been shown to have a direct role in regulating stomatal opening and gene expression (Hetherington and Quatrano, 1991).

In *M. crystallinum*, Chu et al. (1990) found that addition of 1–10 μM ABA to roots or leaves resulted in the induction of enzyme activities for both PEPC and NADP-malic enzyme. However, more recent evidence suggested that the ABA-induction of *Ppc1* transcripts was light- and temperature-dependent (McElwain et al., 1992; Cushman et al., 1993), indicating that ABA cannot be considered the sole mediator of the induction of CAM genes by salt and drought. Due to the antagonistic effects of cytokinins in relation to those elicited by ABA (Letham and Palni, 1983), possible effects of these plant growth regulators during salt adaptation in *M. crystallinum* have also been studied. Addition of 10 μM 6-benzylaminopurine (BAP) to the roots of hydroponically grown *M. crystallinum* plants resulted in nocturnal acidification and accumulation of malate together with increases in PEPC and NADP-malic enzyme, responses associated with the establishment of CAM (Edwards et al., 1996). These results are supported by BAP-induced increases in both protein and mRNA levels for the CAM-specific isoform of PEPC (Thomas et al., 1992b). However, cytokinin levels upon salt-stress did not show a significant change (Thomas et al., 1992b). From these and similar results, Thomas and Bohnert (1993) have suggested that the effects elicited by the exogenous application of cytokinin to *M. crystallinum* may not be directly involved in inducing salt-stress-specific responses. These authors proposed that cytokinins may participate in nitrogen recycling or biosynthesis. The role of other plant growth regulators in CAM induction seems unlikely since neither

Table 1. Initial rates of Na^+-dependent H^+ efflux and ATP-dependent H^+ transport (%F min^{-1} mg $protein^{-1}$) from tonoplast vesicles isolated from leaves of juvenile and adult *M. crystallinum* plants treated for 10–14 days

	Na^+/H^+ Antiport		V-ATPase	
System	Juvenile	Adult	Juvenile	Adult
Control	200	161	534	693
200 mM NaCl	220 (110)	215 (135)	882 (116)	1166 (168)

Three- or six-week old soil grown plants were watered daily with appropriate solutions. Vesicle pH changes were monitored by quenching of quinacrine fluorescence upon addition of 200 mM Na^+ (antiport) or 1.5 mM ATP (V-ATPase) in the presence of 30 μg protein and 50 mM Cl^-. Values in brackets represent the percent of activity with respect to that of control. *F*, fluorescence intensity. Results are from 3 replicates with SD not exceeding 15%.

the auxin 2,4-dichlorophenoxyacetic acid (2,4-D), a gibberellin nor methyl jasmonate supplied through the roots affected the expression of CAM-related genes (Edwards et al., 1996).

ION TRANSPORT MECHANISMS INVOLVED IN VACUOLAR Na^+ ACCUMULATION

For salt-tolerance a plant must be able to effectively distribute Na^+ and maintain cellular water homeostasis. Thus, the study of mechanisms involved in the transport of water and Na^+ within the cell are of prime importance when considering traits essential for salt-tolerance.

A major adaptive difference between glycophytes (plants that do not tolerate salt) and halophytes is the ability of the latter to accumulate Na^+, whereas glycophytes are effectively salt-excluders. As there is little difference between levels of cytoplasmic Na^+ toxicity in salt-tolerant plants as compared to salt-sensitive plants (Osmond and Greenway, 1972), halophytes must efficiently exclude Na^+ from the cytoplasm. The presence of activity associated with a vacuolar membrane (tonoplast) Na^+/H^+ antiporter, or exchanger, involved in vacuolar Na^+ sequestration has been well documented in several plant species (reviewed in Barkla et al., 1994). Secondary active transport of Na^+ into the vacuole via an antiport would be energized by the proton motive force generated by the activity of the tonoplast H^+-ATPase (V-ATPase) and/or H^+-pyrophosphatase (V-PPase). As a fundamental mechanism in salt-tolerance an active antiport would function to maintain cytoplasmic Na^+/K^+ ratios, avoid cytoplasmic Na^+ toxicity, and in parallel, vacuolar Na^+ would serve as an osmoticum necessary for cellular water homeostasis. Evidence for a role of the tonoplast Na^+/H^+ antiport and V-ATPase in plant salt-tolerance has been provided from several studies demonstrating induction of antiport and/or V-ATPase activity upon exposure to NaCl (Blumwald and Poole, 1987; Garbarino and Dupont, 1988; Matsumoto and Chung, 1988; Nakamura et al., 1992; Zingarelli et al., 1994), and NaCl-induction of transcripts for subunits of the V-ATPase (Narasimhan et al., 1991; Binzel et al., 1995; Perera et al., 1995). Biochemical and molecular studies aimed at conclusively identifying the Na^+/H^+ antiport protein or gene have so far been unsuccessful, although preliminary studies suggest the association of a 170 kD polypeptide with the antiport in *Beta vulgaris* (Barkla and Blumwald, 1991).

In *M. crystallinum,* a particularly efficient salt-includer (see Introduction), salt-induced Na^+/H^+ antiport activity has been documented (Barkla et al., 1995). The increase in antiport activity was independent of plant age but induction was larger in adult plants (greater than 6 weeks old) than in juvenile plants (3 weeks old) (Table 1). The NaCl-induced increase in Na^+/H^+ antiport activity was correlated to a similar increase in V-ATPase hydrolytic and proton-pumping activities in both juvenile and adult plants (Rockel et al., 1994; Barkla et al., 1995; Tsiantis et al., 1996; Table 1) and in *M. crystallinum* cell suspensions (Vera-Estrella et al., 1999). It is interesting to note that activity of the Na^+/H^+ antiport are much higher in juvenile plants as compared to adult plants and this observation may explain the smaller induction of this enzyme in salt-stressed juvenile plants. Although no co-ordinate increase in V-ATPase subunits has been documented in *M. crystallinum*, changes in stoichiometry of this multimeric enzyme during the adaptation to salt-stress, which in turn could regulate enzyme activity, appear likely. Freeze fracture analysis of tonoplast particles suggested to represent the V-ATPase V_0 sector have been shown to increase in size in tonoplast from salt-stressed leaves (Rockel et al., 1994). Steady-state leaf transcript levels for the proton-conducting 16 kD proteolipid subunit of the V_0 sector of the V-ATPase have been shown to be selectively increased in leaves of salt-treated plants (Löw et al., 1996; Tsiantis et al., 1996). At the level of protein expression, slight increases in protein amounts for subunit E of the V-ATPase were observed in salt-stressed leaves of mature *M. crystallinum* plants, while protein amounts for subunit A remained unchanged (Dietz et al., 1996).

AQUAPORINS-WATER CHANNELS

A group of membrane channel forming proteins aptly named "aquaporins" facilitate the passive movement of water across membranes. Aquaporins belong to the Major Intrinsic Protein (MIP) family of channels conserved in animals, plants and micro-organisms, that also includes glycerol and ion transporters (for reviews see; Maurel, 1997; Schäffner, 1998). In plants, aquaporins fall into two distinct subclasses of the MIP family, the putative plasma membrane intrinsic proteins (PIPs) and the putative tonoplast intrinsic proteins (TIPs) (Yamada et al., 1995). The demonstration of membrane localization of functional aquaporins within these two groups has supported this classification (Johnson et al., 1989; Daniels et al., 1994; Kaldenhoff et al., 1995; Daniels et al., 1996). The water-conducting ability of several of these plant proteins has been demonstrated by expression in *Xenopus* oocytes (Maurel et al., 1993; Daniels et al., 1994; Kammerloher et al., 1994; Yamada et al., 1995; Daniels et al., 1996; Sarda et al., 1997), and using protoplasts transformed with an antisense construct for an *Arabidopsis thaliana* PIP aquaporin homolog (Kaldenhoff et al., 1995).

Water channels are assumed to largely contribute to cellular hydraulic conductivity (L_p), and thus play a significant role in the control of transmembrane water movement during growth, development and stress responses of plants. Prior to the functional identification of plant aquaporins, changes in L_p were shown to be manifested by treatment of maize roots with NaCl (Azaizeh et al., 1992), and as early as 1990, transcripts for a pea shoot MIP homolog were shown to be induced by water deficit (Guerrero et al., 1990). More recently, drought-induced regulation of a functionally characterized aquaporin gene was demonstrated for a TIP homolog in sunflower (Sarda et al., 1997). In *A. thaliana*, transcript amounts for a PIP homolog, *Rd28*, increased within 10 h of dehydration stress (Yamaguchi-Shinozaki et al., 1992). However, the abundance of the Rd28-PIP protein was not altered by water stress (Daniels et al., 1994).

In *M. crystallinum*, several MIP homologs (MipA through MipE) have been identified by differential screening of a root cDNA library constructed from salt-stressed plants

(Yamada et al., 1995). All of these genes were shown to encode proteins that fall into the PIP subfamily of putative aquaporins. Expression of *MipA* and *MipB* in *Xenopus* oocytes led to characteristic increases in water permeability associated with the water channel function of aquaporins (Yamada et al., 1995). *MipB* transcripts were detected solely in roots, whereas transcripts for *MipA* and *Mip*C were detected in both roots and shoots. Upon salt-stress, mRNA levels for *MipA*, *MipB* and *MipC* decreased within 6 h of stress initiation, continuing through 30 h of treatment. However, by 72 h of stress, transcript levels returned to those detected prior to exposure to NaCl (Yamada et al., 1995). It is possible that the response to salt and/or drought of MIP transcripts in *M. crystallinum*, differs from that observed for sunflower and *A. thaliana* because the latter plants are glycophytes and salt-excluders. *M. crystallinum* is halophilic and selectively accumulates Na^+. As an adaptive mechanism, *M. crystallinum* may, in the short term, respond to salt-stress by rapidly reducing plasma membrane water permeability through the down-regulation of PIP aquaporins. At this time, non-limiting water permeability at the tonoplast could effectively maintain cytoplasmic water homeostasis (Maurel et al., 1997) until sufficient cellular Na^+ was accumulated to compensate for the increase in extracellular osmotic potential. Intrinsic differences in tonoplast and plasma membrane osmotic water permeabilities have been demonstrated by stopped-flow spectrophotometry in membrane vesicles isolated from wheat roots (Niemietz and Tyerman, 1997) and tobacco cell suspensions (Maurel et al., 1997). In both studies tonoplast water permeabilities were significantly higher than those observed for the plasma membrane, 7-fold higher in wheat and close to 100-fold higher in tobacco cell suspensions. These results suggest that in plants, water transport across the plasma membrane is much more tightly regulated than that across the tonoplast and suggests important implications for PIP regulation under environmental stress.

COMPATIBLE SOLUTES

With the accumulation of vacuolar Na^+ in *M. crystallinum,* it is necessary to compensate the osmotic potential in the cytoplasm with the accumulation of compatible solutes. This allows the cell to reduce cellular water potential while avoiding deleteriously high cytoplasmic Na^+ concentrations. Compatible solutes are a group of low molecular weight compounds that can accumulate to high concentrations within the cell without inducing negative effects on metabolic processes (Bohnert et al., 1995). An array of compounds have been categorized as compatible solutes, including proline, betaines, and sugar alcohols, and these are thought to be largely confined to the cytosol (McCue and Hanson, 1990; Tarczynski *et al.*, 1993; Thomas *et al.*, 1995; Hanson *et al.*, 1994). Evidence for a role of these compounds in hydroxy radical scavenging was demonstrated by Smirnoff and Cumbes (1989). These authors showed that sorbitol, mannitol, *myo*-inositol and proline prevented the *in vitro* hydroxylation of salicylate and denaturation of malate dehydrogenase, with glycine betaine being ineffective. Results suggested that these compounds may be involved in the protection of membranes during stress conditions (Smirnoff and Cumbes, 1989). The non-reducing sugar alcohols may store excess carbon under environmental stress conditions (Vernon *et al.*, 1993), and the accumulation of proline has been proposed to protect enzymes and polyribosomes (Hare and Cress, 1997).

In *M. crystallinum*, the salt-induced accumulation of the compatible solutes mannitol, proline, fructans, ononitol and pinitol have been reported (Bohnert *et al.*, 1995). Proline accumulation has been shown to occur in leaves of salt-stressed adult *M. crystallinum* plants, as well as in cell suspension cultures (Adams et al., 1992; Thomas et al.,

1992a). However, Deming and Winter (1986) found that proline is more concentrated in the chloroplasts of leaves from salt-treated *M. crystallinum* plants as compared to the total leaf sap, indicating this amino acid does not play a large role in cytoplasmic osmotic adjustment of plants under salt-stress.

It has been argued that stress-induced cytoplasmic proline accumulation is insufficient to account for pronounced biophysical changes (Hare and Cress, 1997), as not all plants can resist osmotic stress even when proline is accumulated (Delauney and Verma, 1993). Kishor et al. (1995) reported that overexpression of the *Vignia* pyrroline-5-carboxylate synthase gene (P5CS, a key enzyme of proline biosynthesis) in tobacco plants, only increased proline production by 2-fold following imposition of stress and although transgenic plants grew better, biomass was only doubled as compared to control stressed plants. Moreover, Liu and Zhu (1997) produced a salt-hypersensitive mutant from *A. thaliana* (*sos1*) that accumulates more proline than wild type plants under salt-stress, suggesting that proline content is not necessarily a limiting factor for salt tolerance in higher plants. Rather, proline accumulation may be a response sufficient for survival under moderately saline stress conditions, but under severe stress, more specialized mechanisms may be required (Adams et al., 1992).

The major compatible solute contributing to osmotic adjustment in salt-stressed *M. crystallinum* is the cyclic sugar alcohol, pinitol. In plants exposed to NaCl, pinitol increases to become the major low molecular weight carbon compound in the cytosol and chloroplasts (Paul and Cockburn, 1989). The accumulation of pinitol in the plant mirrors the gradient for Na^+ accumulation, with the highest amounts of both solutes being deposited in the youngest tissues (Adams et al., 1998). Vernon and Bohnert (1992) reported the cloning and molecular characterization of a polypeptide homologous to methyltransferases. Expression of this protein in *E. coli* and activity assays identified this enzyme as L-*myo*-inositol methyltransferase (IMT1), a key enzyme in the pinitol biosynthesis pathway (Fig. 2). Vernon et al. (1993) have shown that induction of *Imt1* transcripts was specific for salt-stress, indicating the ability of the plant to distinguish between salt- and drought-stress; increases in IMT1 protein paralleled increases in pinitol concentration. Salt-stress induction of IMT1 protein is independent of plant age, being observed at all stages of development, including seedlings, juvenile plants and adult plants (Nelson et al. 1998, Fig. 3), and is observed in both roots and leaves (Nelson et al. 1998). Another member of the pinitol biosynthesis pathway, INPS (*myo*-inositol 1-phosphate synthase; Fig. 2), is present in all cell types and following salinity stress, is induced in leaves and transiently induced in roots (Nelson et al., 1988).

To understand the importance of pinitol biosynthesis in plant adaptation to salt-stress, *M. crystallinum Imt1* was expressed in tobacco (Vernon et al., 1993; Sheveleva et al., 1997). Transgenic plants appeared phenotypically normal, exhibited IMT1 enzyme activity, and accumulated high levels of ononitol, however, pinitol was not detected. The absence of pinitol was thought to be due to the lack of the epimerase responsible for the conversion of ononitol to pinitol present in *M. crystallinum*. Plant transformation appeared to provide better protection under drought and salt-stress conditions (Sheveleva et al., 1997).

Early studies employing cell suspension cultures generated from hypocotyls of *M. crystallinum*, failed to detect pinitol in salt-stressed cell suspensions, and it was suggested that the accumulation of this polyol may require tissue organization for expression (Adams et al., 1992; 1998; Thomas et al., 1992a). However, in cell suspensions generated from leaves of adult plants, increased levels of pinitol correlated to increased amounts of IMT protein in the cell sap of salt-treated cell suspensions of *M. crystallinum* (Vera-Estrella *et al.*, 1998). These results are consistent with observations made with whole plants, indicating that the accumulation of pinitol upon salt stress does not require tissue organization. However, in contrast to whole plants studies, INPS protein amounts do not in-

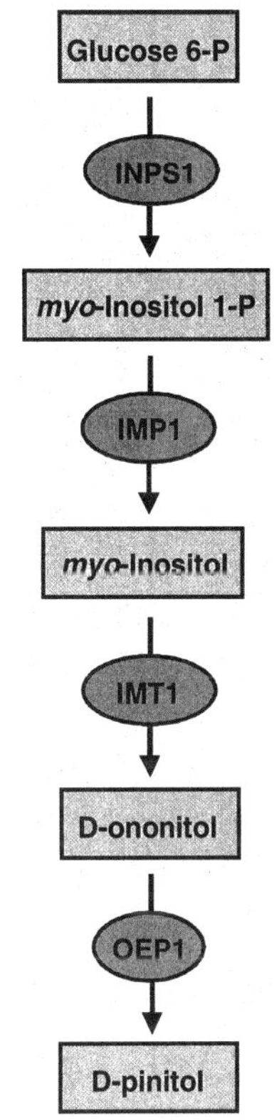

Figure 2. Pinitol biosynthesis pathway. Pinitol biosynthesis starts with the formation of *myo*-inositol-1-phosphate (*myo*-inositol 1-P) from glucose-6-phosphate (glucose 6-P) by inositol 1-P synthetase (INPS1). *myo*-inositol 1-P is dephosphorylated to *myo*-inositol by inositol-1-phosphatase (IMP1). Methylation of *myo*-inositol by inositol O-methyltransferase (IMT1) produces D-ononitol, which is finally epimerased to D-pinitol by ononitol epimerase (OEP1). (Adapted from Bohnert et al., 1995).

crease in salt-stressed *M. crystallinum* cell suspensions (Fig. 4). These cultures should provide a valuable tool for studying signaling and biochemical mechanisms involved in salt recognition and response.

PHYTOREMEDIATION OF SALINE SOILS

In addition to the use of *M. crystallinum* as a potential donor of salt-stress tolerance genes, this halophilic plant also presents characteristics that suggest its use for the recovery of saline soils. Bioremediation exploits the capacity of living organisms for the re-

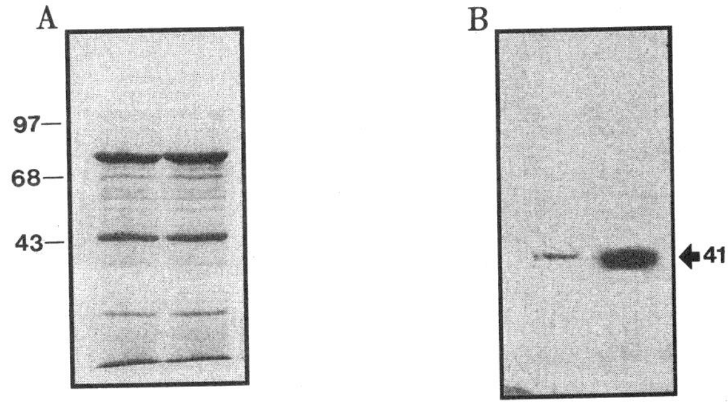

Figure 3. Western blot analysis of IMT. Western blot analysis of total protein (12 μg) using a polyclonal antibody directed against the 41 kD *myo*-inositol methyltransferase (IMT1) protein. Total protein extracts were isolated from leaves of control (lane 1) and 200 mM NaCl (lane 2) 10-day-treated three-week-old *M. crystallinum* plants.

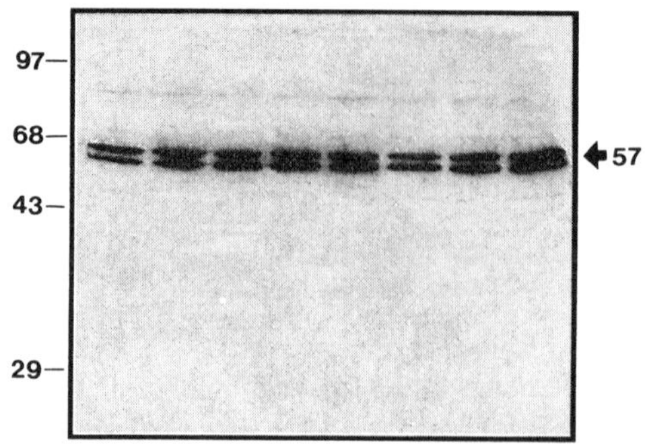

Figure 4. Western blot analysis of INPS Western blot analysis of total protein (12 μg) using a polyclonal antibody directed against the 57 kD *myo*-inositol-phosphate synthase (INPS) protein. Total protein extracts were isolated from control (lane 1), 50 mM NaCl (lane 2), 100 mM NaCl (lane 3), 150 mM NaCl (lane 4), 200 mM NaCl (lane 5), 400 mM NaCl (lane 6) or 400 mM mannitol (lane 7), 4-day-treated *M. crystallinum* cell suspensions.

moval of toxic compounds from contaminated waters or soils. In the case of plants (phytoremediation), the best known application is that related to the removal of heavy metals by the "hyperaccumulators", species that are able to concentrate these metals at higher levels than those found in the soil (Raskin et al., 1994). However, phytoremediation has a major limitation, in that all known hyperaccumulating species have a small biomass, preventing significant heavy metal removal.

The capacity of *M. crystallinum* to accumulate Na^+ to concentrations higher than those found in the growth medium suggests that this plant could be considered as a "Na^+ hyperaccumulator". In contrast to the heavy metal hyperaccumulating plants, *M. crystallinum* can reach a fresh weight of 20 kg (Adams et al., 1992), making it feasible to be employed as a phytoremediator for saline soils.

According to our own observations, *M. crystallinum* plants grown hydroponically in 1 l vessels containing Hoagland's solution, supplied with 200 mM NaCl for 12 weeks, reached a fresh weight close to 150 g and were able to accumulate up to 2.5 g of Na^+, and data from soil grown plants showed decreases of approximately 30% of the initial Na^+ by a single *M. crystallinum* plant in soils with concentrations in the range of 50 mM (unpublished results). From these data and that reported for the maximal fresh weight of 20 kg reached by a plant under unlimited conditions of space (Adams et al., 1992), we can estimate the possible effects of growing *M. crystallinum* in salt-affected soils. Assuming a density of one plant/m^2, we calculated that a field of plants could remove up to 3.3 tons Na^+/hectare, strongly indicating the possibility of employing this plant for the recovery of salt-affected arable lands.

PERSPECTIVES

In this review, key mechanisms involved in conferring salt-tolerance to *M. crystallinum* have been discussed. These include enzymes involved in the synthesis of the compatible solutes; ononitol and pinitol, transport proteins that participate in the vacuolar sequestration of Na^+; and those involved in water-conserving metabolic adaptations. It will now be essential to determine if components of these processes are present in crop plants and what would be the minimum requirement for transformation to obtain the desired activity. One important consideration that must be addressed is how these pathways are regulated. At this time little is understood in this area, it appears there are two distinct effects of Na^+, an ionic and osmotic, and that different pathways are involved for their perception and regulation. More studies are needed in this area before a clear understanding is reached and a salt-tolerant plant breeding program is initiated.

ACKNOWLEDGMENTS

Authors work presented in this review was supported by Grants 3281PN; 25750N from CONACyT; CONACyT-NSF 96–97 program and IN202695 from DGAPA-UNAM.

REFERENCES

Adams, P.; Nelson, D.E.; Yamada, S.; Chmara, W.; Jensen, R.G.; Bohnert, H.J.; Griffiths, H. Growth and development of *Mesembryanthemum crystallinum* (Aizoaceae). *New Phytol.* **1998**, 138, 171–190.

Adams, P.; Thomas, J.C.; Vernon, D.M.; Bohnert, H.J.; Jensen, R.G. Distinct cellular and organismic responses to salt stress. *Plant Cell Physiol.* **1992**, 33, 1215–1223.

Azaizeh, H.; Gunse, B.; Steudle, E. Effects of NaCl and $CaCl_2$ on water transport across root cells of Maize (*Zea mays* L.) seedlings. *Plant Physiol.* **1992**, 99, 886–894.

Barkla, B.J.; Blumwald, E. Identification of a 170-kDa protein associated with the vacuolar Na^+/H^+ antiport of *Beta vulgaris*. *Proc. Natl. Acad. Sci. USA* **1991**, 88, 11177–11181.

Barkla, B.J.; Apse, M.P.; Manolson, M.F.; Blumwald, E. The plant vacuolar Na^+/H^+ antiport. In *Membrane transport in Plants and Fungi: Molecular mechanisms and control*; Blatt, M.R., Leigh, R.A., Sanders, D.A., Eds.; Company of Biologists: Cambridge, **1994**.

Barkla, B.J.; Zingarelli, L.; Blumwald, E.; Smith, J.A.C. Tonoplast Na^+/H^+ antiport activity and its energization by the vacuolar H^+-ATPase in the halophytic plant *Mesembryanthemum crystallinum*. *Plant Physiol.* **1995**, 105, 549–556.

Binzel, M.L.; Dunlap, J.R. Abscisic acid does not mediate NaCl-induced accumulation of 70-kDa subunit tonoplast H^+-ATPase message in tomato. *Planta* **1995**, 197, 563–568.

Blumwald, E.; Poole, R.J.; Salt tolerance in suspension cultures of sugar beet: induction of Na^+/H^+ antiport activity at the tonoplast by growth in salt. *Plant Physiol.* **1987**, 83, 884–887.

Bohnert, H.J.; Nelson, D.E.; Jensen, R.G. Adaptations to environmental stresses. *Plant Cell.* **1995**, 7, 1099–1111.

Chu, C,; Dai, Z.; Ku, M.S.B.; Edwards, G.E. Induction of crassulacean acid metabolism in the facultative halophyte *Mesembryanthemum crystallinum* by abscisic acid. *Plant Physiol.* **1990**, 93,1253–1260.

Cushman, J.C. Characterization and expression of NADP-malic enzyme cDNA induced by salt stress from the facultative CAM plant, *Mesembryanthemum crystallinum*. *Eur. J. Biochem.* **1992**, 208, 259–266.

Cushman, J.C. Molecular cloning and expression of chloroplast NADP-malate dehydrogenase during crassulacean acid metabolism induction by salt stress. *Photosynth. Res.* **1993**, 35, 15–27.

Cushman, J.C.; Bohnert, H.J. Molecular genetics of Crassulacean acid metabolism. *Plant Physiol.* **1997**, 113, 667–676.

Cushman, J.C.; Bohnert, H.J. Transcriptional activation of CAM genes during development and environmental stress. In *Crassulacean acid metabolism. Biochemistry, ecophysiology and evolution;* Winter, K., Smith, J.A.C. Eds. Springer-Verlag: Berlin, **1996.**

Cushman, J.C.; Meyer, G.; Michalowski, C.B.; Schmitt, J.M.; Bohnert, H.J. Salt stress leads to differential expression of phosphoenolpyruvate carboxylase during crassulacean acid metabolism induction in the common ice plant. *Plant Cell.* **1989**, 1, 715–725.

Cushman, J.C.; Vernon, D.M.; Bohnert, H.J. ABA and the transcriptional control of CAM induction during salt stress in the common ice plant. In *Control of plant gene expression;* Verma, D.P.S. Ed. CRC Press; Boca Raton, **1993.**

Daniels, M.J.; Chaumont, F.; Mirkov, T.E.; Chrispeels, M.J. Characterization of a new vacuolar membrane aquaporin sensitive to mercury at a unique site. *Plant Cell.* **1996**, 8, 587–599.

Daniels, M.J.; Mirkov, T.E.; Chrispeels, M.J. The plasma membrane of *Arabidopsis thaliana* contains a mercury-insensitive aquaporin that is a homolog of the tonoplast water channel protein TIP. *Plant Physiol.* **1994**, 106, 1325–1333.

Delauney, A.J.; Verma, D.P.S. Proline biosynthesis and osmoregulation in plants. *Plant J.* **1993**, 4, 215–223.

Demming, B.; Winter, K. Sodium, potassium, chloride and proline concentrations of chloroplasts isolated from the halophyte, *Mesembryanthemum crystallinum*. *Planta.* **1986**, 168, 421–426.

Dietz, K.J.; Arbinger, B. cDNA sequence and expression of subunit E of the vacuolar H^+-ATPase in the inducible Crassulacean acid metabolism plant *Mesembryanthemum crystallinum*. *Biochem. Biophys. Acta* **1996**, 1281,134–138.

Edwards, G.E.; Dai, Z.; Cheng, S.H.; Ku, M.S.B. Factors affecting the induction of crassulacean acid metabolism in *Mesembryanthemum crystallinum*. In *Crassulacean acid metabolism. Biochemistry, ecophysiology and evolution;* Winter, K., Smith, J.A.C. Eds. Springer-Verlag: Berlin, **1996.**

El-Ghareeb, R. Vegetation and soil changes induced by *Mesembryanthemum crystallinum* L. in a Mediterranean desert ecosystem. *J. Arid Environments.* **1991**, 20, 321–330.

Flowers, J.T.; Yeo, R.A. Breeding for salinity resistance in crop plants: where next? *Aust. J. Plant. Physiol.* **1995**, 22, 875–884.

Garbarino, J.; DuPont, F.M. NaCl induced a Na^+/H^+ antiport in tonoplast vesicles from barley roots. *Plant Physiol.* **1988**, 86, 231–236.

Guerrero, F.D.; Jones, J.T.; Mullet, J.E. Turgor-responsive gene transcription and RNA levels increase rapidly when pea shoots are wilted. Sequence and expression of three inducible genes. *Plant Mol. Biol.* **1990**, 15, 11–26.

Hanson, A.D.; Rathinasabapathi, B.; Rivoal, J.; Burnet, M.; Dillon, M.O.; Gage, D.A. Osmoprotective compounds in the Plumbaginaceae: A natural experiment in metabolic engineering of stress tolerance. *Proc. Natl. Acad. Sci. USA* **1994**, 91, 306–310.

Hare, P.D.; Cress, W.A. Metabolic implications of stress-induced proline accumulation in plants. *Plant Growth Regul.* **1997**, 21, 79–102.

Hetherington, A.M.; Quatrano, R.S. Mechanisms of action of abscisic acid at the cellular level. *New Phytol.* **1991**, 119, 9–32.

Johnson, K.D.; Herman, E.M.; Chrispeels, M.J. An abundant, highly conserved tonoplast protein in seeds. *Plant Physiol.* **1989**, 91, 1006–1013.

Kaldenhoff, R.; Kölling, A.; Meyers, J.; Karmann, U.; Ruppel, G.; Richter, G. The blue light-responsive *AthH2* gene of *Arabidopsis thaliana* is primarily expressed in expanding as well as in differentiating cells and encodes a putative channel protein of the plasmalemma. *Plant J.* **1995**, 7, 87–95.

Kammerloher, W.; Fisher, U.; Piechottka, G.P.; Schaffner, A.R. Water channels in the plant plasma membrane cloned by immunoselection from a mammalian expression system. *Plant J.* **1994**, 6, 187–199.

Kishor, P.; Hong, Z.L.; Miao, G.H.; Hu, C.; Verma, D. Overexpression of delta-pyrroline-5-carboxylate synthase increases proline production and confers osmotolerance in transgenic plants. *Plant Physiol.* **1995**, 108, 1387–1394.

Letham, D.S.; Palni, L.M.S. The biosynthesis and metabolism of cytokinins. *Ann. Rev. Plant Phyiol.* **1983**, 34, 163–197.

Liu, J.; Zhu, J-K. Proline accumulation and salt-stress-induced gene expression in a salt-hypersensitive mutant of *Arabidopsis. Plant Physiol.* **1997**, 114, 591–596.

Löw, R.; Rockel, B.; Kirsch, M.; Ratajczak, R.; Hörtensteiner, S.; Martinoia, E., Lüttge, U.; Rausch, T. Early stress effects on the differential expression of vacuolar H^+-ATPase genes in roots and leaves of *Mesembryanthemum crystallinum. Plant Physiol.* **1996**, 110, 259–265.

Matsumoto, H.; Chung, G.C. Increase in proton-transport activity of tonoplast vesicles as an adaptive response of barley roots to NaCl stress. *Plant Cell Physiol.* **1988**, 29, 1133–1140.

Maurel, C. Aquaporins and water permeability of plant membranes. *Annu. Rev. Plant Physiol. Plant Mol. Biol.* **1997**, 48, 399–429.

Maurel, C.; Tacnet, F.; Güclü, J.; Guern, J.; Ripoche, P. Purified vesicles of tobacco cells vacuolar and plasma membranes exhibit dramatically different water permeability and water channel activity. *Proc. Natl. Acad. Sci. USA.* **1997**, 94, 7103–7108.

Maurel, C.; Reizer, J.; Schroeder, J.I.; Chrispeels, M.J. The vacuolar membrane protein γ-TIP creates water specific channels in *Xenopus* oocytes. *EMBO J.* **1993**. 12, 2241–2247.

McCue, K.F.; Hanson, A.D. Drought and salt tolerance: towards the understanding and application. *Trends Biotechnol.* **1990**, 8, 358–362.

McElwain, E.F.; Bohnert, H.J.; Thomas, J.C. Light moderates the induction of phosphoenolpyruvate carboxylase by NaCl and abscisic acid in *Mesembryanthemum crystallinum. Plant Physiol.* **1992**, 99, 1261–1264.

Nakamura, Y.; Kasamo, K.; Shimosato, N.; Sakata, M.; Ohta, E. Stimulation of the extrusion of protons and H^+-ATPase activities with the decline in pyrophosphatase activity of the tonoplast in intact mung bean roots under high-NaCl stress and its relation to external levels of Ca^{2+} ions. *Plant Cell Physiol.* **1992**, 33, 139–149.

Narasimhan, M.L.; Binzel, M.L.; Perez-Prat, E.; Chen, Z.; Nelson, D.E.; Singh, N.K.; Bressan, R.A.; Hasegawa, P.M. NaCl regulation of tonoplast ATPase 70-kilodalton subunit mRNA in tobacco cells. *Plant Physiol.* **1991**, 97, 562–568.

Nelson, D.E.; Rammesmayer, G.; Bohnert, H.J. The regulation of cell-specific inositol metabolism and transport in plant salinity tolerance. *Plant Cell.* **1998**, 10, 1–14.

Niemietz, C.M.; Tyerman, S.D. Characterization of water channels in wheat root membrane vesicles. *Plant Physiol.* **1997**, 115, 561–567.

Osmond, C.B.; Greenway, H.; Salt response of carboxylation enzymes from species differing in salt tolerance. *Plant Physiol.* **1972**, 49, 260–263.

Ostrem, J.A.; Vernon, D.M.; Bohnert, H.J. Increased expression of a gene coding for NAD:glyceraldehyde-3-phosphate dehydrogenase during the transition from C_3 photosynthesis to crassulacean acid metabolism in *Mesembryanthemum crystallinum. J. Biol. Chem.* **1990**, 265, 3497–3502.

Paul, M.J.; Cockburn, W. Pinitol, a compatible solute in *Mesembryanthemum crystallinum.* L. *J. Exp. Bot.* **1989**, 40, 1093–1098.

Perera, I.Y.; Li, X.; Sze, H. Several distinct genes encode nearly identical 16 kDa proteolipids of the vacuolar H^+-ATPase from *Arabidopsis thaliana. Plant Mol. Biol.* **1995**, 29, 227–244.

Raskin, I.; Nanda-Kumar, P.B.; Dushenkov, S.; Salt, D.E. Bioconcentration of heavy metals by plants. *Current Biology* **1994**, 5, 285–290.

Rockel, B.; Ratajczak, R.; Becker, A.; Lüttge, U. Changed densities and diameters of intra-membrane tonoplast particles of *Mesembryanthemum crystallinum* in correlation with NaCl-induced CAM. *J. Plant Physiol.* **1994**, 143, 318–324.

Sarda, X.; Tousch, D.; Ferrare, K.; Legrand, E.; Dupuis, J.M.; Casse-Delbart, F.; Lamaze, T. Two TIP-like genes encoding aquaporins are expressed in sunflower guard cells. *Plant J.* **1997**, 12, 1103–1111.

Schäffner, A.R. Aquaporin function, structure, and expression: are there more surprises to surface in water relations? *Planta* **1998**, 204, 131–139.

Sheveleva, E.; Chmara, W.; Bohnert, H.J.; Jensen, R.G. Increased salt and drought tolerance by D-ononitol production in transgenic *Nicotiana tabacum* L. *Plant Physiol.* **1997**, 115, 1211–1219.

Smirnoff N.; Cumbes Q.J. Hydroxyl radical scavenging activity of compatible solutes. *Phytochemistry* **1989**, 28, 1057–1060.

Szaboles, I. The global problems of salt-affected soils. *Acta Agronomica Hungarica* **1987**, 36, 159–172.

Tarczynski, M.C.; Jensen, R.G.; Bohnert, H.J. Stress protection of transgenic tobacco by production of the osmolyte mannitol. *Science.* **1993**, 259, 508–510.

Thomas, J.C.; Bohnert, H.J. Salt stress perception and plant growth regulators in the halophyte *Mesembryanthemum crystallinum. Plant Physiol.* **1993**, 103, 1299–1304.

Thomas, J.C.; De Armond, R.L.; Bohnert, H.J. Influence of NaCl on growth, proline, and phosphoenolpyruvate carboxylase levels in *Mesembryanthemum crystallinum* suspension cultures. *Plant Physiol.* **1992a**, 98, 626–631.

Thomas, J.C.; McElwain, E.F.; Bohnert, H.J. Convergent induction of osmotic stress-responses. Abscisic acid, cytokinin, and the effects of NaCl. *Plant Physiol.* **1992b**, 100, 416–423.

Thomas, J.C.; Sepahi, M,; Arendall, B.; Bohnert, H.J. Enhancement of seed germination in high salinity by engineering mannitol-expression in *Arabidopsis thaliana. Plant Cell Environ.* **1995**, 18, 801–806.

Tsiantis, M.S.; Bartholomew, D.M.; Smith, J.A.C. Salt-regulation of transcript levels for the c subunit of a leaf vacuolar H^+-ATPase in the halophyte *Mesembryanthemum crystallinum. Plant J.* **1996**, 9, 729–736.

van Schilfgaarde, J. Irrigated agriculture, is it sustainable? In *Agricultural salinity assessment and management.* Tanji, K. Ed. American Society of Civil Engineers; New York, **1990**.

Vera-Estrella R.; Barkla B.J.; Bohnert H.J.; Pantoja, O. Salt-stress in *Mesembryanthemum crystallinum* cell suspensions activates mechanisms similar to those observed in the whole plant. *Plants* **1999**, 207, 426–435.

Vernon, D.M.; Bohnert, H.J. Increased expression of a *myo*-inositol methyl transferase in *Mesembryanthemum crystallinum* is part of a stress response distinct from crassulacean acid metabolism induction. *Plant Physiol.* **1992**, 99, 1695–1698.

Vernon, D.M.; Tarczynski, M.C.; Jensen, R.G.; Bohnert, H.J. Cyclitol production in transgenic tobacco. *Plant J.* **1993**, 4, 199–205.

Yamada, S.; Katsuhara, M.; Kelly, W.B.; Michalowski, C.B.; Bohnert, H.J. A family of transcripts encoding water channel proteins: Tissue-specific expression in the common ice plant. *Plant Cell.* **1995**, 7, 1129–1142.

Yamaguchi-Shinozaki, K.; Koizumi, M.; Urao, S.; Shinozaki, K. Molecular cloning and characterization of 9 cDNAs for genes that are responsive to desiccation in *Arabidopsis thaliana*: Sequence analysis of one cDNA that encodes a putative transmembrane channel protein. *Plant Cell Physiol.* **1992**, 3, 217–224.

Zingarelli, L.; Anzani, P.; Lado, P. Enhanced K^+-stimulated pyrophosphatase activity in NaCl-adapted cells of *Acer pseudoplatanus. Physiol. Plant.* **1994**, 91, 510–516.

8

COMBINATORIAL CHEMISTRY AND ITS APPLICATIONS IN AGRICULTURE AND FOOD

Dominic W. S. Wong and George H. Robertson

Western Regional Research Center, Agricultural Research Service
United States Department of Agriculture
800 Buchanan Street, Albany, California 94710

ABSTRACT

Combinatorial chemistry has become a major focus of research activity in the pharmaceutical industry for development new therapeutic compounds. The same techniques could be potentially applied to benefit agricultural and food research. This article reviews the various procedures used in combinatorial chemistry, outlines some of the strengths and limitations of the various methods, and proposes potential areas in agriculture and food that could be benefited by this technology. These areas include developing new antimicrobial agents, antioxidants, and other additives, creating antigen-binding molecules for the detection or removal of food pathogens or toxicants, engineering food proteins and enzymes for specific functions, and modifying biosynthetic pathways for the production of novel natural products.

INTRODUCTION

The concept of combinatorial chemistry is reshaping the way the pharmaceutical industry develops new therapeutic compounds. Traditional strategies for identifying novel lead compounds often rely on the screening of natural substances for medicinal properties or "rational design" based on the detailed information of the structure-function of the molecules. In contrast, the central idea of combinatorial chemistry is the construction of a chemical library—a pool of all possible variants of the molecule of interest. For example, a simple chemical having 5 substituents randomly attached to a core structure will have a library of 3625 structural variants. A combinatorial library of a pentapeptide will have a diverse population of over a million sequence structures. The enormous repertoire of molecules can be screened with appropriate selection steps for individual molecules that interact with a target

Chemicals via Higher Plant Bioengineering, edited by Shahidi *et al.*
Kluwer Academic / Plenum Publishers, New York, 1999.

and exert a desired biological property. Combinatorial experiments combine the ability to synthesize libraries of vast diverse populations with high-throughput screening methods to find the best fit candidate molecules without the need for prior knowledge of structure-function relationships of the molecules created. In some case, the candidate molecules selected in the initial round of screening can be subjected to mutation and amplification to generate a new population for the next selection. By repeating the process of selection, mutation and amplification, the final population will be increasingly composed of the best-fit molecules. This strategy has been successfully used in the production of antibodies, catalytic antibodies, analogs of hormones/receptors, therapeutic compounds, novel proteins and enzymes. The concept of combinatorial chemistry can be utilized equally well for the benefits of agriculture and food, although the application in this area has not been explored. This article reviews the various procedures used in combinatorial chemistry, and proposes potential applications of this technology in agriculture and food.

THE CHEMICAL APPROACH

Combinatorial libraries are prepared by either chemical or biological approaches (Fig. 1). Chemical approaches are employed to synthesize combinatorial libraries of oligomeric molecules such as peptides and small organic and inorganic molecules. Biological approaches are commonly used for peptides, proteins and nucleic acids. Combinatorial libraries that are produced by chemical synthesis are usually limited in size and diversity, with the exception of the split synthesis procedure for peptides described below. Libraries generated by biological methods that are encoded genetically can potentially attain a population of hundreds of millions of molecules.

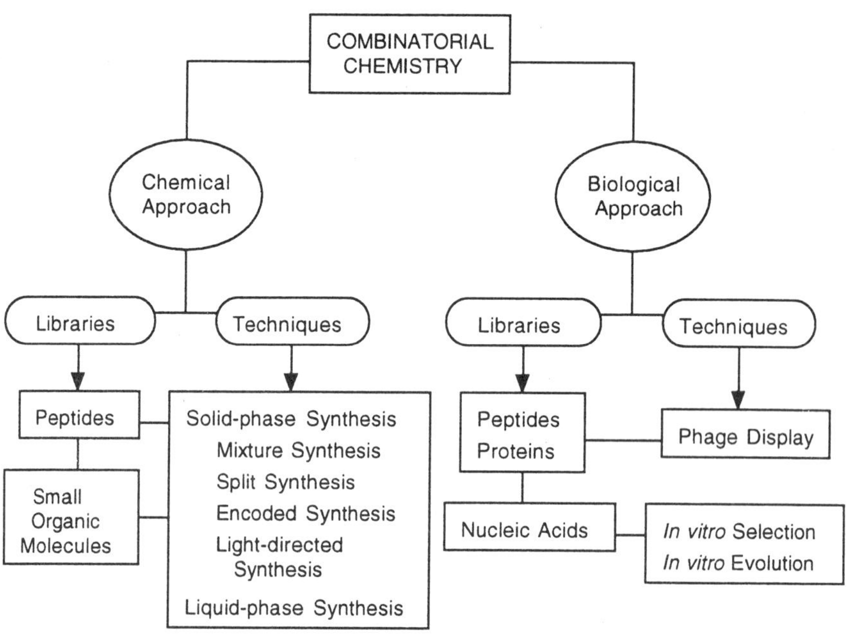

Figure 1. Chemical and biological approaches in creating combinatorial libraries.

MIXTURE SYNTHESIS WITH SPATIAL SEGREGATION

A majority of techniques in the category of chemical approaches employs solid-phase synthesis in which a stepwise coupling of mixtures of chemical units is conducted on a solid support such as polystyrene resin. The early procedures used a random mixture of activated amino acids in standard peptide synthesis by an iterative process. In the multi-pin method (Geysen et al. 1984; Bray et al. 1991), peptides were synthesized simultaneously on polyethylene rods arranged in a Telfon tray with spatially arranged wells similar to a microtiter plate. The procedure was used to synthesize all possible overlapping hexapeptides in parallel, covering the entire amino acid sequence of a coat protein of the foot-and-mouth disease virus. The sequence of the epitopic peptide (selected by ELISA) could be decoded, since each well was indexed to one hexapeptide unit. The size of libraries that can be generated in spatially segregated reaction wells is relatively small (Carter et al. 1992). In the "Tea-Bag" method (Houghten 1985; Houghten et al. 1991), peptide

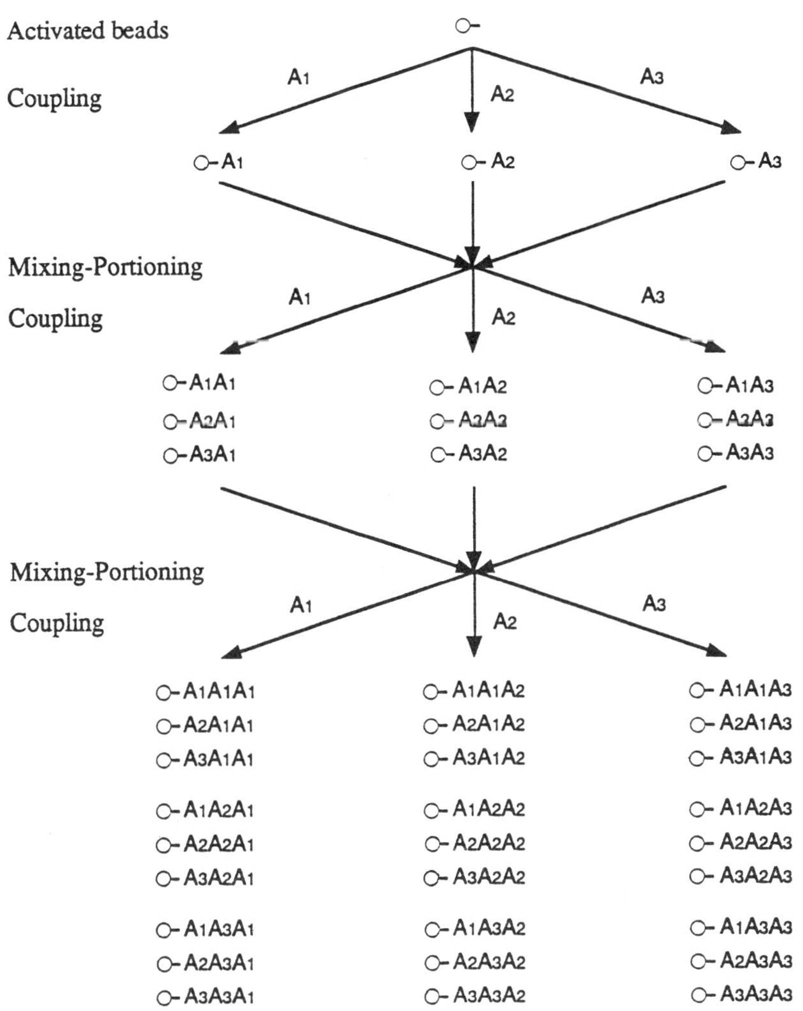

Figure 2. Split synthesis of one-bead, one-structure peptide library (adapted from Lam et al. 1991).

synthesis was carried out on resin beads contained in porous polypropylene packets. Individual packets were carried through the steps in standard solid-phase peptide synthesis. Both methods suffer from the disadvantage of having uneven product distribution due to differential coupling rates of amino acids in the reaction mixture.

SPLIT SYNTHESIS

With the introduction of the split synthesis method by Furka et al. (1991) and later by Lam et al. (1991), the size of a library increases dramatically to several millions of peptide variants. The method consists of cycles of randomization by combining and splitting the pool of resin beads after each coupling step for the next reaction. Figure 2 illustrates the essential steps using only 3 amino acids as building blocks, producing a library of 27 tripeptides. The size of the library expands exponentially with increasing number of building blocks. For a complete library of 19 amino acids with 5 cycles, the number of pentapeptides will be 2.5 million, with each bead carrying a unique peptide sequence. Libraries created in this manner are truly random, and each peptide structure in the library is prepared on a single resin bead. This one-bead, one-peptide technique has since been adapted to semi-automation and automation (Zuckermann et al. 1992; Neimark and Briand 1993; Lebl et al. 1995; Cargill and Maiefski 1996).

ENCODED LIBRARIES BY MOLECULAR TAGS

Based on the framework of these basic strategies, a number of innovative modifications have been reported. Individual molecules in a population of a synthetic library can be labelled with a particular oligonucleotide "tag" so that each chemical structure is specified and readily identified (Brenner and Lerner 1992; Ohlmeyer et al. 1993). The process consists of alternating stepwise synthesis of a peptide and a short oligonucleotide sequence after each mixing and portioning. Active molecules selected can be decoded by amplifying the genetic tag by PCR. The system presumably is applicable to nonpeptides as well. Molecular tags containing alcohols with different hydrocarbon chain lengths have also been used to encode the series of steps in the synthesis of a peptide library (Ohlmeyer et al. 1993; Borchardt and Still 1994). The codes can be analyzed by electron capture capillary gas chromatography, following selection of the active molecules.

LIGHT-DIRECTED SYNTHESIS

Another technique employs photochemical labile protecting groups and photolithography to provide light-directed combinatorial chemical synthesis of biopolymers on a solid support (Fodor et al. 1991; Holmes et al. 1995) (Fig. 3). Chemical linkers having the end groups derivatized with photochemically removable protecting groups (3,4-dimethoxy-6-nitroveratryloxycarbonyl) were attached to a solid substrate. Localized photodeprotection was achieved by directing light through a photolithographic mask to a specific area of the substrate. Chemical coupling in the next step with NVOC-protected amino acids occurred only at those sites which had been activated by illumination (with protective groups removed). The chemical cycle was then repeated except that the illumination was directed to a different region of the substrate through a new mask. The combination of a shifting pattern of photolithographic masking, and the sequence of reactions defines the products and their pre-

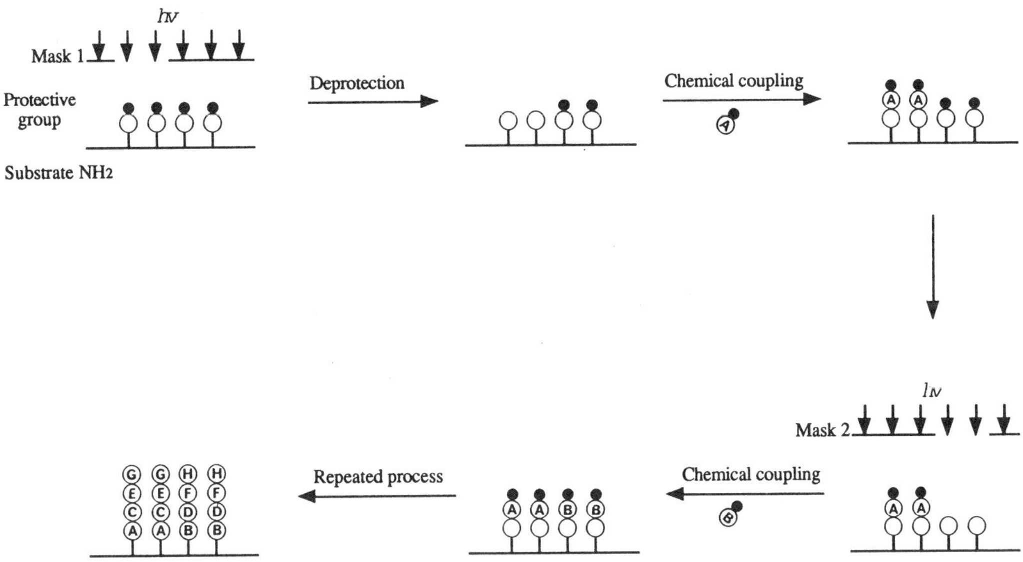

Figure 3. Combinatorial chemical synthesis of biopolymers on a solid support using photoligthographic techniques (Adapted from Fodor et al. 1991).

cise locations in the combinatorial set. For an 8 step 4 masking synthesis, the number of possible combinations is 65,536, all fitted into a 1.6 cm2 with 50 mm square sites.

PEPTIDE LIBRARIES

Synthetic peptide combinatorial libraries have been used in epitope mapping (Geysen et al. 1984; Houghten 1985; Lam et al. 1991; Holmes et al. 1995), development of novel antimicrobial peptides (Houghten et al. 1992), identification of functional antibody fragments (Francisco et al. 1993), discovery of new ligands to receptor proteins and other target macromolecules (Jayawickreme et al. 1994; Lam et al. 1994). Following the identification of an antagonist of eukephalin with a high affinity for m opioid receptors from a combinatorial library of 52 x 106 hexapeptides, Dooley et al. (1993, 1994) synthesized an unnatural all D-amino acid peptide library and isolated an opioid peptide using an iterative selection process. The D-amino acid peptide exhibited high selectivity for m receptors and induction of analgesia in mice. Koppel et al. (1995) identified several tripeptide pharmacophores with potent inhibitor on serotonin uptake from screening a tripeptide cassette library. A tetrameric tripeptide that binds Fc immunoglobulin has been identified through screening of a synthetic peptide library (Fassina et al. 1996). The peptide ligand immobilized onto epoxy-activated resins was utilized for antibody affinity purification mimicking the function of protein A columns.

LIBRARIES OF SMALL ORGANIC MOLECULES

An increasing number of nonpeptide libraries have been constructed for the screening of bioactive compounds. In contrast to peptide synthesis where monomeric units are sequentially added to an elongating chain, the diversity of a library of small organic molecules is defined by the spatial arrangement of the functional building blocks (reactive substituents) on a core molecule. Scaffolding the building blocks at random generates a random set of compounds. Libraries of complex structures can be constructed by a combination of a variety of core molecules and building blocks.

A library of structurally diverse 1,4-benzodiazepine derivatives has been constructed containing a large number of functional groups including phenols, carboxylic acids, amines, amides, and indoles (Bunin and Ellman 1992; Bunin et al. 1994). The library was screened using competitive labelling assay for bioactivity in the binding of cholecystokinin A receptor, and the structure-function relationship of the active derivatives was confirmed. A number of combinatorial libraries consisting of cyclic and heterocyclic core molecules has been synthesized. Using all-cis cyclopentane anhydride as the template, Patek et al. (1994) synthesized cyclopentanes with an array of substitutions of diverse functional secondary and tertiary amides with regioselectivity. Along the same line, a library of piperazinediones has been constructed by reductive alkylation with reducing agents followed by cyclization (Gordon and Steele 1995). A combinatorial library of phosphodiester oligomers was constructed using pyrrolidine phosphoramidite monomer with N-acetylation by a wide variety of carboxylic acids (Hebert et al. 1994). The combinatorial approach has been extended to the design of polymers for specialty applications. A library of polyarylates was created by reacting a set of tyrosine-derived diphenols carrying a variable pendent chain to a set of diacids with systematic variation in backbone structures (Brocchini et al. 1997). Cellular adhesion and proliferation were tested on the polymers and found to be correlated to surface properties. Using light-directed chemical synthesis, Cho et al. (1993) developed a library of oligocarbamates, analogs of peptides with a carbamate backbone. The library was selected for binding affinity to a monoclonal antibody raised against phosphonate hapten, a transition state analogue for ester hydrolysis.

The use of combinatorial techniques for the production of new solid-state materials was reported by Xiang et al. (1995). The library consists of different stoichiometrics and sequences of depositions of individual oxides of barium, bismuth, calcium, copper, lead, strontium, and yttrium in the construction of thin films. The array of sites containing combinations of oxides were sintered at 840°C for the metals to fuse into new materials. A new class of materials whose resistance varies with applied magnetic field has been identified. Sites containing BiSrCaCuOx and YBa2Cu3Ox were identified to be superconducting. Using a similar but automated synthesis of thin-film libraries, Danielson et al. (1997) identified a new red phosphor with a quantum efficiency comparable to those of existing commercial phosphors.

Recently, liquid-phase combinatorial synthesis of both peptide and non-peptide (arylsulfonamides) libraries have been achieved by using a soluble polymer, polyethylene glycol monomethyl ether, instead of solid beads (Han et al. 1995). The polymer is soluble in aqueous and organic solvents, and purification can be easily accomplished by crystallization. Most importantly, the synthesis may be carried out under homogeneous conditions eliminating some of the drawbacks of solid-phase synthesis, such as differential coupling and unequal access to reactive sites. Direct synthesis and screening of chemical libraries of small compounds, such as free in solution have also been tested (Carell et al. 1995).

THE BIOLOGICAL APPROACH

In the biological approach, peptide or protein variants in a library are expressed and displayed on the surface of filamentous phage particles by gene fusion. Other biological systems, including *E. coli* and yeast, have also been used and reported.

THE PHAGE DISPLAY TECHNIQUE FOR PEPTIDES AND PROTEINS

Filamentous bacteriophage produces coat proteins, two of which are the pIII and pVIII, which anchor in the membrane with the N-terminal domain exposed on the surface. Both proteins are synthesized with post-translational cleavage of the corresponding signal peptide. The idea of inserting a foreign DNA into gene *III* or *VIII*, to be expressed as fusion proteins on the surface of a phage particle, was first described by Smith (1985) and the procedure has been widely adopted as the method of choice in generating combinatorial libraries of peptides and proteins (Smith and Scott 1993).

The general scheme involves the creation of a library containing all combinations of DNA sequences which encode the peptide or protein variants of interest by codon mutation. The DNA fragments are inserted downstream of the signal peptide sequence of the pIII gene. A number of phage vectors have been designed with cloning sites and antibiotic resistance markers to facilitate cloning and screening (Parmley and Smith 1988). Infection of *E. coli* host with the recombinant DNA will give progeny phage particles displaying the peptide or protein variants as fusion proteins on the surface (Fig. 4). Each phage particle displays a single variant in multiple (up to 5) copies. The pool of phage particles can be screened by various detection methods depending on the specific application. For example, a hexapeptide library was screened for binding interactions to a specific antibody (V3 loop of HIV-1 gp120 protein) by immobilized biotinylated antibody complexed to streptavidin-coated oxirane-derivatized beads (Christian et al. 1992; Straptavidin coating can be immobilized on polystyrene petri plates, in a method known as "biopanning" (Parmley and Smith 1988)). Peptide variants retained by affinity binding to the immobilized antibodies were recovered and identified.

In several studies, phagemid vectors, consisting of plasmid replication origin, an antibiotic resistance marker, and the phage intergenic regions, are used. In this type of system, a helper phage needs to be coinfected to supply the phage genes to facilitate replication and packaging. A high percentage of the phage particles generated in this man-

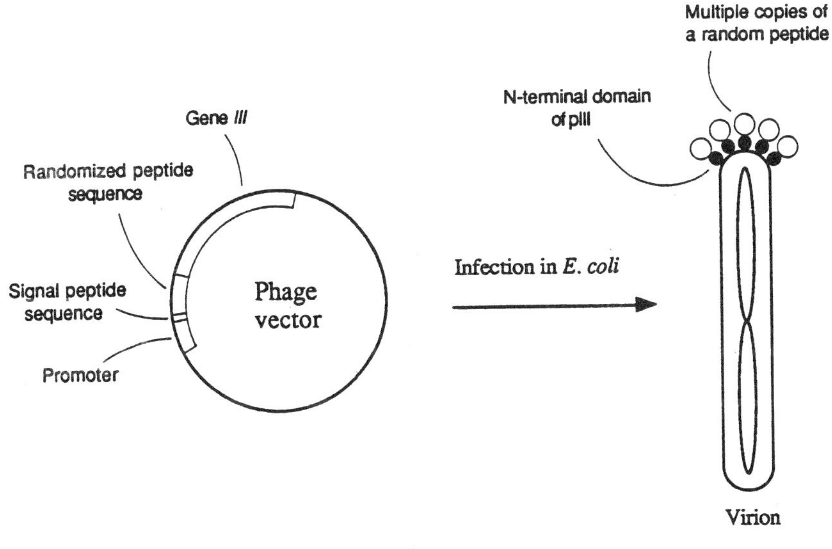

Figure 4. Constructing libraries of peptides displayed by filamentous phage.

ner display one copy of the peptide or protein variants, due to suppression of the transcription of the gene fusion protein (Lowman et al. 1991; Matthews and Wells 1993).

APPLICATIONS OF PHAGE DISPLAY

Applications in the use of the phage display technique encompass the following areas: generation of monoclonal Fab fragments, and catalytic antibodies (Huse et al. 1989; Clackson et al. 1991; Janda et al. 1994), hormones and hormone analogs with enhanced receptor affinity (Lowman et al. 1991; Cull et al. 1992), identification of antigenic determinants of proteins (Pinilla et al. 1993), isolation of molecules with binding specificity to bioactive compounds (Cwirla et al. 1990; Blond-Elguindi et al. 1993; Dedman et al. 1993), and peptide mimetics of carbohydrate ligands for sugar-binding proteins (Hoess et al. 1993; Oldenburg et al. 1992).

The technique has also been extended to include random mutagenesis of proteins and enzymes to probe for structure-function properties, to alter substrate specificity, to enhance thermal/pH stability, and to engineer novel catalytic activities. Procedures for producing and analyzing complete mutant libraries of protein coding domains have been described (Hutchison et al. 1991; Wells and Lowman 1992). A random library of myoglobin was used to screen for O_2 and CO binding characteristics by Huang and Box (1994) and the results yielded information on a new set of specific pathways in myoglobin-ligand binding. Janda et al. (1994) isolated functional Fab molecules from a combinatorial antibody library that catalyzes the hydrolysis of thioester. Schatz (1993) used combinatorial techniques to identify a minimal functional substrate for the *E. coli* biotin holoenzyme which catalyzes the transfer of biotin from biotinyl-5'-adenylate to a unique lysine residue in a limited number of specific proteins. By screening a peptide library of >109 clones, a 13-residue substrate was identified for enzymatic biotinylation. Widersten and Manervick (1995) constructed a library of mutant glutathione transferases by random mutation of 10 amino acids at the substrate binding site. Using transition state analogs for affinity selection, mutants were isolated with enhanced binding activity towards a number of substrates. Graham et al. (1993) constructed a serine protease library by random substitution of a number of residues at the substrate-binding site, using an a-lytic protease mutant (Met190Ala) as the parent molecule. The alterations in substrate specificity, cleavage preference, and catalytic activity, were then used to systematically analyze the structure-function relationships of the subsites of the enzymes. Replacement of Met213 at the S1 subsite with a His residue was found to increase substantially the specificity toward cleavage at the P1 His or Met in synthetic amide substrates. Recently, Urabe's group has created a library of artificial proteins of 141 amino acid residues and 20% were expressed in *E. coli* as soluble proteins. Two of the artificial proteins were found to possess low but distinct esterase activity (Prijambada et al. 1996; Yomo and Urabe 1997). The results substantiate the contention that the unique sequence and conformation of a natural enzyme evolves through natural selection for activity function.

Many of the examples in the literature utilize selection procedures initially based on complexation between protein/enzyme mutants and the target molecules (substrate, inhibitor, etc.) For screening enzyme mutants targeted for improved catalysis, it is desirable to have a system that can be performed directly on the basis of catalytic turnover. Avalle et al. (1997) used suicide inhibitor linked to a biotin moiety to select active b-lactamase mutants from a mixed library of β-lactamase mutants and penicillin-binding proteins. The former enzyme hydrolyzes the antibiotic substrate going through an acyl-enzyme formation, while the latter protein forms a stable acyl-enzyme without further hydrolysis. Janda et al. (1997) constructed a phage library of antibodies raised against haptens that were designed to mimic a glycosidic bond cleavage in its transition state. Catalytic antibodies were selected by a mechanism-based procedure. In this particular system, the suicide inhibitor, *o*-, *p*-(difluoromethyl)aryl-β-D-glucosides, was coupled to BSA via a disulfide bridge. Antibodies that catalyzed glycosidic cleavage would generate a reactive quinone methide species alkylating any nucleophilic side chains at the active site, thereby covalently trapping the catalytic phage clones.

COMBINATORIAL BIOSYNTHESIS

A novel use of combinatorial approaches has been the reshuffling of genes to effect changes in the biosynthetic pathways in microorganisms for the production of new products. In essence, this represents a restructuring of the building blocks of complex polymers by manipulating gene composition of relating multifunctional enzyme complex. This strategy has been explored in the biosynthesis of polyketides, an important class of compounds that includes many antibiotics and anticancer agents, using a module of polyketide synthase complex in a *Streptomyces coelicolor* host-vector system (McDaniel et. al. 1995; Marsden et al. 1998). The synthase gene cluster consists of ketosynthase, acyltransferase, chain length determining factor, acyl carrier protein, ketoreductase, and cyclase, that together catalyze cycles of Claisen condensation, reduction, and cyclization from malonyl CoA starter units. Structural variability of the polyketides arises from the variation of substituents in the starter units, the extent of reduction, and the location of intramolecular ring-closure. The chemical diversity of the products is related to the modular organization of the synthase gene cluster. A large repertoire of polyketides can be generated by the manipulation (rearrangement, deletion, addition) of the polyketide synthase gene complex.

IN VITRO EVOLUTION OF RNA ENZYMES

There has been a marked progress in using the combinatorial approach in engineering RNA molecules with novel binding properties, and a wide range of catalytic activities that are not found in natural ribozymes. In the construction of nucleic acid libraries, RNA pools can be generated by in vitro transcription of DNA prepared by PCR using randomized oligomers. A selection step is applied to the population of a random library of RNA to screen and isolate individual variants that possess the desired property, followed by amplification of the molecules. The heterogeneous population is eventually reduced to the molecule that best fits the requirement by iterative selection and amplification. Various selection constraints can be applied to direct the evolutionary path of the molecules in each generation. Mutations can be introduced after each selection to generate a new pool of the surviving molecules isolated from the previous selection (Fig. 5).

An in vitro system was developed for rapid performance of amplification, mutation and selection of RNA by Joyce (1989). In the general scheme, cDNA obtained from reverse transcription of the RNA pool was used to synthesize the second DNA strands using an oligoprimer with a T7 promoter in a 5' overhang sequence. Partial digestion with T7 gene 6 5'->3' exonuclease under controlled conditions produces ssDNA templates for DNA polymerase extension to generate mutant strands. For polymerase extension, two primers were used—one primer hybridized to the 3' end, and the other primer (carrying random mutation region of mismatch bases) hybridized to the upstream sequence (Fig. 5).

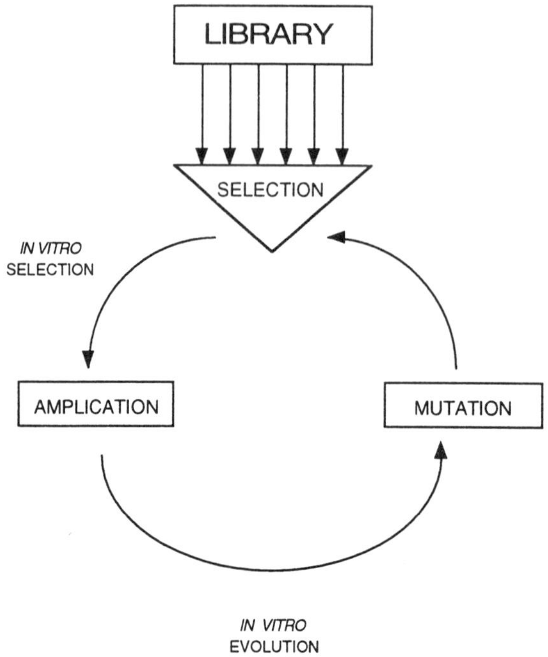

Figure 5. Procedure for mutagenesis and amplification of RNA (adapted from Joyce 1989, Joyce and Inoue 1989).

The resulting dsDNA was then used to transcribe RNA for a second round of selection, mutation, and amplification.

In vitro selection of high affinity RNA ligands to specific target DNA or protein has been an area of continuous focus (Ellington and Szostak 1990). RNA recognition of DNA and protein may have potential therapeutic applications. Ribozymes with unique catalytic activities have been produced, and a list of reactions catalyzed by synthetic RNA enzymes include—cleavage of single-stranded DNA, ligation, polynucleotide kinase activity, N-alkylation, and cleavage of amide bond. Recently, the activity of a group I ribozyme for DNA cleavage was enhanced by 100-fold versus to that of RNA, after a total of 63 generations of evolution, resulting in an average of 28 mutations relative to wild-type enzyme (Tsang and Joyce 1996).

Several research groups have extended the concept of in vitro evolution to make single-stranded DNAs that fold up into molecules with catalytic activity. The first DNA enzyme (deoxyribozyme) was produced by Breaker and Joyce (1994) that, in combination with metal ions, catalyzes the cleavage of RNA. Later work from the same group succeeded in creating a self-cleavaging DNA (Carmi et al. 1996). Sen and Li (1996) isolated a 24-nucleotide DNA strand that catalyzes the insertion of a metal ion into a porphyrin ring. In this catalyst, the active site is likely located between the upper and middle guanine quartets held by hydrogen bonds.

POTENTIAL APPLICATIONS TO AGRICULTURE AND FOOD

Despite the rapid growth in the application of combinatorial chemistry in the pharmaceutical industry, there has been little attention to its use in agriculture and food. The

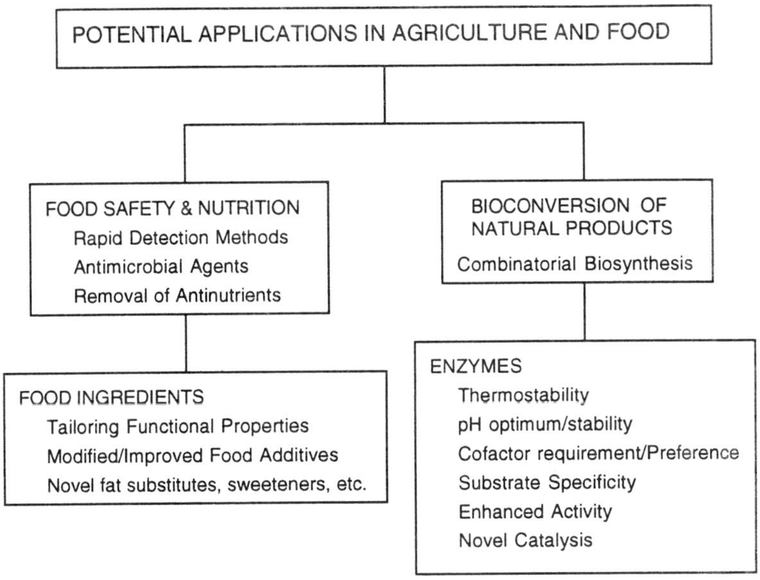

Figure 6. A list of potential applications of combinatorial chemistry in agriculture and food.

concept of combinatorial chemistry should have wide and profound implications in the utilization of agricultural resources and in the production of better foods (Fig. 6). The following highlights some of the areas, initially proposed at a UJNR joint protein panel meeting (Wong et al. 1995), that could be benefited by this technology.

1. Food safety and nutrition. The combinatorial approach is most applicable to the development of rapid detection methods for food pathogens. Many of the current detection methods rely on antibody-antigen reaction or DNA hybridization coupled with enzyme-linked colorimetric assay. Functional Fab fragments have been produced by generating and screening libraries of the immunoglobulin gene repertoire as already mentioned. Ligands of less complex structures can be created to have high affinity to specific surface proteins in pathogenic organisms. High affinity ligands can be produced to form tight complexes with antinutrients or toxicants rendering them unavailable for digestion or absorption. Antimicrobial activity of existing additives can be improved by selection of libraries of structural variants of the chemicals in use. A series of hexapeptides, isolated from by iterative selection of a synthetic peptide library, have been shown to possess various degrees of antimicrobial activity against Pseudomonas aeruginoa, Escherichia coli, Staphylococcus aureus, Streptococcus sanguis, and Candida albicans (Blondelle et al. 1995).
2. Bioconversion of natural products. Biosynthesis of natural products can be engineered using combinatorial approaches. The strategy has been exemplified in the biosynthesis of polyketides using a Streptomyces coelicolor host-vector system as described above. This concept can be used in general applications in engineering biosynthetic routes in the production of new compounds. It offers a viable approach to manipulate the organization of endogenous gene clusters to effect alterations in the nature and quality of natural products.

3. Food ingredients. Chemical and physical properties of food components can be tailored for food formulation and processing. New ingredients may be generated with unique functionalities. Food additives, such as antioxidants and emulsifiers, can be modified or improved by creating libraries of diverse structural variants of existing chemicals and selecting for desired properties (Nokihara et al. 1997). Combinatorial chemistry offers an expansion of the pools of candidate compounds for screening and selecting for unique properties and applications. It is a particularly attractive alternative for identifying novel fat substitutes, low-calorie sweeteners, preservatives, and flavoring compounds, in contrast to the traditional strategies of screening natural substances and computer modeling for structural designs of new compounds.
4. Modification of proteins and enzymes. With the advent of gene technology, tailoring proteins and enzymes for specific functions has become a reality. Combinatorial chemistry offers an additional powerful tool to this rapidly advancing field. In the last decade, the structures of a number of enzymes important to food systems and food processing have been studied in fine detail. The list includes, among others, a-amylase, b-amylase, glucoamylase, cellobiohydrolase, endoglucanase, lipase, phospholipase, lipoxygenase, glucose oxidase, peroxidase, catalase, and xylose isomerase. It now becomes practical to generate combinatorial mutagenesis at a localized site or domain of a protein to achieve alteration of its functional properties. Enzyme reactions are involved in every aspect from the growth, storage, processing of raw materials to production of food products, and their consumption. In nearly all the industrial enzymes currently used there are properties awaiting improvement. For example, in the case of xylose isomerase, there is intense interest to develop mutants with increased thermostability, lower pH optimum, altered metal cation preference, and shift in substrate specificity from xylose to glucose (Wong 1995). With the concerns of the general public on the use of chemicals, the utilization of enzymes in industrial processing will definitely increase.

CONCLUSION

The technology of combinatorial chemistry offers a route to synthesize libraries of vast repertoire, that when coupled with sensitive and fast screening methods and integrated with mutagenesis, will expedite the creation and isolation of molecules of interest. The capability of exploring the structural and functional limits of molecular diversity in a population of chemical compounds makes combinatorial approach a powerful tool. It has been used primarily to synthesize and identify bioactive compounds and macromolecules aimed for pharmaceutical and therapeutic applications. Similar combinatorial strategies for new applications in agriculture and food have not been available. The techniques should be extended to include a broad spectrum of areas including food safety and nutrition, development of product ingredients, and processing and bioconversion of natural products. Combinatorial chemistry is a whole new way of searching for the right molecule.

Reference to a company and/or products is only for the purposes of information and does not imply approval or recommendation of the product to the exclusion of others which may also be suitable. All programs and services of the U.S. Department of Agriculture are offered on a nondiscriminatory basis without regard to race, color, national origin, religion, sex, age, marital status, or handicap.

REFERENCES

Avalle, B., Vanwetswinkel, S., and Fastrez, J. In vitro selection for catalytic turnover from a library of b-lactamase mutants and penicillin-binding proteins. *Bioorg. Medicinal Chem. Lett.* **1997**, *7*, 479–484.

Blond-Elguindi, S., Cwirla, S. E., Dower, W. J., Lipshutz, R. J., Sprang, S. R., Sambrook, J. F., Gething, M.-J. H. Affinity panning of a library of peptides displayed on bacteriophages reveals the binding specificity of BiP. *Cell*, **1993**, *75*, 717–728.

Blondelle, S. E., Takahashi, E. Dinh, K. T., and Houghten, R. A. The antimicrobial activity of hexapeptides derived from synthetic combinatorial libraries. *J. Appl. Bacteriol.* **1995**, *78*, 39–46.

Borchardt, A., and Still, W. C. Synthetic receptor binding elucidated with an encoded combinatorial library. *J. Am. Chem. Soc.* **1994**, *116*, 373–374.

Bray, A. M., Maeji, N. J., Valerio, R. M., Campbell, R. A., and Geysen, H. M. Direct cleavage of peptides from a solid support into aqueous buffer. Application in simultaneous multiple peptide synthesis. *J. Org. Chem.* **1991**, *56*, 6659–6666.

Breaker, R. R., and Joyce, G. F. A DNA enzyme that cleaves RNA. *Chem. Biol.* **1994**, *1*, 223–229.

Brenner, S., and Lerner, R. A. Encoded combinatorial chemistry. *Proc. Natl. Acad. Sci. USA.* **1992**, *89*, 5381–5383.

Brocchini, S., James, K., Tangpasuthadol, V., and Kohn, J. A combinatorial approach for polymer design. *J. Am. Chem. Soc.* **1997**, *119*, 4553–4554.

Bunin, B. A., and Ellman, J. A. A general and expedient method for the solid-phase synthesis of 1,4-benzodiazepine derivatives. *J. Am. Chem. Soc.* **1992**, *114*, 10997–10998.

Bunn, B. a., Plunkett, M. J., and Ellman, J. A. The combinatorial synthesis and chemical and biological evaluation of a 1,4-benzodiazepine library. *Proc. Natl. Acad. Sci. USA* **1994**, *91*, 4708–4712.

Carell, T., Winter, E. A., Sutherland, A. J., Rebek, J., Jr., Dunayevskiy, Y. M., and Vouros, P. New promise in combinatorial chemistry: synthesis, characterization, and screening of small-molecule libraries in solution. *Chem. Biol.* **1995**, *2, 171–183.*

Cargill, J. F., and Maiefski, R. R. Automated combinatorial chemistry on solid phase. *Laboratory Robotics and Automation* **1996**, *8*, 139–148.

Carmi, N., Schultz, L. A., and Breaker, R. R. In vitro selection of self-cleaving DNAs. *Chem. Biol.* **1996**, *3*, 1039–1046.

Carter, J. M., Vanalbert, S., Lee, J., Lyon, J., and Deal, C. Shedding light on peptide synthesis. *Bio/Technology* **1992**, *10*, 509–513.

Cho. C. Y., Moran. E. J., Cherry. S. R., Stephans. J. C., Fodor. S. P. A., Adams. C. L., Sundaram. A., Jacobs. J. W., and Schultz, P. G. An unnaturnal biopolymer. *Science* **1993**, *261*, 1303–1305.

Christian, R. B., Zuckermann, R. N., Kerr, J. M., Wang, L., Malcolm, B. A. Simplified methods for construction, assessment and rapid screening of peptide libraries in bacteriophage. *J. Mol. Biol.* **1992**, *27*, 711–718.

Clackson, T., Hoogenboom, H. R., Griffiths, A. D., and Winter, G. Making antibody fragments using phage display libraries. *Nature* **1991**, *352*, 624–628.

Cull, M. G., Miller, J. F., and Schatz, P. J. Screening for receptor ligands using large libraries of peptides linked to the C terminus of the lac repressor. *Proc. Natl. Acad. Sci. USA* **1992**, *89*, 1865–1869.

Cwirla, S. E., Peters, E. A., Barrett, R. W., and Dower, W. J. Peptides on phage: a vast library of peptides for identifying ligands. *Proc. Natl. Acad. Sci. USA* **1990**, *87*, 6378–6382.

Danielson, E., Golden, J. H., McFarland, E. W., Reaves, C. M., Weinberg, W. H., and Wu, X. D. A combinatorial approach to the discovery and optimization of luminescent materials. *Nature* **1997**, *389*, 944–948.

Dedman, J. R., Kaetzel, M. A., Chan, H. C., Nelson, D. J., and Jamieson, G. A. Jr. Selection of targeted biological modifiers from a bacteriophage library of random peptides. *J. Biol. Chem.* **1993**, *268*, 23025–23030.

Dooley, C. T., Chung, N. N., Schiller, P. W., and Houghten, R. A. Acetalins: Opioid receptor antagonists determined through the use of synthetic peptide combinatorial libraries. *Proc. Natl. Acad. Sci. USA* **1993**, *90*, 10811–10815.

Dooley, C. T., Chung, N. N., Wilkes, B. C., Schiller, P. W., Bidlack, J. M., Pasternak, G. W., and Houghten, R. A. An all D-amino acid opioid peptide with central analgesic activity from a combinatorial library. *Science* **1994**, *266*, 2019–2022.

Ellington, A. D. and Szostak, J. W. In vitro selection of RNA molecules that bind specific ligands. *Nature* **1990**, *346*, 818–822.

Fodor, S. P. A., Read, J. L., Pirrung, M. C., Stryer, L., Lu, A. T., and Solas, D. Light-directed, spatially addresable parallel chemical synthesis. *Science* **1991**, *251*, 767–773.

Francisco, J. A., Campbell, R., Iverson, B. L., and Georgiou, G. Production and fluorescence-activated cell sorting of Escherichia coli expressing a functional antibody fragment on the external surface. *Proc. Natl. Acad. Sci. USA* **1993**, *90*, 10444–10448.

Furka, A., Sebestyen, F., Asgedom, M., and Dibo, G. General method for rapid synthesis of multicomponent peptide mixtures. *Int. J. Peptide Protein Res.* **1991**, *37*, 487–493.

Geysen, H. M., Meloen, R. H., and Barteling, S. J. Use of peptide synthesis to probe viral antigens for epitopes to a resolution of a single amino acid. *Proc. Natl. Acad. Sci. USA* **1984**, *81*, 3998–4002.

Gordon, D. W., and Steele, J. Reductive alkylation on a solid phase: synthesis of a piperazinedione combinatorial chemistry. *Bioorg. Medicinal Chem. Lett.* **1995**, *5*, 47–50.

Graham, L. D., Haggett, K. D., Jennings, P. A., Le Brocque, D. S., and Whittaker, R. G. Random mutageneis of the substrate-binding site of a serine protease can generate enzymes with increased activities and altered primary specificies. *Biochemistry* **1993**, *32*, 6250–6258.

Han, H., Wolfe, M. M., Brenner, S., and Janda, K. D. Liquid-phase combinatorial synthesis. *Proc. Natl. Acad. Sci. USA.* **1995**, *92*, 6419–6423.

Hebert, N., Davis, P. W., DeBaets, E. L., and Acevedo, O. L. Synthesis of N-substituted hydroxyprolinol phosphoramidites for the preparation of combinatorial libraries. *Tetrahedron Letts.* **1994,** *35*, 9509–9512.

Hoess, R., Brinkmann, U., Handel, T., and Pastan, I. Identification of a peptide which binds to the carbohydrate-specific monoclonal antibody B3. *Gene*, **1993**, *128*, 43–49.

Holmes, C. P., Adams, C. L., Kochersperger, L. M., Mortensen, R. B., and Aldwin, L. A. The use of light-directed combinatorial peptide synthesis in epitope mapping. *Biopolymers* **1995**, *37*, 199–211.

Houghten, R. A. General method for the rapid solid-phase synthesis of large numbers of peptides: Specificity of antigen-antibody interaction at the level of individual amino acids. *Proc. Natl. Acad. Sci. USA* **1985**, *82*, 5131–5135.

Houghten, R. A., Appel, J. R., Blondelle, S. E., Cuervo, J. H., Dooley, C. T., and Pinilla, C. The use of synthetic peptide combinatorial libraries for the identification of bioactive peptides. *BioTechniques* **1992**, *13*, 412–421.

Houghten, R. A., Pinilla, C., Blondelle, S. E., Appel, J. R., Dooley, C. T., and Cuervo, J. H. Generation and use of synthetic peptide combinatorial libraries for basic research and drug discovery. *Nature* **1991**, *354*, 84–86.

Huang, X., and Boxer, S. G. Discovery of new ligand binding pathways in myoglobin by random mutagenesis. *Structural Biology* **1994**, *1*, 226–229.

Huse, W. D., Sastry, L., Iverson, S. A., Kang, A. S., Alting-Mees, M., Burton, D. R., Benkovic, S. J., and Lerner, R. A. Generation of a large combinatorial library of the immunoglobulin repertoire in phage lambda. *Science* **1989**, *246*, 1275–1281.

Hutchison, C. A. III, Swanstrom, R., and Loeb, D. D. Complete mutagenesis of protein coding domains. *Meth. Enzymol.* **1991**, *202*, 356–390.

Janda, K. D., Lo, L-C., Lo, C.-H. L., Sim, M.-M., Wang, R., Wong, C.-H., and Lerner, R. A. Chemical selection for catalysis in combinatorial antibody libraries. *Science* **1977**, *275*, 945–948.

Janda, K. D., Lo, C.-H. L., Li, T., Barbas, C. F. III, Wirsching, P., and Lerner, R. A. Direct selection for a catalytic mechanism from combinatorial antibody libraries. *Proc. Natl. Acad. Sci. USA* **1994**, *91*, 2532–2536.

Jayawickreme, C. K., Graminski, G. F., Quillan, J. M., and Lerner, M. R. Creation and functional screening of a multi-use peptide library. *Proc. Natl. Acad. Sci. USA* **1994**, *91*, 1614–1618.

Joyce, G. F. Amplification, mutation and selection of catalytic RNA. *Gene* **1989**, *82*, 83–87.

Lam, K. S., Salmon, S. E., Hersh, E. M., Hruby, V. J., Kazmierski, W. M., and Knapp, R. J. A new type of synthetic peptide library for identifying ligand-binding activity. *Nature* **1991**, *354*, 82–84.

Lam, K. S., Zhao, S.-G., Wade, S., Krchnak, V., and Lebl, M. Identification of small peptides that interact specifically with a small organic dye. *Drug Development Research* **1994**, *33*, 157–160.

Lebl, M., Krchnak, V., Sepetov, N. F., Seligmann, B., Strop, Felder, S., and Lam, K. S. One-bead-one-structure combinatorial libraries. *Biopolymers* **1995**, *37*, 177–198.

Lowman, H. B., Bass, S. H., Simpson, N., and Wells, J. A. Selecting high-affinity binding proteins by monovalent phage display. *Biochemistry* **1991**, *30*, 10832–10838.

Marsden, A. F. A., Wilkinson, B., Cortes, J., Dunster, N. J., Staunton, J., and Leadlay, P. F. Engineering broader specificity into an antibiotic-producing polyketide synthase. *Science* **1998**, *279*, 199–202.

Matthews, D. J., and Well, J. A. Substrate phage: selection of protease substrates by monovalent phage display. *Science* **1993**, *260*, 1113–1117.

McDaniel, R., Ebert-Khosla, S., Hopwood, D. A., and Khosla, C. Rational design of aromatic polyketide natural products by recombinant assembly of enzymatic subunits. *Nature* **1995**, *375*, 549–554.

Neimark, J., and Briand, J.-P. Development of a fully automated multichannel peptide synthesizer with integrated TFA cleavage capacity. *Pept. Res.* **1993**, *6*, 219–228.

Nokihara, K., Yasuhara, T., Muramoto, K., Ando, E., and Wray, V. **1997.** Studies on peptides exhibiting antioxidative activity: Construction of a peptide library and screening. Peptide Chemistry 1996, ed. C. Kitada, Protein Research Foundation, Osaka, Japan.

Ohlmeyer, M. H. J., Swanson, R. N., Dillard, L. W., Reader, J. C., Asouline, G., Kobayashi, R., Wigler, M., and Still, W. C. Complex synthetic chemical libraries indexed with molecular tags. *Proc. Natl. Acad. Sci. USA* **1993,** *90,* 10922–10926.

Oldenburg, K. R., Longanathan, D., Goldstein, I. J., Schultz, P. C., and Gallop, M. A. 1992. Peptide ligands for a sugar-binding protein isolated from a random peptide library. *Proc. Natl. Acad. Sci. USA* **1992,** *89,* 5393–5397.

Parmley, S. F., and Smith, G. P. Antibody-selectable filamentous fd phage vectors: affinity purification of target genes. *Gene* **1988,** *73,* 305–318.

Patek, M., Drake, B., and Lebl, M. All-cis cyclopentane scaffolding for combinatorial solid phase synthesis of small non-peptide compounds. *Tetrahedron Letts.* **1994,** *35,* 9169–9172.

Pinilla, C., Appel, J. R., and Houghten, R. A. Synthetic peptide combinatorial libraries (SPXLs): identification of the antigenic determinant of b-endorphin recxognized by monoclonal antibody 3E7. *Gene* **1993,** *128,* 71–76.

Prijambada, I. D., Yomo, T., Tanaka, F., Kawama, T., Yamamoto, K., Hasegawa, A., Shima, Y., Negoro, S., and Urabe, I. Solubility of artifical proteins with random sequences. *FEBS Lett.* **1996,** *382,* 21–25.

Yomo, T., and Urabe, I. Properties of artificial proteins with random sequences. Enzyme Engineering XIV Conference, Beijing, China, Oct. 12–17, 1997.

Schatz, P. J. Use of peptide libraries to map the substrate specificity of a peptide-modifying enzyme: 13 residue consensus peptide specifies biotinylation in Escherichia coli. *Bio/Technology* **1993,** *11,* 1138–1143.

Smith, G. P. Filamentous fusion phage: novel expression vectors that display cloned antigens on the virion surface. *Science* **1985,** 228, 1315–1317.

Smith, G. P., and Scott, J. K. Libraries of peptides and proteins displayed on filamentous phage. *Meth. Enzymol.* **1993,** *217,* 228–257.

Tsang, J., and Joyce, G. F. Specialization of the DNA-cleaving activity of a group I ribozyme through in vitro evolution. *J. Mol. Biol.* **1996,** *262,* 31–42.

Wells, J. A., and Lowman, H. B. Rapid evolution of peptide and protein binding properties in vitro. *Curr. Opin. Biotechnol.* **1992,** *3,* 355–362.

Widersten, M., and Mannervik, B. Glutathione transferases with novel active sites isolated by phage display from a library of random mutants. *J. Mol. Biol.* **1995,** *250,* 115–122.

Wong, D. W. S. **1995.** Food Enzymes: Structure and Mechanism. Chapman and Hall, New York.

Wong, D. W. S., Pavlath, A. E., and Robertson, G. H. **1995**. Combinatorial approach in generating RNA and DNA enzymes. Proc. UNJR Protein Panel Meeting, Tsukuba Science City, Japan.

Xiang, X.-D., Sun, X., Briceno, G., Lou, Y., Wang, K. A., Chang, H., Wallace-Freedman, W. G., Chen, S.-W., and Schultz, P. G. A combinatorial approach to materials discovery. *Science* **1995,** *268,* 1738–1740.

Zuckermann, R. N., Kerr, J. M., Siani, M. A., Banville, S. C., and Santi, D. V. Identification of highest-affinity ligands by affinity selection from equimolar peptide mixtures generated by robotic synthesis. *Proc. Natl. Acad. Sci. USA* **1992,** *89,* 4505–4509.

9

BIOLOGICAL NITROGEN FIXATION AND FUTURE CHALLENGES OF AGRICULTURE

The Endophytic Connection

Federico Sánchez,* Luis Cárdenas, and Carmen Quinto

Departamento de Biología Molecular de Plantas
Instituto de Biotecnología
Universidad Nacional Autónoma de México
Apartado Postal 510-3, Cuernavaca Morelos, 62271, México

ABSTRACT

Feeding the growing global population, anticipated to be 8 billion by the year 2020, is one of the most important recent challenges of agriculture. The increase in cereal grain yield, to cope with this demand, directly implies a dramatic increase in the use of nitrogen-based fertilizers and agrochemicals. Some of these intensive agricultural practices have progressive detrimental effects on the environment. This review is focused on some novel insights gained into the understanding of associative and symbiotic interactions of plants with nitrogen-fixing organisms that makes Biological Nitrogen Fixation (BNF) a viable answer to this compelling dilemma.

INTRODUCTION

Food production must be markedly increased during the next decades, since human population is expected to increase by about 95 million annually. It is estimated that there will be 8 billion people by the year 2020 (Chrispeels and Sadava, 1994).

Both striking increases in cereal grain yields in the developed countries and "green revolution" success in agriculture in the developing world, during the last few decades, is a direct consequence of dramatic increases in the use of nitrogen-based fertilizers (Vance,

* Corresponding author; telephone 52-73291653; fax 52-73-136600; e-mail: federico@ibt.unam.mx

Chemicals via Higher Plant Bioengineering, edited by Shahidi *et al.*
Kluwer Academic / Plenum Publishers, New York, 1999.

1998) with concomitant negative effects on the environment. However, there is not a simple answer to this predicament; the fundamental problem of feeding the growing population and the progressive environmental degradation of the ecosystems associated with some intensive agricultural practices demands effective solutions to these problems.

After photosynthesis, biological nitrogen fixation (BNF) is probably the most important metabolic process on Earth. Whereas photosynthesis supplies carbon and oxygen necessary for life on our planet, BNF results in the reduction of nitrogen gas (approximately 80% of our atmosphere) into ammonia. Although all living organisms require reduced ammonia for the synthesis of essential biochemicals such as nucleic and amino acids, only certain prokaryotes are capable of "fixing nitrogen" due to the synthesis of the enzyme nitrogenase which catalyzes the conversion of atmospheric nitrogen into ammonia.

The association between grasses and endophytic diazotrophs and the interaction between legume crops and endosymbiotic nitrogen-fixing bacteria give to some plants the selective ability to grow in nitrogen-poor soils. These remarkable associations and symbiotic interactions are especially important for natural ecosystems, forestry and for sustainable and intensive agriculture. The use of BNF, as compared to fertilizer application, has the advantage of putting all the N into the plant and spares the use of fossil fuel, which are, and will continue to be in the future, major points to consider for the protection of the environment. On the other hand, most food production in the developing countries of the tropics is based on small farms, often family operated, especially in the Central and South America, and Africa where the use of fertilizer and agro-chemicals is minimal. Improving of nitrogen-fixing traits for grasses and cereal-associations through molecular tools and genetic manipulations for better inocula, and breeding for high yield grain legume crops in such circumstances, provides excellent opportunities for the production of green manure, and high protein food sources, at the same time, helping to protect the soil organic matter and enhance soil fertility (Rolfe et al., 1998b).

ENDOPHYTIC DIAZOTROPHIC ASSOCIATIONS WITH CEREALS

Research on nitrogen fixation in cereals started with the pioneering work of Döbereiner (von Bülow and Döbereiner, 1975) and the isolation and description of bacteria of the genus *Azospirillum* (Tarrand et al., 1978). This genus comprises free-living N_2 fixing rhizosphere bacteria that have been isolated from different soil types and from the roots of numerous wild and cultivated plants all over the world. Field trials, carried out at multiple locations, have shown that under certain environmental and soil conditions, inoculation with *Azospirillum* has beneficial effects on plant yields (see the review by Okon and Labandera-González, 1994). However, in general it is not N_2 fixation, but bacterial phytohormone biosynthesis, that has been proposed for being responsible for the observed plant growth promotion upon *Azospirillum* inoculation, since proliferation of root mass is always beneficial to the plant in absorbing extra nutrient (Okon and Vanderleyden, 1997). Regarding direct plant growth promotion, the ability of *Azospirillum brasilense* to produce indol-3-acetic acid (IAA) is now intensively studied. IAA biosynthesis in *A. brasilense* has proven to be very complex, since at least three biosynthetic pathways are operative (Costacurta and Vanderleyden, 1995). This property is exploited with the commercialization of inoculants such as Azogreen by the Lipha Company in France.

One of the restrictions with the inoculation of diazotrophic bacteria, like *Azospirillum,* is that they only fix nitrogen under nitrogen-limiting, microaerobic conditions. Therefore, their limited ability to secrete ammonium ion in the presence of reduced sources of nitrogen is a drawback. Recent findings that could facilitate the "evolution" of an effective nitrogen-fixing association towards a successful endosymbiosis between diazotrophs and cereals are: 1) The isolation of a spontaneous *A. brasilense* Sp7 mutant (Sp7-S) impaired in flocculation which has reduced exopolysaccharide production. Such a mutant presents a modified colonization pattern and superior nitrogen fixation rate, in association with wheat roots, under conditions in which the nitrogenase activity of free-living *Azospirillum* organisms was inhibited by an excess of oxygen (Katupitiya et al., 1995; Pereg-Gerk et al., 1998). 2) The generation of engineered *A. brasilense* strains with a truncated NifA regulatory protein (a transcriptional activator of the nitrogenase structural genes) and other genetic modifications which allow these mutated bacteria, to excrete ammonium and still fix nitrogen in the presence of ammonium (Arsene et al., 1996; de Zamaroczy et al., 1996). 3) *Azospirillum irakenses* is the only *Azospirillum* species which displays a clear xylanase and pectinolytic activity. By means of this activity, *A. irakense*, in contrast to the other *Azospirullum* species, is able to grow and fix nitrogen on media with pectin or polygalacturonate (PGA) as the sole source of carbon. This property could explain why this bacteria, which is isolated from rice roots, can penetrate and be intracellularly established in high numbers as compared to other *Azospirillum* species that lack these lytic enzymes (van de Broek et al., 1998). It is clear that the aforementioned process of facilitated evolution, which eventually would lead to more effective nitrogen-fixing, endophytic associations, involves the stepwise genetic improvement of both the prospective microsymbionts and the cereal host. In this regard, clear evidence is provided that crack-entry at the points of emergence of lateral roots or of 2,4-D induced para-nodules in wheat (Zeman et al., 1992) is the most significant route of enhanced endophytic colonization. Furthermore, using this laboratory model of para-nodulated wheat, there is now evidence that the endophytic ability of *Azospirilum brasilense* (Sp7-S) can be genetically controlled (Pereg-Gerk et al., 1998)

In the last ten years an increasing number of examples describing systemic colonization of graminous crops by diazotrophic bacteria of the genus *Azoarcus* (Reinhold-Hurek et al., 1993a), *Herbaspirillum* (Baldani et al., 1996) and *Acetobacter* (Cavalcante and Döbereiner, 1988), have been reported. Recent insights gained with biochemical and genetical approaches, and the use of molecular tools developed for studying the synergistic interactions, clearly indicate that these bacteria stimulate growth by fixing nitrogen on their host plants.

Most strains belonging to the genus *Azoarcus* spp. are root-associated diazotrophs capable of invading the roots of both Kallar grass (*Leptochloa fusca* (L.) Kunth) and rice seedlings, especially in the zone of elongation, and differentiation just above the root tips (Hurek et al., 1994). *Azoarcus* isolates are grouped into five species (Reinhold-Hurek et al., 1993b; Hurek and Reinhold-Hurek, 1995). Three species which have been isolated from the root interior of Kallar grass have not yet been detected in the soil, thus suggesting that they are closely associated with plants. However, recent data indicate that they can be isolated from the sclerotia of a basidomycete which is related to the phytopathogen *Ustilago hordei* found in the soil of the former Kallar grass field which had been recently planted with rice. This fungi probably serves as a niche for survival of *Azoarcus* in the soil and as a source for reinfection of plants (Hurek et al., 1997).

Recent findings indicate that strains of *Herbaspirillum* and *Acetobacter,* in particular *A. diazotrophicus,* colonize specific cultivars of sugarcane as endophytes and make a significant contribution to the nitrogen economy of the plants (Fuentes-Ramírez et al., 1993; Caballero-Mellado and Martínez-Romero, 1994; Boddey et al., 1998). This significant property could even rise to a novel type of endophytic interaction, one in which, like the Rhizobium/Legume symbiosis, the plant partner supplies photosynthate to support bacterial nitrogen fixation and the bacterium provides fixed nitrogen for plant growth. Early reports indicated that *A. diazotrophicus* occurs endophytically only in sugarcane and in a few other vegetatively propagated sucrose-rich plants (Döbereiner, 1993; Caballero-Mellado et al., 1995). On the other hand, recent published data have demonstrated that this bacteria is capable of colonizing both the rhizosphere and tissues of coffee plants propagated through seeds (Jiménez-Salgado et al., 1997). Moreover, maize plants can be colonized in high numbers by *A. diazotrophicus*, albeit without exerting beneficial or detrimental effects on the growth of maize (Caballero-Mellado et al., 1998). This exciting result broadens the host plant spectrum and the potentiality of introducing beneficial traits into plants by endophytic bacteria. This is because the bacteria is capable of colonizing the internal tissues of plants not considered natural hosts until now. Obviously a more detailed study is required because colonization and plant growth promotion by *A. diazotrophicus* could also be dependent on the plant host genotype.

It is interesting to note that, as in the case of *Azoarcus*, which has been detected in association with facultative parasitic fungi, *A. diazotrophicus* occurs inside Vesicular Arbuscular Micorrhiza (VAM) (Paula et al., 1993).

ENDOSYMBIOTIC NITROGEN-FIXING INTERACTIONS WITH LEGUMES

Several families of bacteria associate intimately with plants to establish symbiotic interactions whereby carbon from the plant is supplied to the bacteria which in turn furnishes the plant with reduced nitrogen. The most agronomically important of these symbiotic associations is between members of the legume family (Fabaceae) and the bacterial family Rhizobiaceae, which consists of *Azorhizobium*, *Bradyrhizobium*, *Rhizobium*, and *Sinorhizobium*. The symbiotic association with rhizobia is restricted to legumes with one interesting exception, *Parasponia* in the Ulmaceae. The central question is what makes the legumes unique in that only they can associate with these nitrogen-fixing bacteria? Is there a specific receptor for recognizing the rhizobia or rhizobial gene products? Why aren't all plants capable of establishing an effective symbiosis with nitrogen-fixing bacteria?

The interaction between *Rhizobium* and its legume host is very specific. Generally one species of *Rhizobium* can infects only one or a few closely related species of legumes. This specificity resides in the signaling molecules that both organisms use. The host legume releases flavonoids into the soil, with different plant species producing diverse flavonoids. These compounds stimulate the coordinated expression of bacterial genes which are required for nodulation (*nod* genes). The *nod* genes, in turn, encode for enzymes involved in the synthesis of small substituted lipochitin oligosaccharides (LCOs), signal molecules that are potent plant morphogens with unique structure, the so called "Nod factors" (Lerouge et al., 1990; Fisher and Long, 1992; Spaink, 1996). Purified Nod factors have been shown to induce plant responses at concentrations as low as 10^{-12} M (Lerouge et al., 1990; Spaink et al., 1991; Truchet et al., 1991). The responses reported for the purified Nod factors include root-hair deformation (Heidstra et al., 1994), root hair curling (Relic et al., 1993), formation of pre-infection threads (van Brussel et al., 1992), cell divisions in the root cortex and in some legumes genuine nodule primordia (see Mylona et al., 1995). At the membrane level, it has been demonstrated that Nod factors are able to induce depolarization of the root-hair membrane potential in a specific way (Ehrhardt et al., 1992; Felle et al., 1995). In addition, Nod factors induce the transcription of the plant early nodulin (ENOD) genes, such as ENOD40 and ENOD12, which are related to the infection process (Horvath et al., 1993; Bauer et al., 1994; Crespi et al., 1994; Journet et al., 1994; Vijn et al., 1995; Fang and Hirsch, 1998). Nevertheless, rhizobia are required for the formation of an infection thread, which facilitates rhizobial entry into host cells (Callaham and Torrey, 1981). Thus, Nod factors could be considered a primary morphogenetic signal for nodulation because it triggers on a compatible host some of the earliest stages of nodule development. This is probably one of the most significant discoveries of the last ten years in this field.

The symbiotic interaction culminates in the formation of nitrogen-fixing nodules (Brewin, 1991; Hirsch, 1992; Mylona, et al., 1995, Hirsch and LaRue, 1997). The number of nodules and their morphology is determined by the plant host, and not by the bacterium (Dobritsa and Novik, 1992; Pawlowski and Bisseling, 1996). The number of nodules on the host root is autoregulated. This mechanism is abolished in some legume mutants displaying the so-called supernodulation phenotype (see Caetano-Anollés and Gresshoff, 1991). As a consequence of the increased nodule number, the aerial parts of supernodulated mutants display reduced growth (Hansen et al., 1990; 1992), indicating that in such conditions rhizobial colonization can become parasitic.

In tropical areas, a number of naturally occurring plant-*Rhizobium* interactions, which could play an important role in maintenance and improving of soil fertility, remain largely unexploited. Lately, the isolation of presumptive diazotrophic endophytic bacteria from rice seedlings has been reported. *Rhizobium leguminosarum* bv. *trifolii* has been found as an endophytic bacteria growing in *Oryza* roots (Rolfe et al., 1998a). It is yet to be found how the bacteria reach the intracellular stage in these tissues and adapt to selective rice varieties. Nevertheless, these studies undoubtedly will open the possibility for further inoculation approaches to improve the productivity of rice crops.

Coinoculation trials in legume crops with both strains of *Azospirillum* and *Rhizobium* (Penmetsa et al., 1997; Plazinski and Rolfe, 1985), or endomycorrhizic microorganisms and *Rhizobium* (Makonese et al., 1998), have proven to have a synergetic effect on nitrogen fixation and plant growth. These beneficial and complementary interactions between endophytic and endosymbiotic microorganisms deserve further attention and deeper fundamental understanding.

In contrast to what has been unraveled about the bacterial partner, very little is known about the plant genes important for establishing a nitrogen-fixing symbiosis. Several Nod- (non-nodulating) and Fix- (non-nitrogen-fixing) legume mutants have been isolated, suggesting that multiple genes are required for nodule development (Mylona et al., 1995; Hirsch and LaRue, 1997). However, so far the identity of the genes that are disrupted in these mutants and the proteins encoded by these genes remain unknown (Geurts et al., 1997)

Ever since the chemical identification of Nod factor, a signal molecule generated by rhizobia that triggers plant morphogenesis, there has been great interest in finding the receptor with which Nod factor or *Rhizobium* interacts. In part, the interest lies in the fact that the process of nodulation and subsequent nitrogen fixation initiated by the interaction with rhizobia appears (with one exception) to be restricted to legumes. Thus, the quest for finding a potential receptor that interacts with Nod factor/*Rhizobium* is attractive not only for understanding the mechanics and evolution of nodule formation and nitrogen-fixing ability in legumes, but also for speculating as to whether the introduction of such a receptor into a non-legume plant would enable such a plant to nodulate in response to rhizobia.

Nevertheless, conferring nitrogen-fixing ability to a non-leguminous plant is not going to be easy based as these appears to involve many steps even for the initial stages of nodulation. It is likely that of the various separate complementation groups of identified non-nodulating mutants one may represent a mutation in a potential receptor.

Additional evidence which indicates that some non-nodulating mutants are blocked in perceiving signal molecules, due to a mutation in a receptor, is the fact that several Nod^- indeterminate nodule-forming mutants, such as alfalfa MN1008 and various pea mutants, are also Myc-, i.e. incapable of establishing a mycorrhizal association with phosphate-acquiring fungi (Duc et al., 1989; Bradbury et al., 1991; Sagan et al., 1995). Indeed, Nod factors are related to a class of molecules that are more typical of fungal cell walls than of Gram-negative bacterial surfaces (Bonfante-Fasolo et al., 1990). In addition, it has been reported that alfalfa mycorrhizal roots contain transcripts for ENOD2 and ENOD40 just like *Rhizobium*-inoculated alfalfa roots (Wyss et al., 1990; van Rhijn et al., 1997). This sharing of downstream gene products strongly suggests that the signal transduction pathways for nodulation and mycorrhizal formation are conserved (Hirsch and Kapulnik, 1998). LaRue and Weeden (1994) proposed that *Rhizobium*'s ability to associate with plants evolved from VAM-fungi, and that Nod factor could be the link. This is a very interesting hypothesis which needs to be explored with genetic and molecular techniques.

CONCLUDING REMARKS

It is now clear that there is a great potential to include in agriculture one of the most important processes on Earth, biological nitrogen fixation, as valuable tool, together with plant breeding, genetic engineering and biotechnology. To use the associations between endophytic diazotrophs with cereals and the synergistic interaction of endosymbionts such as VAM-fungi and *Rhizobium* as effective and environmentally safe inoculants to promote crop yield. There is still a great deal to learn and to improve through fundamental research and coordinated extension. These efforts should be supported and encouraged, in order to change the actual concept of traditional and intensive agricultural practices and to cope with the increasing demand for food production in order to feed the growing population, a paradox that could eventually lead to a matter of survival or extinction.

REFERENCES

Arsene, F.; Kaminski, P.A.; Elmerich, C. Modulation of NifA activity by PII in *Azospirillum brasilense*: evidence for a regulatory role of the NifA N-terminal domain. J. Bacteriol. **1996**, 178, 4830–4838.

Baldani, J.I.; Pot, B.; Kirchhof, G.; Falsen, E.; Baldani, V.L.; Olivares, F.L.; Hoste, B.; Kersters, K.; Hartmann, A.; Gillis, M.; Döbereiner, J. Emended description of Herbaspirillum; inclusion of (Pseudomona) rubrisubalbicans, a milk plant pathogen, as Herbaspirillum rubrisubalbicans comb. nov.; and classification of a group of clinical isolates (EF goup I) as Herbaspirillum species 3. Int. J. Syst. Bacteriol. **1996**, 46, 802–810.

Bauer, P.; Crespi, M.D.; Szecsi, J.;Allison, L.A.; Schultze, M.; Ratet, P.; Kondorosi, E.; Kondorosi, E. Alfalfa *ENOD12* genes are differentially regulated during nodule development by Nod factors and *Rhizobium* invasion. Plant Physiol. **1994**, 105, 585–592.

Boddey, R.M.; Giller, K.E.; Cadish, G.; Alves, B.J.R.; Urquiaga, S. Contribution of biological nitrogen fixation to tropical agriculture: actual and potential. In Biological Nitrogen Fixation for the 21st Century. Elmerich, C., Kondorosi, A., Newton, W.E. (eds), 599–603. Kluwer Academic Publishers. The Netherlands. **1998**.

Bonfante-Fasolo, P.; Faccio, A.; Perotto, S.; Schubert, A. Correlation between chitin distribution and cell wall morphology in the mycorrhizal fungus *Glomus versiforme*. Mycol. Res. **1990**, 94, 157–165.

Brewin, N.J. Development of the legume root nodule. Annu. Rev. Cell Biol. **1991**, 7, 191–226.

Bradbury, S.M.; Peterson, R.L.; Bowley, S.R. Interactions between three alfalfa nodulation genotypes and two Glomus species. New Phytol. **1991**, 119, 115–120.

Caballero-Mellado, J.; Martínez-Romero, E. Limited genetic diversity in the endophytic sugarcane bacterium *Acetobacter diazotrophicus*. Appl. Environ. Microbiol. **1994**, 60, 1532–1537.

Caballero-Mellado, J.; Fuentes-Ramírez, L.E.; Reis, V.M.; Martínez-Romero, E. Genetic structure of *Acetobacter diazotrophicus* populations and identification of a new genetically distant group. Appl. Environ. Microbiol. **1995**, 61, 3008–3013.

Caballero-Mellado, J.; Martínez-Romero, E.; Estrada de los Santos, P.; and Fuentes-Ramírez, L.E. Maize colonization by *Acetobacter diazotrophicus*. In Biological Nitrogen Fixation for the 21st Century. Elmerich, C., Kondorosi, A., Newton, W.E. (eds), 381–382. Kluwer Academic Publishers. Dordrecht, The Netherlands. **1998**.

Caetano-Anollés, G.; Gresshoff, P. M. Plant genetic control of nodulation. Annu. Rev. Microbiol. **1991**, 45, 345–382.

Callaham, D.A.; Torrey, J.G. The structural basis for infection of root hairs of *Trifolium repens* by *Rhizobium*. Can. J. Bot. **1981**, 59, 1647–1664.

Cavalcante, V.A.; Döbereiner, J. A new-acid tolerant nitrogen fixing bacterium associated with sugarcane. Plant Soil. **1988**, 108, 23–31.

Chrispeels M.J.; Sadava, D.E. Human population growth: lessons from demography. In Plants, Genes, and Agriculture. 1, 1–24. Jones and Bartlett Publishers International. Boston, MA. **1994**.

Costacurta, A.; Vanderleyden, J. Synthesis of phytohormones by plant-associated bacteria. Crit. Rev. MIcrobiol. **1995**, 21, 1–18.

Crespi, M.D.; Jurkevitch, E.; Poiret,.M.; d'Aubenton-Carafa, Y.; Petrovics, G.; Kondorosi, E.; Kondorosi, A. enod40, a gene expressed during nodule organogenesis, codes for a non-translatable RNA involved in plant growth. EMBO J. **1994**, 13, 5099–5112.

De Zamaroczy, A..; Paquelin, A.; Peltre, G.; Forchhammer, K.; Elmerich, C. Coexistence of two structurally similar but functionally different P II proteins in Azospirillum brasilense. J. Bacteriol. **1996**, 178, 4143–4149.

Döbereiner, J. History and new perpectives of diazotrophs in asociation with non-leguminous plants. Symbiosis **1993**, 13, 1–13.

Dobritsa, S.V.; Novik, S.N. Feedbaack regulation of nodule formation in *Hippophae rhaminoides*. Plant Soil. **1992**, 144, 45–50.

Duc, G.; Trouvelot, A.; Gianinazzi-Pearson, V.; Gianinazzi, S. First report of non-mycorrhizal plant mutants (Myc-) obtained in pea (Pisum sativum L.) and fababean (Vicia faba L.). Plant Sci. **1989**, 60, 215–222.

Ehrhardt, D. W.; Atkinson, E. M.; Long, S. R. Depolarization of alfalfa root hair membrane potential by *Rhizobium meliloti* Nod factors. Science. **1992**, 256, 998–1000.

Fang, Y; Hirsch, A. Studying early nodulin gene *ENOD40* expression and induction by nodulation factor and cytokinin in transgenic alfalfa. Plant Physiol. **1998**, 116, 53–68.

Felle, H. H.; Kondorosi, E.; Kondorosi, A.; Schultze, M. Nod signal-induced plasma membrane potential changes in alfalfa root hairs are differentially sensitive to structural modifications of the lipochitooligosaccharide. Plant J. **1995**, 7, 6, 939–947.

Fisher, R.F.; Long, S.R. *Rhizobium*-plant signal exchange. Nature. **1992**, 357, 655–660.

Fuentes-Ramírez, L.E.; Jiménez-Salgado, T.; Abarca-Ocampo, I.R.; Caballero-Mellado, J. *Acetobacter diazotrophicus*, an indolacetic acid producing bacterium isolated from sugarcane cultivars of México. Plant Soil. **1993**, 154, 145–150.

Geurts, R.; Heidstra, R.; Hadri, A.E.; Downie, A.J.; Franssen, H.; van Kammen, A; Bisseling. T. Sym2 of pea is involved in a nodulation factor-perception mechanism that controls the infection process in the epidermis. Plant Physiol. **1997**, 115, 351–359.

Hansen, A.P.; Peoples, M.B.; Brown, P.H.; Carroll, B.J.; Gresshoff, P.M. Nitrogen partioning during early development of supernodulating soybean (*Glycine max* (L.) Merrill) mutants and their wild-type parent. J. Exp. Bot. **1990**, 1239–1244.

Hansen, A.P.; Yoneyama, T.; Kouchi, H. Short-term nitrate effects of hydroponically-grown soybean cv. Bragg and its supernodulating mutant: II. Distribution and respiration of recently fixed ^{13}C-labeled photosynthate. J. Exp. Bot. **1992**, 43, 9–14.

Heidstra, R.; Geurts, R.; Franssen, H.D.; Spaink, H.P.; van Kammen, A.; Bisseling, T. Root-hair deformation activity of nodulation factors and their fate on Vicia sativa. Plant Physiol. **1994**, 105, 787–797.

Hirsch, A.M. Developmental biology of legume nodulation New Phytol. **1992**, 122, 211–237.

Hirsch, A.M.; LaRue, T.A. Is the legume nodule a modified root or stem or an organ *sui generis*? Crit. Rev. Plant Sci. **1997**, 16, 4, 361–392.

Hirsch, A.M.; Kapulnik, Y. Signal transduction pathways in mycorrhizal associations: Comparisons with the Rhizobium-legume symbiosis. Fungal Genet. Biol. **1998**, 23, 205–212.

Horvath, B.; Heidstra, R.; Lados, M.; Moerman. M.; Spaink. H.P.; Promé, J.-C.; Van Kammen, A.; Bisseling, T. Lipo-oligosaccharides of *Rhizobium* induce infection-related early nodulin gene expression in pea root hairs. Plant. J. **1993**, 4, 727–733.

Hurek, T.; Reinhold-Hurek, B.; Van Montagu, M.; Kellenberger, E. Root colonization and systemic spreading of Azoarcus sp. strain BH72 in grasses. J. Bacteriol. **1994**, 176, 1913–1923.

Hurek, T.; Reinhold-Hurek, B. Identification of grass-associated and toluene degrading diazotrops Appl. Environ. Microbiol. **1995**, 61, 2257–2261.

Hurek, T.; Wagner, B.; Reinhold-Hurek, B. Identification of N_2-fixing plant-and fungus-associated Azoarcus species by PCR-bases genomic fingerprints. Environ. Microbiol. **1997**, 63, 4331–4339.

Jiménez-Salgado, T.; Fuentes-Ramírez, L.E.; Tapia-Hernández, A.; Mascarua-Esparza, M.A.; Martínez-Romero, E.; Caballero-Mellado, J. *Coffea arabica* L., a new host plant for *Acetobacter diazotrophicus* and isolation of other nitrogen-fixing acetobacteria. Appl. Environ. Microbiol. **1997**, 63, 3676–3683.

Journet, E.P.; Pichon, M.; Dedie, A.; de Billy, F.; Truchet, G.; Barker, D.G. *Rhizobium meliloti* nod factors elicit cell-specific transcription of the ENOD 12 gene in transgenic alfalfa. Plant J. **1994**, 6, 241–249.

Katupitiya, S.; Millet, J.; Vesk, M.; Viccars, L.; Zeman, A.; Lidong, Z.; Elmerich, C.; Kennedy, I.R. A mutant of *Azospirillum brasilense* Sp7 impaired in flocculation with a modified colonization pattern and superior nitrogen fixation in association with wheat. Appl. Environ. Microbiol. **1995**, 61, 1987–1995.

LaRue, T.; Weeden, N.F. The symbiosis genes of the host. In: Proc. First European Nitrogen Fixation Conference. Kiss, G.B. and Endre, G., Eds. 147–151. Officina Press, Szeged, **1994**.

Lerouge, P.; Roche, P.; Faucher, C.; Maillet, F.; Truchet, G.; Promé, J.-C.; Dénarié, J. Symbiotic host-specificity of *Rhizobium meliloti* is determined by a sulfated and acylated glucosamine oligosacchaaride signal. Nature. **1990**, 344, 781–784.

Makonese, F.T.; Mpepereki, S.; Kasasa, P.; Mafongoya, P. Response of soybean (Glycine max) to co-inoculation with rhizobia and mycorrhizae. In Biological Nitrogen Fixation for the 21st Century. Elmerich, C., Kondorosi, A., Newton, W.E. (eds), 633. Kluwer Academic Publishers. Dordrecht, The Netherlands. **1998**.

Mylona, P.; Pawlowski, K.; Bisseling, T. Symbiotic Nitrogen Fixation. The Plant Cell. **1995**, 7, 869–885

Okon, Y.; Labandera-González, C.A. Agronomic applications of *Azospirillum*. An evaluation of 20 years worldwide field inoculation. Soil Biol. Biochem. **1994**, 26, 1551–1601.

Okon, Y.; Vanderleyden J. Root-associated *Azospirillum* species can stimulate plants. ASM News. **1997**, 63, 7, 366–370.

Pawlowski, K.; Bisseling, T. Rhizobial and actinorhizal symbiosis: what are the shared features?. The Plant Cell. **1996**, 8, 1899–1913.

Paula, M.A.; Siqueira, O.; Döbereiner, J. Ocorrencia de fungos micorrixicos vesiculoarbusculares e de bacterias diazotroficas na cultura da batata-doce. Rev. Bras. Ci. Solo, Campinas **1993**, 17, 349–356.

Penmetsa, R.V.; Cook, D.R. A legume ethylene insensitive mutant hyperinfected by its rhizobial symbiont. Science. **1997**, 275, 527–530.

Pereg-Gerk, L.; Paquelin, A.; Gounon, P.; Kennedy, I.R.; Elmerich, C. A transcriptional regulator of the Lux-R-UhpA family, FlcA, controls flocculation and wheat root surface colonization by *Azospirillum brasilense* Sp7. Mol. Plant Microbe Interac. **1998**, 11, 3, 177–187.

Plazinski, J.; Rolfe, R.G. Influence of *Azospirillum* strains on the nodulation of clovers by *Rhizobium* strains. Appl. Environ. Microbiol. **1985**, 49, 984–989.

Reinhold-Hurek, B.; Hurek, T.; Claeyssens, M., van Montagu, M. Cloning, expression in Escherichia coli, and characterization of cellulolytic enzymes of *Azoarcus* sp., a root-invading diazotroph. J. Bacteriol. **1993a**, 175, 21,7056–7065.

Reinhold-Hurek, B.; T. Hurek.; Gillis, M.; Hoste, B.; Vancanneyt, M.; Kesters, K.; De Ley J. *Azoarcus* gen. nov., nitrogen-fixing proteobacteria associated with roots of Kallar grass (*Leptochloa fusca* (L.) Kunth), and description of two species, *Azoarcus indigens*, sp.nov. and *Azoarcus communis* sp. nov. Int. J. Syst. Bacteriol. **1993b**, 43, 574–584b.

Relic, B.; Talmont, F.; Kopcinska, J.; Golinowski, W.; Promé, J.C.; Broughton W.J. Biological activity of *Rhizobium* sp. NGR 234 Nod-factors on *Macroptilium atropurpureum*. Mol Plant-Microb Interact. **1993**, 6, 764–774.

Rolfe, B.G.; McIver, J.M.; Anest, K.; Weinman, J.J.; Djorjevic, M.A.; Yanni, Y.G.; Squartini, A.; Dazzo, F.B. *Rhizobium leguminosarum* bv *trifolii* and its interaction with rice plants. In Biological Nitrogen Fixation for the 21st Century. Elmerich, C., Kondorosi, A., Newton, W.E. (eds), 665–666. Kluwer Academic Publishers. Dordrecht, The Netherlands. **1998a**.

Rolfe, B.G.; Verma, D.P.S.; Potrykus, I.; Dixon, R.; McCully, M.; Sautter, C.; Denarie, J.; Sprent, J.; Reinhold-Hurek, B.; Vanderleyden, J.; Ladha, K.; Dazzo, F.B.; Kennedy, I.; Cocking, E.C. Round Table: Agriculture 2020: 8 billion people. In Biological Nitrogen Fixation for the 21st Century. Elmerich, C., Kondorosi, A., Newton, W.E. (eds), 685–692. Kluwer Academic Publishers. Dordrecht, The Netherlands. **1998b**.

Sagan, M.; Morandi, D.; Tarenghi, E.; Duc, G. Selection of nodulation and mycorrhizal mutants in the model plant *Medicago truncatula* (Gaertn.) after g-ray mutagenesis. Plant Sci. **1995**, 111, 63–71.

Spaink, H.P.; Sheeley, M.; van Brussel, A.A.N.; Glushka, J.; York, W.S.; Tak, T.; Geiger, O.; Kennedy, E.P.; Reinhold, V.N.; Lugtenberg, B.J.J. A novel highly unsaturated fatty acid moiety of lipo-oligosaccharide signals determines host specificity of *Rhizobium*. Nature. **1991**, 354, 125–130.

Spaink, H.P. Regulation of plant morphogenesis by lipo-chitin oligosaccharides. Crit. Rev. Plant Sci. **1996**, 15, 559–582.

Tarrand, J.J.; Krieg, N.R.; Döbereiner, J. A taxonomic study of the Spirillum lipoferum group, with descriptions of a new genus, Azospirillum gen. nov. and two species, *Azospirillum lipoferum* (Beijerinck) comb. nov. and *Azospirillum brasilense* sp. nov. Can. J. Microbiol. **1978**, 24, 967–980.

Truchet, G.; Roche, P.; Lerouge, P.; Vasse, J.; Camut, S.; de Billy, F.; Promé, J.-C.; Dénarié, J. Sulphated lipo-oligosaccharide signals of *Rhizobium meliloti* elicit root nodule organogenesis in alfalfa. Nature. **1991**, 351, 670–673.

von Bülow, J.F.W.; Döbereiner, J. Potencial for nitrogen fixation in maize genotypes in Brazil. Proc. Natl. Acad. Sci. USA. **1975**, 72, 6, 2389–2393.

van Brussel, A.A.N.; Bakhuizen, R.; Van Spronsen, P.C.; Spaink, H.P.; Tak, T.; Lugtenberg, B.J.J.; Kijne, J.W. Induction of pre-infection thread structures in the leguminous host plant by mitogenic lipo-oligosaccharides of *Rhizobium*. Science **1992**, 257, 70–71.

Vance, C. Legume symbiotic nitrogen fixation: agronomic aspects. In the Rhizobiaceae. Spaink, H.P. Kondorosi, A., and Hoykaas, P. 509–530 (eds) Kluwer Academic Publishers. Dordrecht, The Netherlands. **1998**.

van de Broek, A.; Bekri, A.M.; Dossselaere, F.; Faure, D.; Lambrecht. M.; Okon, Y.; Costacurta, A.; Prinsen, E.; De Troch P.; Desair, J.; Keijers, V.; Vanderleyden, J. Azospirillum-plant root association: genetics of IAA biosynthesis and plant cell wall degradation. In Biological Nitrogen Fixation for the 21st Century. Elmerich, C., Kondorosi, A., Newton, W.E. (eds), 377–378. Kluwer Academic Publishers. Dordrecht, The Netherlands. **1998**.

van Rhijn, P.; Fang, Y.; Galili, S.; Shaul, O.; Atzmon, N.; Winiger, S.; Eshead, Y.; Lum, M.; Li, Y.; To, V.; Fujishige, N. Kapulnik, Y.; Hirsch, A.M. Expression of early nodulin genes in alfalfa mycorrhizae indicates that signal transduction pathways used in forming arbuscular mycorrhizae and *Rhizobium*-induced nodules may be conserved. Proc. Natl. Acad. Sci. USA. **1997**, 94, 5467–5472.

Vijn, I.; Martínez-Abarca, F.; Yang, W.C.; das Neves, L.; van Brussel, Van Kammen, A.; Bisseling, T. Early nodulin expression during Nod factor induced processes in *Viciae sativa*. Plant J. **1995**, 8, 111–119.

Wyss, P.; Mellor, R.B.; Wiemken, A. Vesicular-arbuscular mycorrhizas of wild-type soybean and non-nodulating mutants with *Glomus mosseae* contain symbiosis-specific polypeptides (mycorrhizins), immunologically cross-reactive with nodulins. Planta. **1990**, 182, 22–26.

Zeman, A.M., Tchan, Y.T., Elmerich, C., Kennedy, I.R. Nitrogenase activity in wheat seedlings bearing para-nodules induced by 2,4-dichlorophenoxyacetic acid (2,4-D) and inoculated wir *Azospirillum*. Res. Microbiol. **1992**, 143, 847–855.

MOLECULAR STRATEGIES TO IMPROVE THE NUTRITIONAL QUALITY OF LEGUME PROTEINS

Benito O. de Lumen,[1*] Alfredo F. Galvez,[1] M. Jamela Revilleza,[2] and Deanne C. Krenz[1]

[1]Division of Nutritional Sciences and Toxicology
University of California, Berkeley, California 94720-3104
[2]Institute of Chemistry
University of the Philippines at Los Banos College
Laguna, Philippines

ABSTRACT

Legume proteins are relatively deficient in the sulfur amino acids. Among the different strategies to increase the sulfur amino acid contents of legumes, we have chosen to increase the biosynthesis of endogenous, non-abundant sulfur-rich proteins we have identified in soybean seed. We identified and isolated an 8 kDa sulfur-rich protein from the albumin fraction of soybean seed. Subsequently, a cDNA designated Gm2S-1, has been isolated and characterized. The Gm2S-1 cDNA encodes a pre-proprotein that contains a signal peptide and a precursor protein that undergoes post-translational processing to yield the mature 8 kDa protein containing 7.8% methionine and 7.8% cysteine. The Gm2S-1 gene is an ideal candidate for overexpression to improve the nutritional quality of soybean and other legumes.

INTRODUCTION

Nutritional Rationale and Goals

Methionine is the first limiting essential amino acid in legumes because the major storage proteins in legumes, the globulins, contain low amounts of this amino acid. Cyste-

* Corresponding author

Chemicals via Higher Plant Bioengineering, edited by Shahidi *et al.*
Kluwer Academic / Plenum Publishers, New York, 1999.

ine, although not an essential amino acid, is included with methionine because it has a sparing effector, methionine when added to the diet. This relative deficiency of legumes in sulfur-containing amino acids, combined with its high lysine content, makes legumes nutritionally complementary with cereals which are first limiting in lysine and relatively high in methionine. Consequently, the overall protein quality of cereal-legume mixtures is better than that of either of these protein sources alone.

Supplementation of legumes with free methionine improves their value in supporting animal growth and increases the efficiency of utilization of dietary protein (Bressani and Elias 1968). In the feeds of poultry and swine where soybean is the main protein source, methionine is used as a supplement. Little practical use of supplementation in human diets has been made despite clinical studies with humans demonstrating their effectiveness.

The rapid accumulation of knowledge about gene expression and regulation in plants has led to the development of strategies to improve the nutritional quality of plant proteins through genetic engineering. The strategies for legume proteins have been discussed in a recent review and the reader is referred to this (de Lumen and Uchimiya, 1997). Briefly, these are:

a. introduction of methionine residues into major seed storage proteins.
b. transfer of heterologous genes encoding methionine-rich proteins (NW) from other species.
c. manipulation of key enzymes in biosynthetic pathways of essential amino acids.
d. increasing the level of endogenuous, non-abundant NW in the target legume.

What would be the nutritional target of genetic engineering to enhance the protein quality of legumes? The sulfur amino acid requirements of humans and animals vwy with age, with infants requiring the most (42 mg/g) and adults the least, and growing pigs requiring more (48 mg/g) than human infants (de Lumen et al., 1997; FAO, 1985). It is recommended that the essential amino acid requirement pattern of 2–5 year old children (25 mg/g) be used to evaluate protein quality for all ages, except those for infants (FAO, 1985). Accordingly, the methionine+cysteine content of legumes has to be increased to approximately 25 mg/g protein to meet the requirements for humans except infants, to 42 mg/g protein for human infants, and to 48 mg/g protein for swine. The digestibility of legume protein, which varies from 72 to 98%, also has to be considered. Taking soybean as an example, its average methionine+cysteine content of 22 mg/g protein (Liener, 1978) has to be increased by 14% for humans, 91% for infants, and 118% for swine. Values would have to be higher when protein digestibility is taken into consideration and for other beans whose methionine+cysteine contents are generally lower than those of soybean. In any case, animal feeding trials need to be carried out to establish the actual nutritional value of transgenic seeds.

ENHANCING THE BIOSYNTHESIS OF ENDOGENOUS METHIONINE-RICH PROTEINS

The strategy we have chosen in our laboratory to improve the nutritional quality of soybean protein is to enhance the biosynthesis of non-abundant, endogenous MRP. This could have the following advantages: (a) increasing the level of an endogenous protein is unlikely to be deleterious to the seed, (b) as a homologous protein, the NW is expected to be processed and transported correctly and is less subject to proteolytic degradation, (c) the gene encoding MRP, as a homologous gene, could lead to a more stable integration into the host genome, and (d) the finding that the transgenic Brazil nut MRP in soybean is allergenic to humans (Nordlee et al, 1996), adds significance to the search for non-allergenic MRP. Although it remains to be proven, the soybean MRP that we have isolated might have the additional advantage of non-allergenicity since it belongs to a molecular weight (MW) class which has not been shown to be allergenic.

Our laboratory has established that the MRP are localized in the albumin (water soluble) fraction soybean protein using a method based on the specific alkylation of the thioether moiety in methionine with [1-^{14}C]-iodoacetic acid (de Lumen and Kho, 1987; Kho and de Lumen, 1988). Resolution of total protein by 2D-SDS-PAGE identifed a 10.8 kDa protein that has 12. 1% methionine and although the N-terminus was sequenced, the insufficiency of NW resulted in some ambiguity of the sequence (George and de Lumen, 1991). Enrichment of the albumin fraction for the low molecular weight components by heparin affinity-chromatography followed by 2D-SDS-PAGE revealed three 8 kDa MRP with different isolectric points (PI) (Revilleza et al, 1996). The N-terminus of the most abundant and most acidic of the three MRP, previously called 2D- 1, was sequenced and the encoding cDNA, now termed Gm2S-1, has been cloned and characterized (Galvez et al, 1997). This is the first MRP cDNA ever cloned from a legume and is distinct from the cDNA coding for a cysteine-rich but methionine-poor albumin cloned from pea (Higgins et al, 1986).

FEATURES OF TBE GM2S-1. DNA AND PROTEIN

The Gm2S-1 cDNA containing the open reading frame of the full length protein is found within the 1.5 kb EcoRl-BamHl fragment near the 3' end of a 2.5 kb clone, isolated from a library prepared from midmaturation mRNA of soybean seed (Galvez et al, 1998). Because of the chimeric nature of the 2.5 kb cDNA clone, the transcription initiation site of the Gm2S-1 coding region was not resolved although the ATG start site was present. Using a modified cRACE protocol (Maruyama et al, 1995), the transcription initiation site was established and is located 17 bp apart from the ATG translation start site.

The Gm2S-1 cDNA encodes a 158 amino acid protein. The first 21 amino acids corresponds to a signal peptide sequence which is characteristically hydrophobic while the rest of the molecule is hydrophilic (Figure 1). A search revealed that the next 43 amino acid segment (SSU), which we now call lunasin, has 100% homology with an aspartic acid-rich peptide isolated from dried mature soybean seed (Odani et al, 1987). Post-translational processing should therefore include cleavage sites after 21-alanine and 64-aspartic acid to release the SSU. Based on the N-terminal sequence of the Gm2S-1 protein (Revilleza et al, 1996), there is a putative proteolytic cleavage site between 81-asparagine and 82-glutamic acid giving rise to a 77 amino acid polypeptide that corresponds to the mature 8 kDa Gm2S-1 protein (LSU) and a 17 amino acid linker peptide between the SSU (lunasin) and the LSU (8 kDa MRP).

Except for the 21 amino acid N-terminus that comprises the hydrophobic signal peptide, the pro-Gm2S-1 protein is mainly hydrophilic (Galvez et al, 1998). The proteolytic cleavage sites corresponding to the asparagine residues at position 65 and 81 are located at the most hydrophilic regions of the Gm2S-1 precursor protein.

Multiple sequence alignment analysis with other 2S albumins (Figure 2) showed that like Gm2S-1, the albumins from Brazil nut, *Arabidopsis, and Brassica napus* all contain a 21–22 amino acid signal peptide and undergo two proteolytic cleavages to generate ma-

Signal Peptide ↓ **Small subunit (lunasin)**

MTKFTILLISLLFCIAHTCSASKWQSPARYVRKQLQGVNL

↓ **Linker**

SPCEKHIMEKIQGRGDDDDDDDDDNHILRTMGGRINYI

↓ **Large subunit (8 kDa MRP)**

RRNEGKDEDEEEGHMQKCCTEMSELRSPKCQCKALQKI

MENQSEELEEKQKKKMEKELINLATMCRFGPMIQCDLSSDD

Figure 1. Complete protein sequence encoded by the Gm2S-1 cDNA clone. The protein consists of a hydrophobic signal peptide, a small subunit (SSU) now called lunasin, a linker peptide and the large subunit (LSU) 8 kDa MRP. The arrows indicate proposed posttranslational proteolytic cleavage sites.

ture heterodimeric proteins. Except for the identity of lunasin with the previously reported soybean aspartic acid-rich peptide, the rest of Gm2S-1 protein shows no significant homology with the other 2S albumins. However, the locations of cysteine residues in both lunasin and LSU are conserved among the different plant species, indicating that the Gm2S-1 is related to other members of the 2S albumin family on structure rather than sequence homology.

Using a PCR fragment that contains the Gm2S-1 coding region as a probe, a transcript of about 700 bp size is detected as early as 3 weeks after flowering which persists up to 7 weeks but is completely absent in the mature seed. Maximal expression is observed at 5–6 weeks after flowering (Figure 3). The 700 bp Gm2S-1 transcript is present only in the cotyledon (5 weeks after flowering) and not in any other tissue (pod, leaf, root, and stem).

To compare the levels of Gm2S-1 mRNA and the 8 kDa Gm2S-1 protein at different stages of seed development, the low molecular weight-enriched albumin fraction of seed proteins, isolated at different times after flowering, were resolved in ID-SDSPAGE and tagged for methionine residues with [1-^{14}C]-iodoacetate. Detectable levels of the 8 kDa MRP fraction are seen at 5 weeks after flowering and maintained up to maturity (Figure 4). The appearance of the maximal level of Gm2S-1 mRNA at 5 weeks after flowering coincides with the initial detection of the 8 kDa NW fraction. The 8 kDa MRP fraction, as noted earlier, constitutes three differently charged proteins in 2D-SDS-PAGE, in which Gm2S-1 is the most abundant and acidic (Revilleza et al, 1996).

Multiple DNA restriction fragments hybridized to the Gm2S-1 PCR probe in *HindIII, Eco RV, Pvu II, NcoI,* and *XbaI* digests of soybean genomic DNA (Galvez et al, 1998). The Gm2S-1 cDNA, like the *Arabidopsis* 2S albumin multi-gene family (Krebbers et al, 1988), may not contain any introns. If this was true, the multiple fragments suggest the presence of at least two copies of the Gm2S-1 gene in the soybean haploid genome.

AMINO ACID CONWOSITION OF GM2S-1 AND OTHER MRP

The essential amino acid compositions of the soybean 8 kDa MRP is compared with other MRP whose cDNA have been cloned (Table 1). Expression of the sunflower MRP in

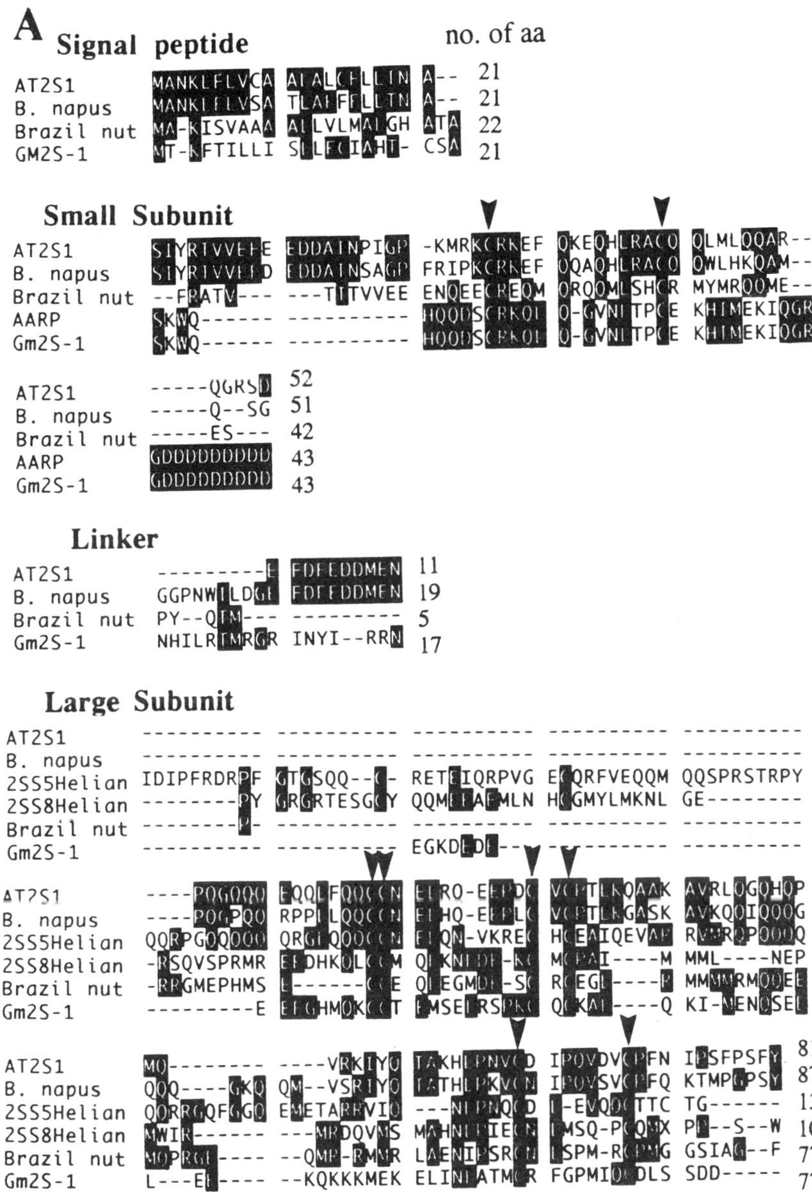

Figure 2. Sequence homology comparison of component polypeptides among the different 2S albumins upon proteolytic cleavage of precursor proteins. **A:** Multiple sequence alignment was done using the Mggins-Sharp algorithm in the MacDNASIS sequence analysis software (Hitachi). Shaded regions indicate 2 or more sequence matches upon alignment. The signal peptide, SSU, and linker polypeptides were compared among the dimeric 2S albumins from *Arabidopsis (AT2S1), Brassica napus (B.* napus), Brazil nut, and Gm2 S-1 including the aspartic acid-rich polypeptide that has a 100% sequence homology to the small subunit of Gm2S-1. Comparisons among the LSU included the monomeric 2S albumins from sunflower (2SS5Helian and 2SS8Helian). Arrows indicate the location of the conserved cysteine residues in the small and large subunits. **B:** Phylogenetic tree showing sequence homology among the LSU sequences as determined from the teggins- Sharp pairwise sequence similarity algorithm in the MacDNASIS sequence analysis program (Etachi).

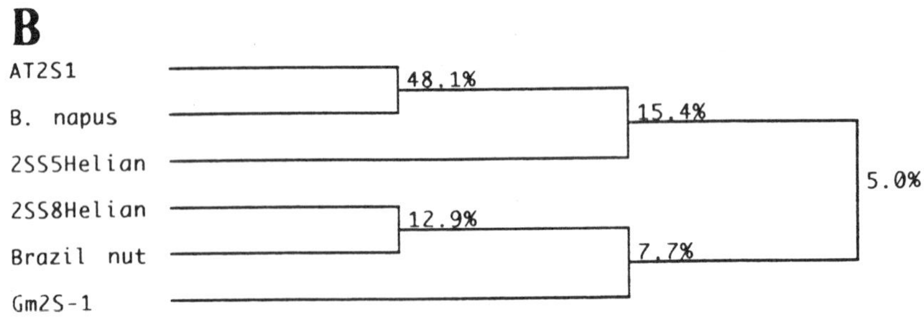

Figure 2. *(Continued)*

the seeds of lupin *(Lupinus angustifolis* L.) resulted in enhanced methionine levels and increased nutritive value of the transgenic seeds (Molvig et al, 1997). The Brazil nut NW has been used in a number of plant species by several laboratories to increase the methionine content in seeds, resulting in varying levels of enhancement (Altenbach et al, 1989, 1992; de Clerq et al, 1990; Guerche et al, 1990; Saalbach et al, 1994). The most successful of these attempts led to an accumulation of Brazil nut NW at levels of up to 8% of the total protein, equivalent to a 26% gain in methionine (Townsend et al, 1992; Nordlee et al, 1994). However, the heterologous Brazil nut NW in transgenic soybean has been found to be allergenic (Nordlee et al, 1996). Soybean is a known source of allergens, although it is widely used for human foods (Shibasaki et al, 1980; Herian et al, 1990; Taylor, 1992). Several allergens from soybean protein with subunit molecular weights from 14 to 60 kDa have been identified (Herian et al, 1990). Earlier studies in Japan have found allergenic activities in the major soy protein components, the 11S-, 7S-, and 2S-globulin fractions (Shibasaki et al, 1980). However, it should be noted from these studies that the low MW albumin fraction from which the Gm2S-1 8 kDa mature protein was isolated, has not been shown to be allergenic. Nevertheless, the issue of Gm2S-1 protein allergenicity still re-

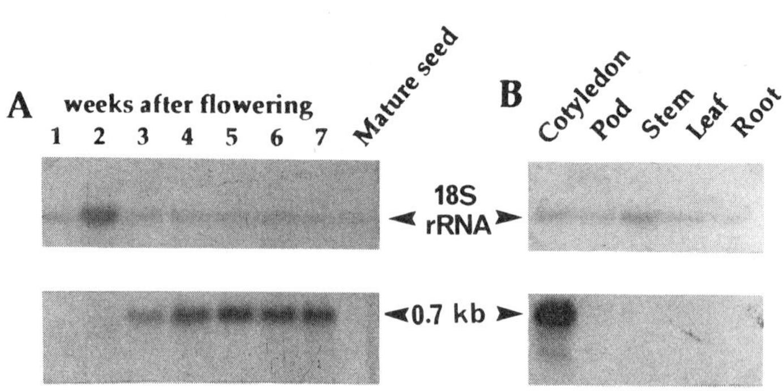

Figure 3. Northern analysis showing temporal **(A)** and seed-specific **(B)** expression of the Gm2S- I CDNA. Approximately 20 µg of total RNA isolated from developing soybean seeds and from other plant tissues were electrophoresed in a 1.5% denaturing agarose gel and transferred to Hybond-N membrane by capillary transfer. The Northern blots were probed with a 32p-labeled PCR fragment containing the full-length Gm2S-1 coding region, and a 326 hp PCR fragment of the 18S rRNA cDNA that was used as an internal control to show equal loading of total RNA in each lane.

Table 1. Essential amino acid composition of methionine-rich proteins[a]

Amino acid	Corn 10 kDa	Corn 15 kDa	Rice 10 kDa	Brazil nut 12 kDa	Sunflower	Soybean
Methionine	22.5	11.3	20.0	18.8	15.5	7.8
Cysteine	3.9	4.4	10.0	7.9	7.8	7.8
Lysine	0.0	0.0	0.9	0.0	3.9	13.0
Tryptophan	0.0	0.0	0.0	0.0	1.0	0.0
Threonine	3.9	2.5	6.4	0.0	0.0	2.6
Isoleucine	2.3	0.6	1.8	1.0	2.9	3.9
Leucine	11.6	10.0	6.4	5.0	8.7	7.8
Phenylalanine	3.9	0.0	2.7	0.0	0.0	1.3
Valine	3.9	2.5	1.8	0.0	1.9	0.0
Histidine	2.3	0.0	0.9	2.0	2.9	1.3

[a] Values, % of total protein, are deduced from sequences of cloned cDNAs (Altenbach and Simpson, 1990; Galvez et al, 1997).

mains to be resolved. Although the Gm2S-1 protein is endogenous to soybean, it is important to test it for allergenicity if it is to be used to enhance the nutritional quality of soybean for human consumption.

In legume-cereal mixtures, lysine is commonly the first limiting essential amino acid. The 8 kDa Gm2S-1 protein contains 13% lysine in contrast with no lysine for all the other NW except that of sunflower. This presents an advantage of using Gm2S-1 for over-expression to increase methionine and lysine as well. It is also interesting to note that at least one of the following essential amino acids are also lacking in some of the MRP including Gm2S-1: tryptophan, threonine, phenylalanine, and valine. If the enhanced synthesis of the NW happens at the expense of other storage proteins, an imbalance of other essential amino acids could occur. However, this remains to be established through amino acid analysis of transgenic seeds and animal feeding studies.

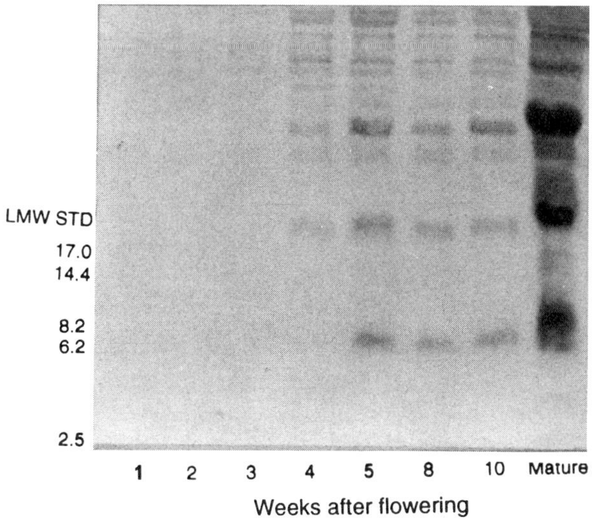

Figure 4. Autoradiograph of low molecular weight albumin fraction with labeled methionine residues at different stages of soybean seed development. One-dimensional SDS-PAGE was carried out on the low molecular weight-enriched albumin fraction isolated from developing cotyledons. The proteins were electroblotted onto Immobilon membrane (Millipore) and the methionine-containing albumins were detected in the membrane by specific alkylation of the thioether moiety with [1-^{14}C]-iodoacetate.

FUTURE EXPERIMENTS

Since the 8 kDa soybean MRP is estimated to make up about 0.5% of the total extractable protein in soybean seed, we calculated that it has to be increased approximately 14-fold over wild type to increase the sulfur amino acid content of soybean to 25mg/g, the requirement for 2–5 year old children (FAO, 1985). It has been demonstrated that modifications in the CaMV 35S constitutive promoter resulted in seedspecific enhanced expression as high as 25- to 40-fold over wild type in tobacco seeds when a segment of the a'-subunit of β-conglycinin (soybean protein) promoter is inserted into the CaMV 35S promoter (Chen et al, 1988). A similar construct will be made by putting the cDNA encoding the signal peptide, the 17 amino acid polypeptide linker and the 8 kDa MRP (*signal peptide-polypeptide linker-8 kDa MRP*) downstream from the modified CaMV 35S promoter in which the β-conglycinin subunit has been inserted in the 5' to 3' orientation at the -90 position (courtesy of Dr. Roger Beachy). The NOS polyadenylation control region will be at the 3' end. If needed, we can add another copy of the modified CaMV 35S to further enhance the expression (Kay et al, 1987). Increasing the number of methionine residues in the 8 kDa protein would reduce the extent of increased overexpression needed to achieve the nutritionally significant increase in sulfur an-fino acid mentioned above. The region between the 4th and 5th cysteine residues and towards the carboxyl end from the 6th cysteine residue are likely sites for addition of methionine residues since they have variable lengths among the different 2S albumins (Galvez et al, 1996). Methionine residues could be inserted in these regions. For example, addition of four additional methionine residues (two between the 4th and 5th cysteine and two after the 6th cysteine) will increase the methionine content of the 8 kDa protein to about 13.0% from 7.8%, and would reduce the required increase from 14-fold to about 9-fold. The constructed gene expression cassettes will be cloned into the pGEM3Zf(+) vector (Promega, Madison, WI) for use in transient gene expression assay (Cho et al., 1995). To hasten the screening of the promoter constructs in a soybean a gene expression cassette using the GUS reporter gene in place of the *signal peptide-polypeptide linker-8 kDa SRP* cDNA will also be made for visual quantitation of expression levels. Once the highest expressing constructs are selected, stable transformation, will be carried out in collaboration with a laboratory with experience in soybean transformation such as the Center for Soybean Tissue Culture and Genetic Engineering, University of Kentucky.

ACKNOWLEDGMENTS

This work was supported in part by USDA NRICGP Grant # 94-02167 and the US Soybean Board. We thank Dr. Robert Goldberg for the cDNA library, Merrill Shum, Cindy Kang, Abigail Jung, Ann Chou, and Gyu Kim for laboratory assistance and Jing-jie Liu and Bill Odegard for helpful discussions.

REFERENCES

Altenbach, S.B.; Kuo, C.C.; Staraci, L.C.; Pearson, K.W.; Wainwright, C.; Georgescu, A.; Townsend, J. Accumulation of a Brazil nut albumin in the seeds of transgenic canola results in enhanced levels of seed protein methionine. *Plant Mol. Biol.* **1992,** *1*, 235–245.

Altenbach, S.B.; Pearson, K.W.; Meeker, G.; Staraci, L.C.; Sun, S.S.M. Enhancement of the methionine content of seed proteins by the expression of a chimeric gene encoding a methionine-rich protein in transgenic plants. *Plant Mol. Biol.* **1989,** *1*, 513–522.

Altenbach, S.B.; Simpson, R.B. Manipulation of methionine-rich protein genes in plant *seeds. TIBTECH* **1990,** *8*, 156–160.

Bressani, R.; Elias, L. G. Processed vegetable protein mixtures for human consumption in developing countries. *Adv. Food Res.* **1968,** *16,* 1–103.

Chen, Z.L.; Pan, N.S.; Beachy, R.N. A DNA sequence element that confers seed-specific enhancement to a constitutive promoter. *EMBO J.* **1988,** *7*, 297–302.

Cho, M.J.; Wilholm, J.M.; Vodkin, L.A. Cassettes for seed-specific expression tested in transformed embryogenic cultures of soybean. *Plant Mol. Biol. Rep.* **1995,** *13*, 255–269.

De Clerq, A.; Vandewiele, M.; Van Damme, J.; Guerche, P.; Van Montagu, M.; Vanderkerchove, J.; Krebbers, E. Stable accumulation of modified 2S albumin seed storage proteins with higher methionine contents in transgenic plants. *Plant Physiol.* **1990,** *94,* 970–979.

de Lumen, B.O.; Uchimiya, H. Molecular strategies to enhance the nutritional quality of legume protein: an update. *AgBiotech. News and Information* **1997,** *9*, 53N-58N.

de Lumen, B.O.; Krenz, D.C.; Revilleza, M.J. Molecular strategies to improve the protein quality of legumes. *Food Tech.* **1997,** *51*, 67–70.

de Lumen, B.O.; Kho, C.J. Identification of methionine containing proteins and quantitation of their methionine contents. *J. Agric. Food Chem.* **1987,** *35*, 688–691.

FAO/WHO/UNU. Energy and protein requirements. VVHO Tech Rep Ser No. 724, **1985,** VVHO Geneva, Switzerland.

Galvez, A.F.; Revilleza, J.M.; de Lumen, B.O.; Krenz DC. Enhancing the biosynthesis of endogenuous methionine-rich proteins (MRP) to improve the protein quality of legumes via genetic engineering. *Proceedings of Food for Health in the Pacific Rim.* 3rd International Conference of Food Science and Technology. Oct 19–23, 1997.

Galvez A.F.; Revilieza, J.M.; de Lumen, B.O. A novel methionine-rich protein from soybean *(Glycine max)* seed: Cloning and characterization of cDNA. (Accession No. AF005030) *Plant Physiol.* **1997,** *114*, 1567.

George, A.A.; de Lumen, B.O. A novel methionine-rich protein in soybean: Identification, amino acid composition and N-terminus sequence. *J . Agric. Food Chem.* **1991,** *39,* 224–227.

Guerche, P.; De Almeida, E.R.P.; Schwarztein, M.A.; Gander, E.; Krebbers, E.; Pelletier,G. Expression of the 2S albumin from *Bertholletia excelsa* in *Brassica napus. Mol. Gen. Genet.* **1990,** *221,* 306–314.

Herian, A.M.; Taylor, S.L.; Bush, R.K. Identification of soybean allergens by immunoblotting with sera from soy-allergic adults. *Int. Arch. Allergy Appl. Immunol.* **1990,** *92*, 193–198.

Higgins, T.J.V.; Chandler, P.M.; Randall, P.J.; Spencer, D.G.; Beach, L.R.; Blagrove, R.J.; Kortt, A.A.; Inglis,A.S. Gene structure, protein structure, and regulation of the synthesis of a sulfur-rich protein in pea seeds. *J. Biol. Chem.* **1986,** *261,* 11124–11130.

Higgins, D.G.; Sharp, P.M. CLUSTAL: a package for performing multiple sequence alignment on a microcomputer. *Gene* **1988,** *73*, 237–244.

Kay, R.; Chan, A.; Daly, M.; McPherson, J. Duplication of a CaMV 35S promoter sequences creates a strong enhancer for plant genes. *Science* **1987,** *236,* 1299–1302.

Kho, C.J.; de Lumen, B.O. Identification and isolation of methionine-cysteine rich protein fraction in soybean seed. *Plant Foods Human Nutr.* **1988,** *38*, 287–296.

Kyte, J.; Doolittle, R.F. A simple method for displaying the hydropathic character of a protein. *J. Biol. Chem.* **1982,** *157*, 105–132.

Liener, I.E. Nutritional value of food protein products. In *Soybeans: Chemistry and Technology Vol. I Proteins.* Smith A.K. and Circle S.J. eds. AVI Publishing, Westport, CT **1978,** pp 203–277.

Maruyarna, I.N.; Rakow, T.L.; Maruyama, H.I. cRACE: a simple method for identification of the 5's end of mRNAs. *Nuc. Acids Res.* **1995,** *23,* 3796–3797.

Molvig, L.; Tabe, L.M.; Eggum, B.O.; Moore, A.E.; Craig, S.; Spencer, D.; Higgins, T.J.V. Enhanced methionine levels and increased nutritive value of seeds of transgenic lupins *(Lupinus angustifolius* L.) expressing a sunflower seed albumin gene. *Proc. Natl. Acad. Sci.* USA **1997,** *94*, 8393–8398.

Nordlee, J.A.; Taylor, S.L.; Townsend, J.A.; Thomas, L.A.; Bush, R.K. Identification of a Brazil nut allergen in transgenic soybeans. *N. Engl. J. Med.* **1996,** *334,* 688–692. Nordlee, J.A.; Taylor, S.L.; Townsend, J.A.; Thomas, L.A.; Townsend, R. Investigations of the allergenicity of Brazil nut 2S seed storage protein in transgenic soybean. *OECD Workshop on Evaluation of New Foods,* **1994,** Oriel College Oxford, UK.

Odani, S.; Koide, T.; Ono, T. Amino acid sequence of a soybean *(Glycine max)* seed polypeptide having a poly (L-aspartic acid) structure. *J. Biol Chem.* **1987,** *262,* 10502–10505

Revilleza, J.M.; Galvez, A.F.; Krenz, D.C.; de Lumen, B.O. An 8 kDa methionine-rich protein from soybean *(Glycine max)* cotyledon: Identification, purification and N-terminal sequence. *J. Agric. Food. Chem.* **1996,** *44,* 2930–2935.

Saalbach, I.; Pickardt, T.; Machemehl, F.; Saalbach, G.; Schieder, O.; Muntz, K. A chimeric gene encoding the methionine-rich 2S albumin of the Brazil nut *(Bertholettia excelsa* H.B.K.) is stably expressed and inherited in transgenic grain legumes. *Mol. Gen. Genet.* **1994,** *242,* 226–236.

Shibasaki, M.; Susuki, S.; Tajima, S.; Nemoto, H.; Kuruome, T. Allergenicity of major component proteins of soybean. *Int. Archs. Allergy Appl. Immun.* **1980,** *61,* 441–448.

Townsend, J.A.; Thomas, L.A.; Kulsek, E.S.; Daywalt, M.J.; Winter, K.R.K.; Altenbach, S.B. Improving the quality of seed proteins in soybean. *Proceedings of the 4th Biennial Conference on Molecular and Cellular Biology of Soybean.* July 27–29, 1992, **1992** Iowa State University, Ames IA. p 4.

11

MOLECULAR FARMING OF INDUSTRIAL PROTEINS FROM TRANSGENIC MAIZE

Elizabeth E. Hood,* Ann Kusnadi,[1] Zivko Nikolov,[1,2] and John A. Howard

[1]Department of Food Science and Human Nutrition
[2]Department of Agricultural and Biosystems Engineering
Iowa State University, Ames, Iowa 50101

ABSTRACT

Recombinant egg white avidin and bacterial B-glucuronidase (GUS) from transgenic maize have been commercially produced. High levels of expression were obtained in seed by employing the ubiquitin promoter from maize. The recombinant proteins had activities that were indistinguishable from their native counterparts. We have illustrated that downstream activities in the production of these recombinant proteins, such as stabilizing the germplasm and processing for purification, were accomplished without any major obstacles. Avidin (A8706) and GUS (G2035) are currently marketed by Sigma Chemical Co.

INTRODUCTION

Plant biotechnology was developed over the last 10 years with the major objective of generating better agricultural products. The targeted traits to engineer included insect, disease or herbicide resistance. The tools of biotechnology, however, are now being applied to other industries such as production of human and animal health products, the food and feed industries and production of industrial enzymes. Because of the many applications of biotechnology, examples abound for expression of foreign genes in plants (Benfey and Chua, 1989; Fisk and Dandekar, 1993). One application on which we have focused is the use of transgenic plants for large-scale production of proteins for industrial, pharmaceutical, veterinary and agricultural use, a system which offers the potential to be one of the most economical approaches (Kusnadi et al., 1997; Austin et al., 1994; Krebbers et al.,

* ProdiGene, Inc., 1500 Research Parkway, College Station, TX 77845, (409)862-4062, fax (409)862-8538, (email) eha105@aol.com

1993; Pen et al., 1993A; Whitelam et al., 1993). Advantages of plant systems include the low cost of growing a large biomass, easy scale-up (increase of planted acreage), natural storage organs (tubers, seeds), and established practices for efficient harvesting, transporting, storing, and processing of the plant (Kusnadi et al., 1997; Whitelam et al., 1993). Recombinant proteins can be targeted to seeds which allows stable storage of the recombinant proteins for an extended period (Holzman, 1994; Pen et al., 1994b). Plants offer advantages over other production systems because some proteins may be used without extensive purification, since for many applications, plant material is used directly as a food source or feedstock.

In our studies, we have generated transgenic maize plants that express the chicken *avidin* gene and the *E. coli β-glucuronidase* gene. Avidin is a glycoprotein found in avian, reptilian and amphibian egg white and is used primarily as a diagnostic reagent. The protein avidin is composed of four identical subunits, each 128 amino acids long, the amino acid sequence of which has been known since 1971 (Delange and Huang, 1971). The cDNA of the chicken oviduct *avidin* gene was documented by Gope et al. (1987) and a genomic clone was isolated by Keinanen et al. (1988). More recently, Keinanen et al. (1994) reported on a family of closely related *avidin* genes from chicken. The customary source for commercial production of avidin has been chicken egg white.

β-glucuronidase (GUS) is a hydrolase that catalyzes the cleavage of terminal glucuronic acid that is β-linked to mono-, oligo-, or polyaccharides or phenols. The *β-glucuronidase* gene has been cloned and sequenced and encodes an enzyme which has great utility in plant molecular biology (Jefferson et al., 1986; Jefferson, 1987; Jefferson et al., 1987), being among the most commonly used reporter genes. The protein is composed of four identical subunits, each having a molecular weight of 68 kDa on SDS-PAGE. As an enzyme, β-glucuronidase can act on a variety of substrates, such as indole-β-glucuronide and methylumbelliferone β-glucuronide (MUG). The extensive use of GUS is partially due to its versatility in that the enzyme can be detected by both cytochemical and spectrophotometric assays. In addition, fluorometric detection of this enzyme using MUG as a substrate is highly sensitive.

Avidin and β-glucuronidase are demonstration products for the process of molecular farming of foreign proteins from transgenic plants. Though they have viable markets, these proteins are distinct from each other in a variety of ways and thus their greatest value is that they are examples of the flexibility of the protein production system in plants. Avidin and GUS have quite different physical characteristics--GUS is a bacterial enzyme, is large—68 kDa monomeric subunit mass, is not glycosylated and has an acidic isoelectric point. In contrast, avidin is small--approximately 17 kDa monomeric subunit mass, is encoded by a eukaryotic gene, is glycosylated and is quite basic. Successful production of these unique industrial proteins from transgenic maize demonstrates the versatility of the plant production system.

Using plants as factories for the production of high value proteins requires an efficient system not only for generating transformants with high expression, but ensuring the integrity of the transgenic protein. In addition, one must be able to effectively grow, harvest and process the production material. We have characterized transgenic plants that express high levels of chicken egg white avidin and b-glucuronidase in maize seed and compared the biochemical properties of the transgenic proteins to the native proteins. We also have begun to examine the variability of expression in different growing environments as well as the stability of the transgenic material through several generations. Finally, data on the storage, processing, extraction and purification of these proteins from field-grown transgenic material are presented. Thus, we describe the first commercial production of foreign proteins using a plant system. This system takes advantage of trans-

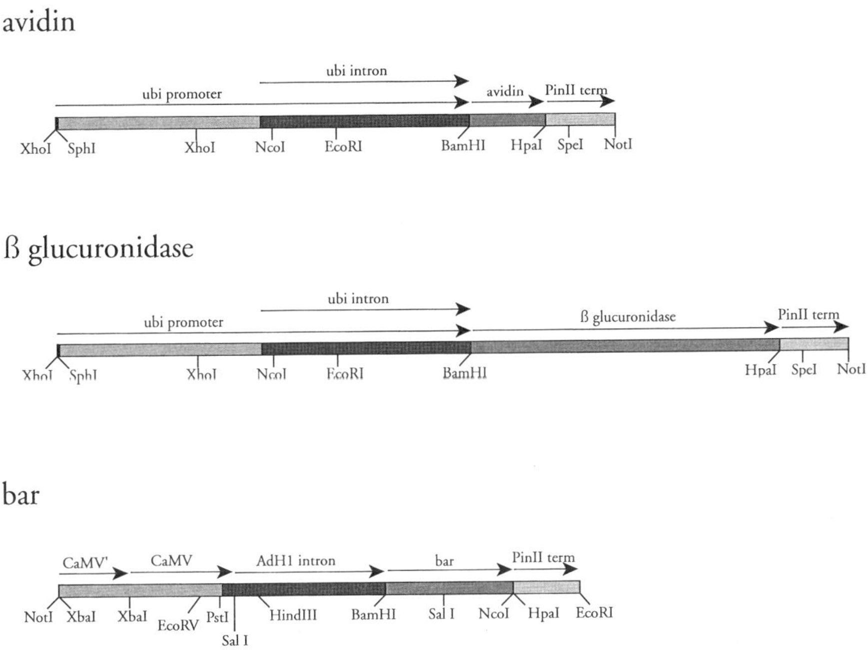

Figure 1. Plant transcription units for maize expression of *avidin, b-glucuronidase* and *bar*. A. The chicken oviduct *avidin* gene was fused upstream to a barley alpha amylase signal sequence, driven by the maize ubiquitin promoter with its nontranslated first exon and intron, and fused downstream to the PinII terminator. B. The *E. coli b-glucuronidase* gene was driven by the maize ubiquitin promoter with its nontranslated first exon and intron, and fused downstream to the PinII terminator. C. The *Streptomyces bar* gene was driven by the tandem CaMV promoter followed by the AdH1 intron from maize, and fused downstream to the PinII terminator.

genic plants, chimeric genes, tissue culture technology, plant breeding and a production and processing system. We have coordinated these factors to establish a viable paradigm for production of high value proteins in a crop plant. These commercial products are currently being marketed by Sigma Chemical Company.

CHARACTERIZATION OF THE AVIDIN PRODUCTION LINE

Construction of Expression Cassettes Containing the *Avidin* and *Bar* Genes

A gene optimized for expression of avidin protein in maize and a maize optimized DNA sequence encoding the barley alpha amylase signal sequence (Rogers, 1985) were generated by annealing and ligating overlapping, complementary synthetic oligonucleotides. The signal sequence was included based on the prediction that higher levels of avidin expression could be obtained if avidin protein was targeted to the extracellular matrix. The expression of avidin was driven by the maize ubiquitin promoter (Christensen et al., 1992; Cornejo et al., 1993) and was terminated with the PinII terminator from potato (An et al., 1989) (Figure 1A). The ubiquitin promoter is considered a constitutive promoter that shows limited tissue specificity, and in the transgenic plants generated in this study, it drives high levels of expression in many tissues, particularly in the embryo.

The expression of the gene which encodes resistance to the herbicide, bialaphos, (*bar*; White et al., 1992), was driven by the 35S promoter derived from Cauliflower Mosaic Virus sequences (Nagy et al., 1985) and was terminated with the PinII terminator from potato (An et al., 1989; Fig. 1C; see Hood et al., 1997).

Generation of *Avidin*-Expressing Transgenic Plants

Particle bombardment of regenerable embryogenic calli was performed to generate transformed calli that were selected on bialaphos for 10 weeks. Individual events were subjected to ELISA analysis to confirm the expression of the *avidin* gene. Plants initially regenerated from bialaphos resistant embryogenic tissue (T0) were used as females in crosses with an untransformed inbred as the pollen donor. Regenerated T0 plants were screened for high expression of avidin in leaf tissue (>0.1% of extracted protein). Our goal, however, was to produce avidin from seed. Therefore, T1 seed from an event showing high leaf expression was screened to confirm high levels of expression also in seed tissue. Avidin expression segregated in the T1 generation, and some seed was negative. However, expression in positive seeds ranged from 2.1–5.7% of aqueous extractable protein (determined by ELISA).

DNA Blot Hybridization Experiments

Five individual T1 plants from our highest expressing event were analyzed for the structure of insert DNA (Figure 2A). The linked *avidin* and *bar* genes were segregating as a single unit among individuals in these populations (see Hood et al., 1997), and it was not surprising to find the genes present in only three of the five plants examined (Fig. 2A). The *avidin* gene copy number was estimated at three to five as determined by single restriction enzyme digests and DNA blot hybridization of genomic DNA (Hood et al., 1997). When double digests using SphI and SpeI were conducted to determine the size of the plant transcription unit, four separate fragments could be identified, none of which was the same size as the original plasmid-derived plant transcription unit (Fig. 2A). The selectable marker gene, *bar*, hybridized to a band of the correct size, but these plants also contained two other transcription unit sizes, one smaller and one larger than the original construct (data not shown). The *avidin* and *bar* genes appear to be inherited as a single linkage unit, and it is not clear which of the inserts is active in transcription and translation. Similar integration patterns were present in all plants examined that were derived from this event.

Cell Biology

In seeds, avidin protein is primarily localized to the embryo, as determined either through tissue printing (Figure 3) or embryo versus endosperm extraction and quantitation of avidin protein (Table 1). Within the embryo, tissue prints reveal that the scutellum is the major sink of avidin, and the embryonic axis accumulates somewhat lesser amounts (Fig. 3B). The endosperm, which represents approximately 80% of the total kernel weight, contributes just 22% of the avidin activity found in whole kernels. An additional 5% of the activity is present in the hulls (pericarp). The embryo, however, contains at least 55% of the avidin protein found in whole dry seed (Table 1). The avidin protein balance indicates that 18% of the total activity is not accounted for and possibly is lost during dry milling.

In situ localization experiments were performed on thin sections of embedded embryos using anti-avidin primary antibodies and TRITC fluorescently labeled secondary an-

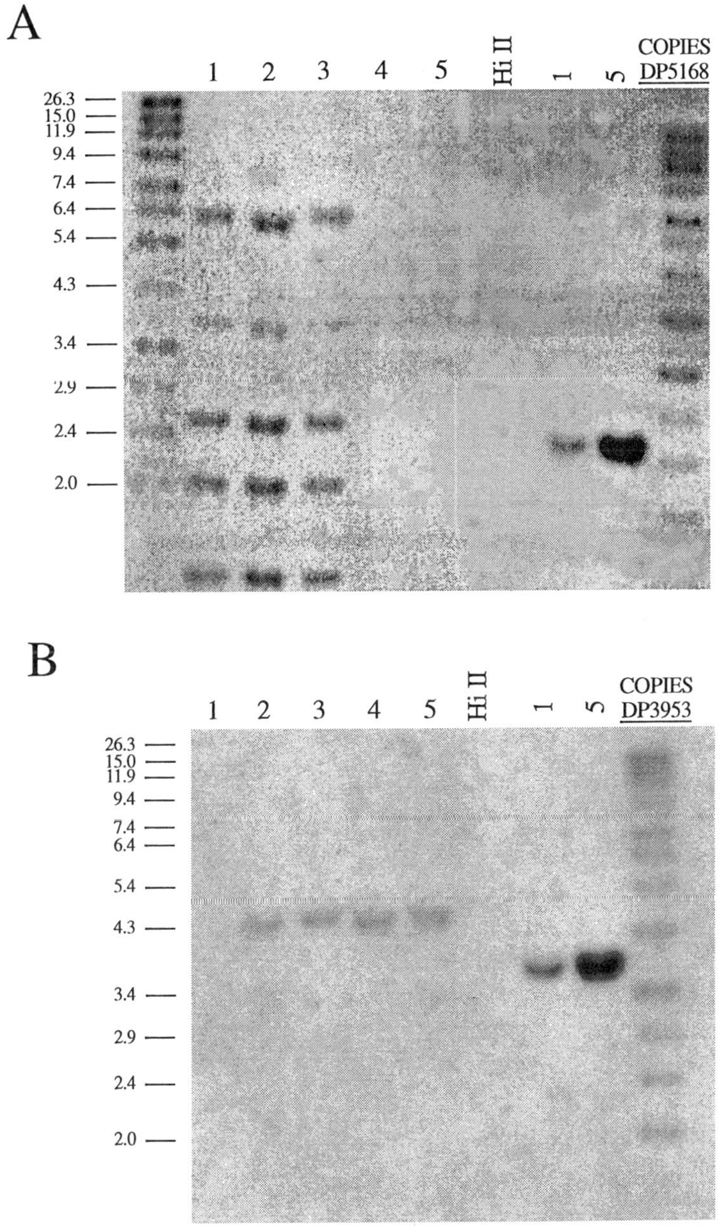

Figure 2. DNA blot hybridization analysis of avidin-containing plants with the avidin gene specific probe and b-glucuronidase-containing plants with the GUS gene specific probe. A. DNA was extracted from five plants (T1 generation). The DNA was digested with SphI and SpeI to generate fragments within the plant transcription unit. The intensity and number of bands hybridizing with the probe indicate the number of copies of the gene in the transgenic plant. Three plants were positive in this hybridization experiment. Lanes 1–5 represent individual plants. Copies 1 and 5 refer to plasmid DNA loaded in one and five copy number reconstructions. B. DNA was extracted from five plants (T3 generation). The DNA was digested with SphI and SpeI to generate a fragment within the plant transcription unit. The blot was probed with the GUS gene. Four plants were positive in this hybridization experiment. Lanes 1–5 represent individual plants. Copies 1 and 5 refer to the plasmid DNA loaded in one or five copy number reconstructions. The intensity and number of the bands indicate the number of copies of the gene in the transgenic plant.

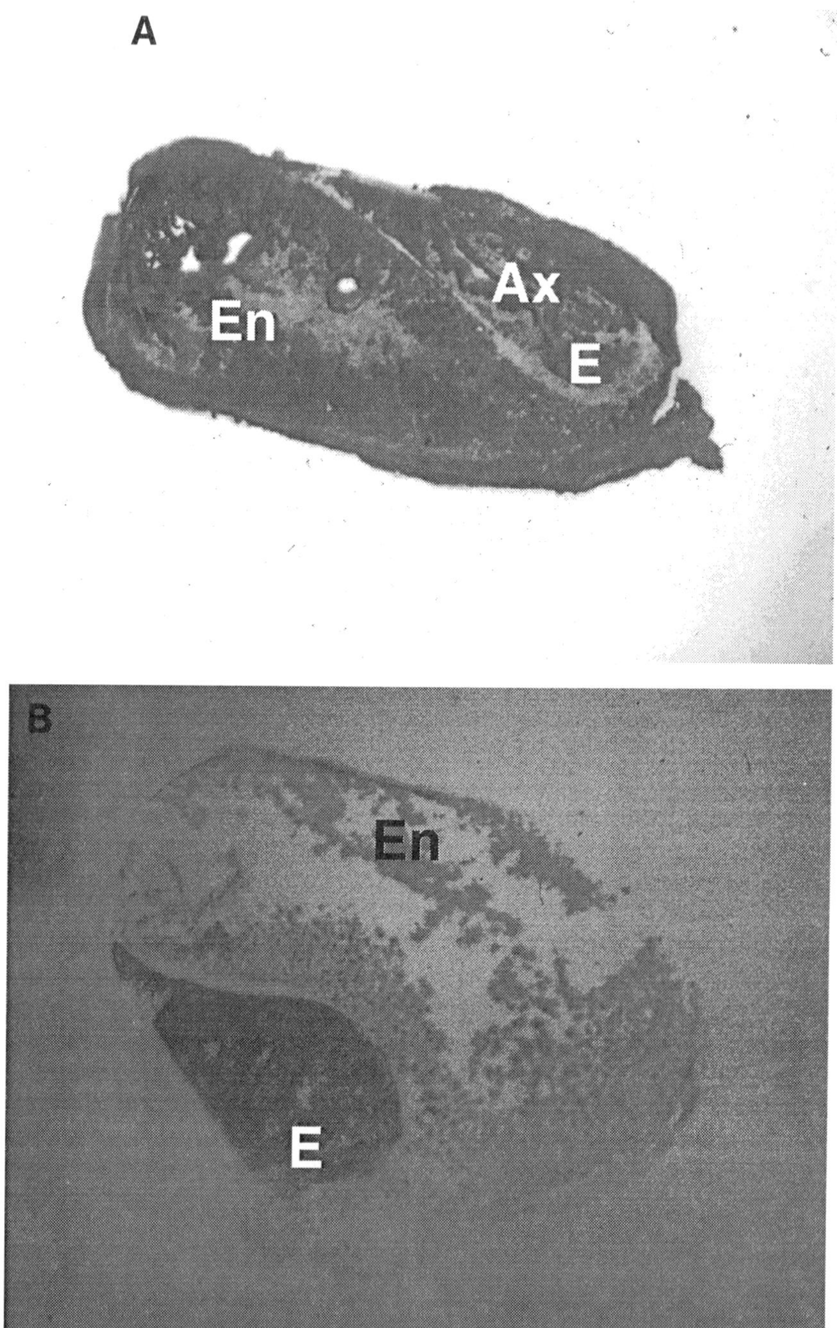

Figure 3. Tissue prints of near mature kernels from an avidin-positive plant. A. Tissue print stained for total protein with India Ink. B. Tissue print incubated with the anti-avidin antibody (ICN, 1:1000 dilution) and an alkaline phosphatase conjugated secondary antibody (Sigma Chemical Co., 1:1000 dilution). Note high concentration of avidin protein in the embryo (E). En, endosperm.

Table 1. Balance of avidin activity from transgenic seed after dry milling

Component	% of total dry weight	% of total avidin activity
Germ	12	55
Endosperm	80	22
Hulls	5	5
Unaccounted		18

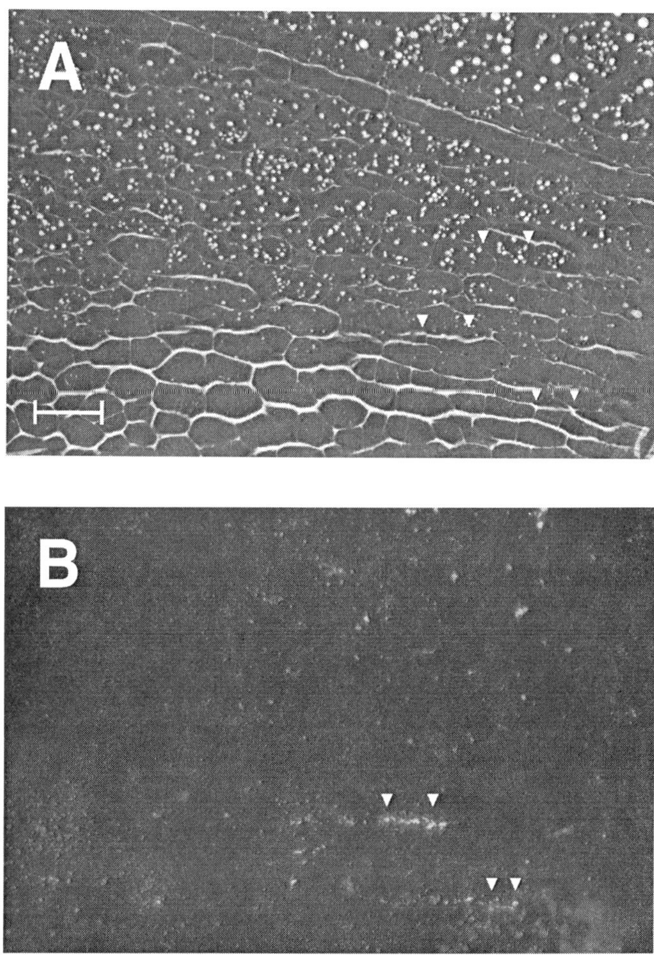

Figure 4. Localization of avidin protein in the embryo at the light microscope level. Two μm plastic-embedded sections were incubated with anti-avidin antibodies and TRITC labeled secondary antibodies (Hood et al., 1997). A. Phase contrast image of embryonic axis. B. Rhodamine (TRITC) fluorescence of same image as in A. Double arrows mark similar locations in each image. Bar represents 10 μm.

Table 2. Correlation of male sterile/limited fertility phenotype with presence of the avidin gene as determined by PCR

PCR Result	# Fertile	# Partially Sterile	# Sterile
+	1	6	32
–	40	0	1

tibodies (Figure 4). Phase contrast images of embedded sections showed excellent preservation of the cellular structure (Fig. 4A). Fluorescence was present in cell walls indicating that avidin is primarily localized to that organelle (Fig. 4B).

Genetics

Because T0 plants regenerated from tissue culture often exhibit problems with fertility, an outcross was performed in this first generation with pollen from an elite Pioneer inbred line. This avidin-expressing line lost resistance in the T1 generation to the herbicide bialaphos, on which it was originally selected in culture. Thus, because field grown material could not be sprayed with herbicide, a different scheme was required to identify transformed plants in the field in subsequent generations. Polymerase chain reaction (PCR) was chosen as an alternative method to select avidin-positive plants. Additionally, however, many of the plants exhibited male sterility, and a 97.5% correlation was observed between presence of the *avidin* gene (by PCR) and the male sterile/limited fertility phenotypes (Table 2). Male sterility also occurred in other avidin-positive lines, making it highly unlikely that this phenotype was a result of foreign gene insertion into a native gene required for pollen development. The male sterile phenotype was subsequently used for scoring of transformed plants in three production fields in 1995 and one in 1996.

Production

A program was designed to scale up the production of seed from this avidin-positive line, the goal of which was to obtain sufficient material from which to purify commercial quantities of recombinant avidin protein. Additionally, a vigorous backcross program was initiated to improve the agronomic character of the tissue culture-produced plants. The phenotype of T3 generation plants was vigorous and seed yield was increased above the T1 generation. Because of the male sterile phenotype, seed production employed vigorous inbred or hybrid lines and fertile sibling plants (Table 3) as pollen bearing parents. Ears were harvested only from plants showing loss of male fertility, and the seed was shelled and bulked.

The data (Table 3) show the levels of avidin present in extracts from seed produced in four different planting locations and over four generations (T1-T4). Quantitation of avidin levels in production seed lots was done by ELISA analysis. Because the *avidin* gene can only be inherited from the female and is hemizygous in those plants, the harvested seed contains a near 50:50 mixture of avidin positive and negative kernels and quantitative estimates incorporate both phenotypes. The expression of avidin was at a level that would yield an estimated 150–300 mg avidin per kg of harvested seed (Table 3), the average being 230 mg per kg of seed. Because the pollen donors utilized to date are

Table 3. Expression level of avidin protein as a percent of aqueous-extractable protein in T1 through T4 generations of maize seed

Location[A]	Avidin as % of extracted protein	Pollen parent	mg Avidin per kg seed
Greenhouse	0.8[B] (T1)	PHN46[G]	80
1	3.0 ± 0.40[C] (T2)	PHN46	300
	1.5 ± 0.16[D] (T3)	3394 +sibs	150
	2.4 ± 0.10[E] (T4)		240
2	1.9 ± 0.23[D] (T3)	3162	190
3	1.8 ± 0.19[D] (T3)	3394 +sibs	180
4	3.0 ± 0.11[F] (T3)	3394 +sibs	300
Average	2.3%		230

[A] Locations: 1. Hawaii winter nursery, March, 1995 (T2); Hawaii isolation field, July, 1995 (T3); Hawaii isolation field, March, 1996 (T4); 2. Texas winter nursery, July, 1995 (T3); 3. Johnston, IA isolation field, October, 1995 (T3); 4. Nebraska isolation field, October, 1995 (T3).
[B] Mean of 10 individuals.
[C] Mean of two bulk populations of 20 seeds each, three replicates each; ± standard error of the mean.
[D] Mean of six populations of 20 seeds and three bulk populations of 50 seeds each; ± standard error of the mean.
[E] Mean of two bulk populations of 10 seeds each; ± standard error of the mean.
[F] Mean of six bulk populations of 20 seeds each; ± standard error of the mean.
[G] PHN46, Pioneer elite inbred; 3394, Pioneer hybrid; sibs, fertile siblings from previous out-cross.

the same in locations where protein levels vary (compare locations 1, 3 and 4 in Table 3), pollen source does not appear to be a factor in variation of protein yield in the seed. The various growing locations were chosen to test whether environmental conditions might affect protein accumulation in the seed. In this context, several observations were made. First, the gene encoding for avidin is not silenced in later generations (at least in T3 or T4). Second, the variation in avidin as percent of extractable protein among locations was not attributable to any obvious genetic or environmental condition, and studies are continuing to determine if similar trends in other years will be seen. The level of avidin expression is slightly lower in T3 generation seed but this range of variation in T3 seed is similar to that observed in seed from a single location (Hawaii) over several generations. We now have produced avidin seed through the T7 generation with no loss of expression.

The *avidin*-expressing transgenic plants produced at least one year with approximately 300–400 kernels, i.e., a 300–400-fold increase. Therefore, 1 kg of transgenic seed should yield approximately 150–200 kg more seed, assuming 50% segregation in the seed source. We produced 160 kg of transgenic avidin seed in 1995 from 1 kg of starting material, greater than a 300-fold increase in total seed.

Processing

Grain drying and processing as currently practiced require an elevated temperature which potentially could destroy avidin activity in the whole kernel as well as in the ground material. Therefore, the effect of temperature on avidin activity was investigated. Avidin in whole kernels was stable and could withstand temperatures up to 50 °C for at least 7 days without loss of activity (data not shown). This allows the seed to be dried using conventional practices which typically operate at temperatures of 41 °C for three days. In addition, our data (Figure 5) indicate that avidin retains its original activity in the processed ground material for up to 95 days at -29 °C and 10 °C . Likewise, no significant change in avidin activity was measured when ground kernels were stored at 25 °C for up to 21 days. However, almost 50% of the initial avidin binding activity was lost when ground

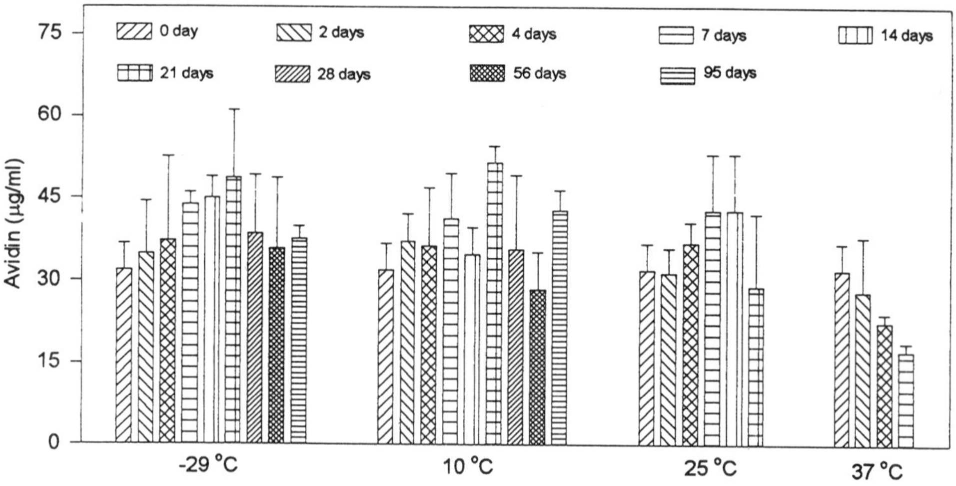

Figure 5. Effect of temperature on avidin stability during storage of ground maize seed. Maize seed was processed by cracking and flaking, then stored at the indicated temperatures. Triplicate samples were extracted and analyzed for avidin binding activity after the times (in days) shown by the shaded bars. Bars represent means, and standard deviations are indicated by the vertical lines.

material was stored at 37 °C for 7 days. The sample-to-sample variability (large standard deviation) was most likely caused by the variability in the activity assay when used on crude seed extracts. In any event, the thermostabilty of avidin in ground and whole kernels allows significant flexibility in terms of storage, processing and transportation conditions to ensure integrity during delivery to the designated purification site.

Extraction and Purification

We have established that equal amounts of avidin protein can be extracted from seed subjected to either cracking and flaking, or to grinding into flour (data not shown). Therefore, the less heat-generating flaking process was selected as the method of choice for seed processing prior to the protein extraction step.

For small-scale purification, avidin is extracted in a sodium bicarbonate/ sodium chloride buffer (pH 11) containing Tween-20 and EDTA. Avidin is affinity purified from this complex mixture by incubating the extract with 2-iminobiotin agarose (Heney and Orr, 1981). As determined by ELISA and western blot analyses, avidin is quantitatively eluted from this column.

Physical Characterization

A comparison of the physical properties of the native and maize-derived avidins was performed. The amino-terminal sequence of the two proteins as determined by Edman degradation (Matsudaira, 1987) was revealed to be identical (Table 4). Egg-white and maize-derived avidins were also compared on stained SDS-polyacrylamide gels and western blots (Figure 6). The protein in maize seed extract and in purified form has a molecular weight of 16,8 kDa (Fig. 6 lanes 1 and 2), approximately 800 Da smaller than the native chicken protein (Gope et al., 1987) (Fig. 6 lanes 3 and 4). However, when proteins

Table 4. Summary of physical characterization data for native and maize-derived avidins

Avidin source	N-Terminal Sequence	MW	Glycosylation	Ki	Antigenic similarity	pI
Egg white	ARKCSLTGKWTNDLGSXMTI	17.6 k Da	yes	3.2 uM	identical	10
Maize-derived	ARKCSLTGKWTNDLGSXMTI	16.8 k Da	yes	3.3 uM	identical	10

from both sources were digested with N-glycosidase A (NGA), the resulting deglycosylated forms were identical with a molecular weight of approximately 12.5 kDa (Figure. 7). These data combined with the N-terminal sequence data (Table 4) strongly suggest that the primary structure of the two avidins is identical.

Based upon avidin's strong biotin-binding property, an analog of biotin, HABA, is utilized for the quantification of avidin activity. Using the HABA binding assay, estimates of total avidin amounts in the fractions eluted from the affinity column (used for purification) were identical to estimates made by ELISA. When competitive binding studies were performed for avidin with biotin and HABA, the native and maize-derived avidins produced identical inhibition constants (Ki, Table 4). The deglycosylated form of chicken egg-white avidin binds biotin as well (Hiller et al., 1987), though it is not known if this is also true for the maize-derived recombinant avidin.

In additional experiments, other physical characteristics of the proteins were addressed (Table 4). Using ouchterlony double diffusion, antigenic identity of native and recombinant avidins was confirmed (data not shown). The isoelectric point (pI) of the native and maize-derived avidins is at least 10, the highest (most basic) value for pI that can be experimentally determined (Table 4). This value correlates well with the calculated value of 10.4.

The end value of a corn crop which is processed for livestock feed is approximately $250 per acre. From our 1994 and 1995 production fields, we recovered 200 to 300 mg

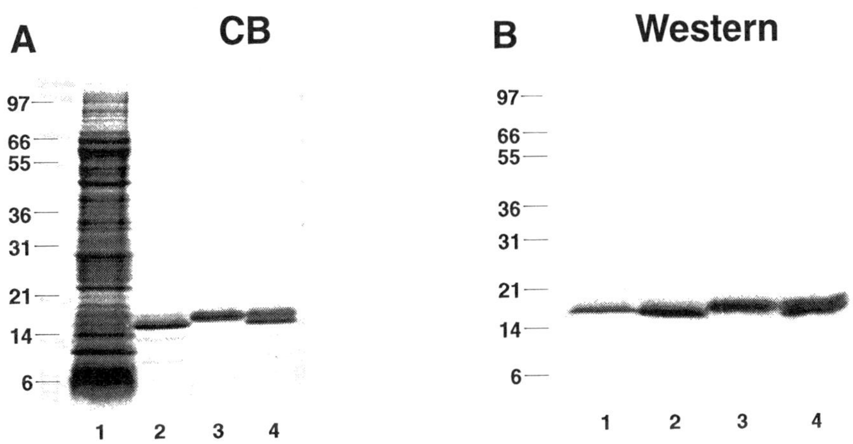

Figure 6. Comparison between recombinant avidin and native avidin. A. Coomassie blue (CB) stained 4–20% polyacrylamide gel of: Lanes: 1. 20 μg avidin-positive maize seed extract; 2. 1 μg of purified recombinant avidin; 3. 1 μg of native avidin; and 4. 0.5 μg each of native and recombinant avidin. B. Western blot (WB) stained with affinity-purified polyclonal antibody raised against avidin. Lanes: 1.20 μg of avidin-positive seed extract; 2. 0.5 μg of recombinant avidin; 3. 0.5 μg of native avidin; and 4. 0.5 μg each of recombinant and native avidin. Molecular weight standards (x 10^{-3}) are indicated on the left.

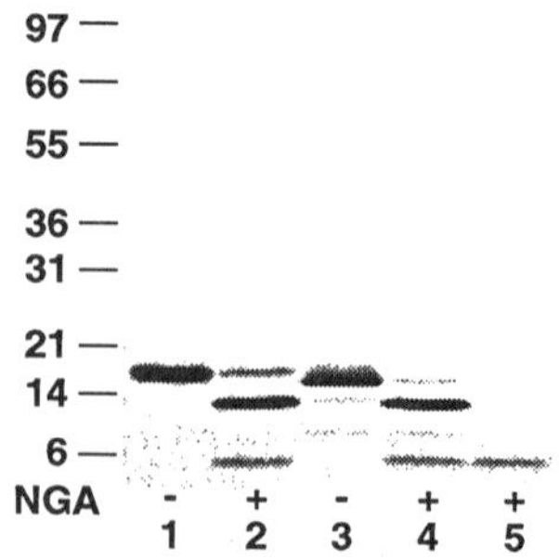

Figure 7. Deglycosylation of avidin. Native avidin (2μg) was incubated overnight at 37°C in the absence (lane 1) or presence (lane 2) of 0.5 munits of N-glycosidase A (NGA). Recombinant avidin (2μg) was also incubated overnight at 37°C in the absence (lane 3) or presence (lane 4) of 0.5 munits of NGA. Lane 5 contains 0.5 munits of NGA incubated overnight at 37°C. The 4–20%polyacrylanide gel was stained with Coomassie blue. Molecular weight standards (x 10^{-3}) are indicated on the left.

avidin protein per kilogram of seed, depending on the production year and location (Table 5). The end value of this material is approximately $2.2 million per acre, based on catalogue price and recovery of product from the raw material. Therefore, the incorporation of the industrial protein, avidin, into the seed adds enormous revenues to be reaped from the corn crop. An additional incentive for this technology is the relative cost of the starting material for processing avidin from corn seed compared to its native source, eggs (Table 6). One ton of eggs at a cost of $1,000 is required to produce an amount of avidin equal to what can be recovered from four bushels of corn (at 2.3% expression level) at a cost of $20.00. It is clear that one advantage of the plant-derived proteins is from the standpoint of starting material costs.

CHARACTERIZATION OF THE β-GLUCURONIDASE PRODUCTION LINE

Several issues have an impact on the commercial production of β-glucuronidase from transgenic plants. First and foremost, though this gene is expressed in hundreds of plants, levels of expression that are necessary for commercial production are 100 to 1000 fold higher than what is histochemically detectable. Additionally, expression must be this high in tissue that is stable, such as in seed, and is easy to store, to transport, and to process for purification. Assuming high levels of expression, this expression must be geneti-

Table 5. Yield and value of avidin from transgenic maize

Year	1994	1995	1995
Location	Hawaii	Hawaii	Texas
Percent male sterile	43	60	51
Percent avidin of extractable protein	3.0	2.0	2.4
Avidin (mg) per kilogram seed	300	240	200
Avidin (g) per acre	469	436	445
Value per acre (catelogue price)	$2.3mm	$2.2mm	$2.2mm

Table 6. Relative value of starting material for avidin purification—eggs versus transgenic maize seed

Source	Egg whites	Corn seed
Volume	1 ton	4 bushels
Cost	$1,000	$20.00

cally stable over several generations. We have addressed all these issues with our commercial line.

Construction of an Expression Cassette Containing the β-*Glucuronidase* Gene

The *Escherichia coli* derived GUS gene (Jefferson et al., 1986) was isolated and cloned into a plasmid derived from pBlueScript SK+ (Stratagene, La Jolla, CA). In this plasmid, the GUS gene was inserted between the maize ubiquitin 5' region, which includes the promoter, the first exon and first intron (Cornejo et al., 1993), and the potato proteinase inhibitor II (PinII) transcription termination region (Fig. 1B; An et al., 1989). The selectable marker expression unit for bialaphos is the same as for the avidin plants (Fig. 1C).

Generation of β-*Glucuronidase* Expressing Transgenic Plants

Regenerable callus lines or immature embryos of Hi-II maize were used as the source tissue for particle bombardment-mediated transformation and were selected on bialaphos for 10 weeks. Resistant calli were subjected to PCR analysis (Jhingan, 1992) or histochemical staining to confirm the presence of the GUS and *bar* genes. Plants initially regenerated from selected embryogenic tissue (T0) were selfed or out-crossed to an elite Pioneer inbred line. Seed from several regenerated T0 plants were screened for high expression using the histochemical or fluorometric assay. From these original transformants, one event expressing high levels of GUS was chosen for further characterization.

DNA Blot Hybridization Experiments

Five individual T3 plants from the identified GUS transformation event were analyzed for their DNA structure on Southern blots (Fig. 2B). Single restriction endonuclease digests showed that the GUS gene was present in four of five sampled plants at one copy (data not shown). When double digests were conducted to determine the size of the plant transcription unit, one band containing the GUS gene fragment was identified and was larger than the original plasmid-derived plant transcription unit (Fig. 2B). For the selectable marker gene, *bar*, a band of the correct size was present in all positive plants but they also contained a smaller band (data not shown). The GUS and *bar* inserts appear to be inherited as a linkage unit.

Cell Biology

Kernels from transgenic plants were fractionated into pericarp, endosperm, and embryo tissue. Enzymatic analysis of recombinant GUS activity in these various tissues illus-

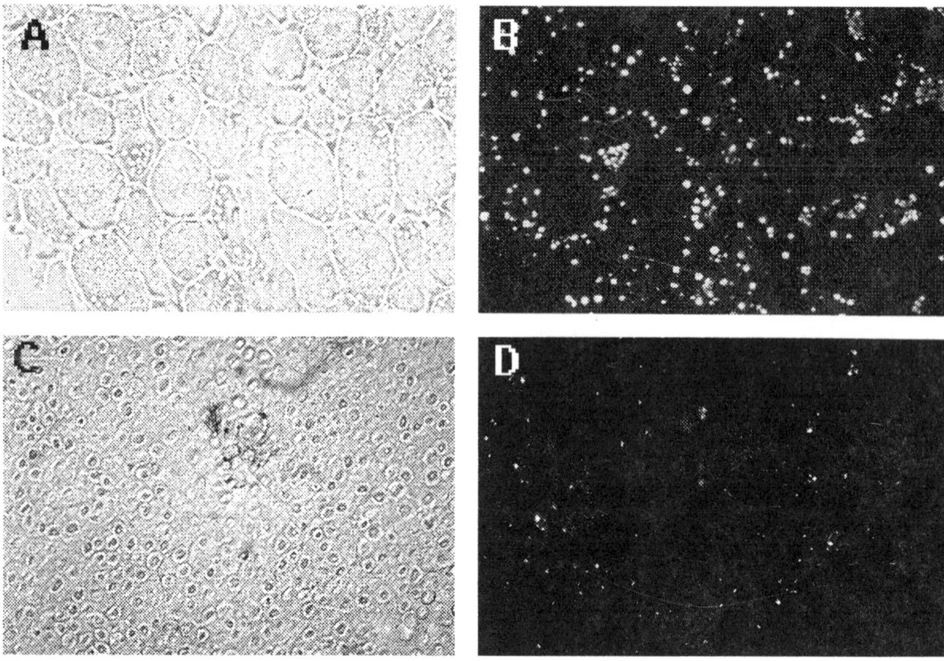

Figure 8. Localization of β-glucuronidase in the seed at the light microscope level. A. Phase contrast and B. fluorescence (TRITC) images of the same 2 μm section of an embryo from GUS positive maize treated with affinity purified anti-β-glucuronidase antibodies and TRITC labeled secondary antibodies. Magnification = 320X. C. Phase contrast and D. fluorescence images of the same β-glucuronidase positive maize endosperm section (5 μm) treated as described above. Transgenic maize kernels containing avidin protein were used as negative controls in these experiments and no label was detected (data not shown). Magnification = 200X.

trated that approximately 93% of the total activity was located in the embryo and the remaining 7% was in the endosperm (Table 7). Staining of the kernel cross-sectional area with X-Gluc confirmed this result (data not shown). Neither the extraction nor the tissue staining showed any GUS accumulation in the pericarp. Because the embryo usually accounts for 10–13% of the dry kernel weight and contains 93% of the β-glucuronidase activity, separation of the embryo from the endosperm before protein extraction could considerably reduce the cost of downstream processing by reducing the amount of total solids and by increasing the concentration of β-glucuronidase in the extract (see below).

To determine the subcellular location of the enzyme, sectioned material was treated with affinity-purified anti- GUS antibodies and detected with fluorescently labeled secondary antibodies (Figure 8) (Witcher et al., 1998). The enzyme was primarily localized to

Table 7. Distribution of β-glucuronidase activity in dry milled fractions of maize seed

Parameter	Embryo	Endosperm
Percent of total seed dry weight	13	80
GUS activity (units/g tissue)	6794	82
Percent of total seed localized GUS activity	93	7

the scutellar tissue (Fig. 8B) with little label in the coleoptile (data not shown). The labeling pattern indicated that the protein is dispersed throughout the cytoplasm, but additionally appears to be associated with a series of small spherical organelles. At high magnification, the X-Gluc stained sections show the enzyme on the surface of these spheres. In addition, an extra-organellar location is consistent with the gene construct because no signal sequence is present that would direct the protein to the intra-organellar pathway.

In endosperm, a similar labeling pattern was seen. Phase contrast images of the endosperm showed good preservation of the cellular structure (Fig. 8C). Diffuse cytoplasmic labeling and an association with larger spherical organelles was observed (Fig. 8D). The sparser label in this tissue reflects the lower concentration of enzyme activity found here compared to embryos. This observation of significantly higher levels of β-glucuronidase in the embryos compared to that of the endosperm is an unexpected result given the fact the constitutive ubiquitin promoter was used in the expression cassette. Quantitative analysis of these results demonstrate that the ubiquitin promoter induced 10–20 fold higher levels of this enzyme in maize seed compared to a number of seed specific promoters (J. Ranch, personal communication). Although there are situations where one would want to limit the expression of a protein to the seed, the ultimate goal for obtaining transgenic maize lines expressing high levels of β-glucuronidase (0.4 to 0.7% of soluble protein) was achieved with the use of the ubiquitin promoter. These expression levels of β-glucuronidase are the highest reported for any plant.

Processing

We have also investigated the storage and heat stability of recombinant GUS in ground and whole kernels, respectively. No significant loss of GUS activity occurred when ground corn was stored at -29, 10 or 25 °C for up to two weeks (Figure 9). When ground samples were stored at 37 °C for one week, 23% of the initial GUS activity was lost. Storing the sample at -29 or 10 °C for up to three months resulted in 14% or 23% loss of GUS activity, respectively. At ambient temperature (25 °C), sample spoilage was observed after three weeks and a loss of 13% of the initial GUS activity was measured. Our recent tests demonstrated that GUS activity in the ground corn samples was preserved for more than a year at 10 °C. Similar storage stability data were obtained for recombinant avidin in transgenic corn (see Fig. 5).

The heating of whole kernels with initial moisture content of 12.6% at 50 °C for up to eight hours or to 70 °C for 1 hour did not affect GUS activity (Kusnadi et al., 1998). Similar results were obtained for avidin as well, indicating that processing of whole seed may stay within these temperatures without affecting the final product.

Extraction of Kernel Fractions

Because the distribution analysis demonstrated that more than 90% of GUS activity accumulated in the embryo (germ), the endosperm and the germ were fractionated by a dry-milling degermination process (Figure 10), and a portion of the germ pool was extracted with hexane to produce a defatted-germ sample. Approximately ten times more GUS was extracted from the ground full-fat germ and defatted-germ than from the kernel samples at either 1:4 or 1:10 solid-to-liquid ratio (Figure 11). Samples were normalized for moisture- and oil-free solids. The same amounts (μg/g solids) of GUS and germ protein, respectively, were extracted from the full-fat and defatted samples at either 1:4 or

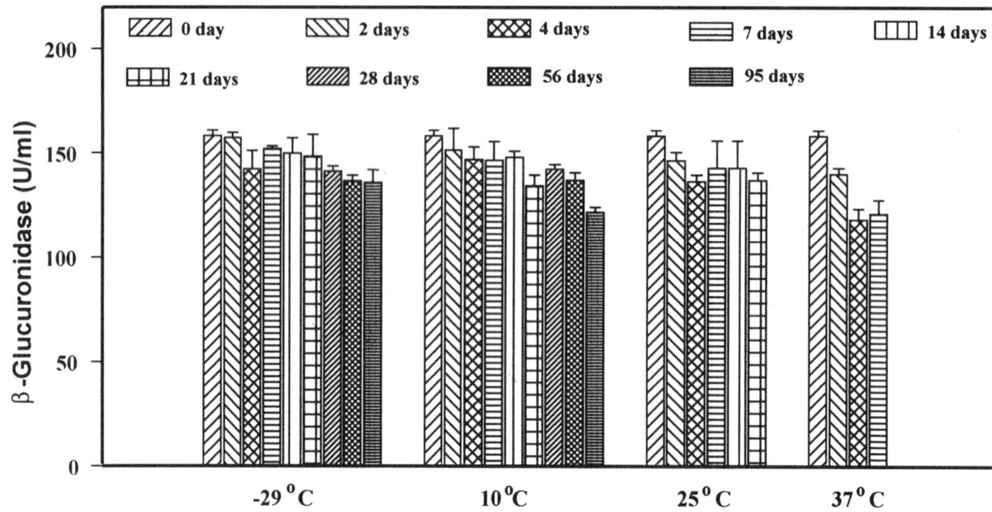

Figure 9. Effect of temperature on β-glucuronidase stability during storage of ground maize seed. Maize seed was processed by cracking and flaking, then stored at the indicated temperatures. Triplicate samples were extracted and analyzed for GUS activity after the times (in days) shown by the shaded bars. Bars represent means, and standard deviations are indicated by the vertical lines.

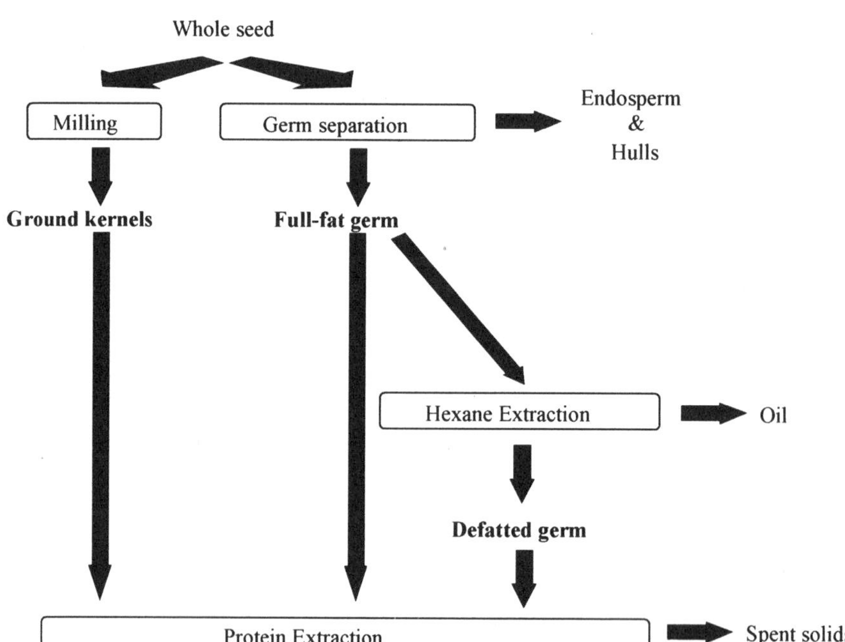

Figure 10. Diagram of processing protocol to produce dry-milled fractions of maize kernels that are subjected to extraction and partial protein purification.

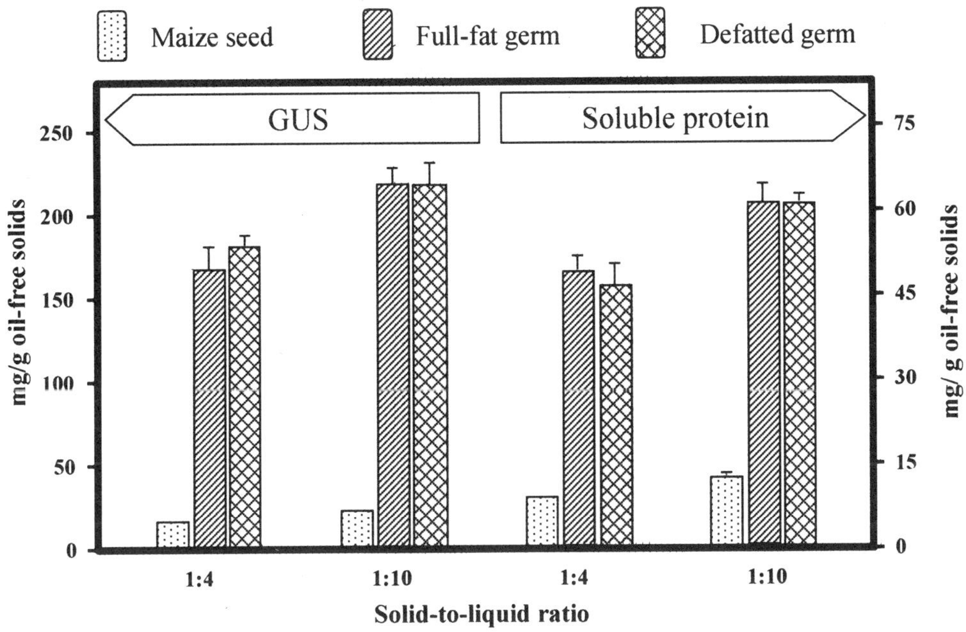

Figure 11. Solid to liquid ratios used to extract protein from dry-milled fractions. Total soluble protein in the extracts was determined and GUS content was estimated with MUG assays. The 1:10 ratio for extraction only produced 30% more protein than the 1:4 ratio, though the volume was increased 2.5 times.

1:10 solid-to-liquid buffer ratio, indicating that the hexane extraction at 60 °C did not affect the amount of extractable GUS activity.

Extractions carried out at a 1:10 solid-to-liquid ratio resulted in 30% more GUS (μg/g solids) than those performed at a 1:4 ratio. Because proportionally more corn protein was also extracted, the GUS concentration per mg of protein in the extract did not change with the solid-to-liquid ratio; the average concentration of GUS in the germ extracts was 3.6 μg/mg and in the kernel extracts was 1.9 μg/mg soluble protein. There was only a two-fold increase in the GUS concentration per mg soluble protein because the germ contains significantly more water and salt-soluble proteins than the endosperm (Wilson, 1987).

Purification and Physical Characterization

In order to validate the commercial protein production system for β-glucuronidase, one must ensure the integrity of the transgenic protein. To characterize the maize produced GUS in detail, the recombinant enzyme was purified from seed extracts by immunoaffinity chromatography and by ion exchange and hydrophobic interaction chromatographies (Table 8). The purified recombinant and native enzymes migrated in an SDS-polyacrylamide gel with a molecular weight of 68 kDa and migrate together when placed in a single lane on the gel. Western blot analysis with affinity-purified polyclonal antibodies against the native enzyme also showed that the proteins migrated at 68,000 Da and identified the recombinant enzyme in the seed extract (Figure 12). Because the molecular weights of the recombinant and native enzymes on native gels were estimated to

Table 8. Purification of recombinant GUS from transgenic maize seed

	Total purification fold	Recovery (%)
Extraction	1	100
IEC (DEAE Sepharose)	8	126
HIC (Phenyl Sepharose)	62	81
IEC (DEAE Sepharose	145	70
SEC (Sephacryl S-300 HR)	915	27

be approximately 300,000 Da (data not shown), they apparently exist as homotetramers (Brot et al., 1978; Gehrmann et al., 1994; Ho, 1991; Rethinaswamy et al., 1994). The isolectric points (pI) of the native and recombinant enzymes were determined to be between 4.8 and 5.0 (Table 9).

The N-terminal amino acid sequence of the maize produced β-glucuronidase, as determined by Edman degradation (Matsudaira, 1987), coincided with that predicted from the transforming gene sequence except for the loss of the initial methionine (Table 9). In order to determine if the initial methionine is removed through a cellular processing event or a deletion in the gene encoding the enzyme, the 5' portion of the transgene was sequenced. The ATG codon for the initial methionine was present, strongly suggesting that

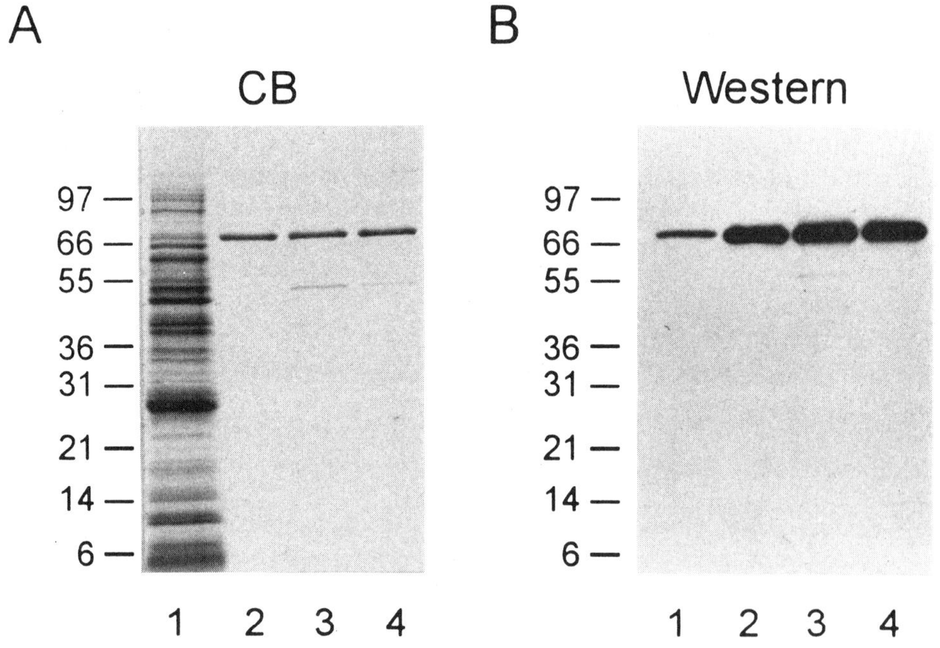

Figure 12. Comparison between recombinant and native GUS. A. Coomassie Blue (CB) stained polyacrylamide gel of 20 μg of β-glucuronidase positive maize extract (lane 1), 1 μg of recombinant β-glucuronidase (lane 2), 1 μg of native β-glucuronidase (lane 3), and 0.5 μg of native and recombinant β-glucuronidases (lane 4). B. An immunoblot of 20 μg of β-glucuronidase positive maize extract (lane 1), 0.5 μg of recombinant β-glucuronidase (lane 2), 0.5 μg of native β-glucuronidase (lane 3), and 0.25 μg of both recombinant and native β-glucuronidase (lane 4). Molecular weight standards (x 10^{-3}) are indicated on the left.

Table 9. Summary of physical properties of β-glucuronidase

Biochemical properties	Native E. coli GUS	Recombinant GUS
Molecular weight	68,000 Da	68,000 Da
Km	0.21 +/- 0.04 nM	0.19 +/- 0.05 nM
Vmax	3.2 x 10^5 +/- 3.3 x 10^4 nmoles/hr	1.5 x 10^5 +/- 3.8 x 10^4 nmoles/hr
PI	4.8-5.0	4.8-5.0
heat stable	yes	yes
Antigenic similarity	yes	yes
N-terminal sequence	native	identical except for initial met

the initial methionine is cleaved from the enzyme through a cellular processing event. This N-terminal methionine processing has been identified in *Escherichia coli* and has also been shown to occur in organelles of animals, yeast, and plants (Clements et al., 1989; Hirel et al., 1989; Schmidt et al., 1992; Termignoni et al., 1991). The extent of the N-terminal methionine cleavage has been demonstrated to be governed by the side-chain length of the penultimate amino acid (Hirel et al., 1989). The analysis of the native and recombinant enzymes on SDS-PAGE, western blots, and isoelectric focusing gels in combination with the N-terminal sequence data indicate that the primary structure of the two proteins is identical except for the N-terminal methionine.

Data for the physical characterization of the native and recombinant enzymes are presented in Table 9. Scatchard analysis of kinetic data revealed an apparent K_m of 0.21 mM and a V_{max} of 2.2 x 10^5 nM MUG/hr for the native enzyme. For the recombinant β-glucuronidase, Scatchard analysis revealed an apparent K_m of 0.20 mM and a V_{max} of 2.9 x 10^5 nM MUG/hr. The result of this kinetic analysis of the recombinant enzyme from maize illustrates that this protein is functionally very similar to the native enzyme. Additionally, both enzymes had an inhibition constant, K_i of 0.12 mM D-saccharic acid 1,4 lactone. These results combined with the kinetic analysis, thermal stability, and enzyme activity over a broad range of pH (data not shown), demonstrate that the native and the recombinant β-glucuronidases are functionally equivalent.

CONCLUSIONS

Several important factors which could impact the production of recombinant proteins in plants are illustrated by these examples. First, the expression cassette must be efficient to drive accumulation of high levels of the recombinant protein in the seed. In this example, the ubiquitin promoter, rather than a seed specific promoter, was employed for high levels of protein expression. The levels of avidin expression (2.3%) and β-glucuronidase expression (0.4 to 0.7 %) in our production lines are the highest reported for any plant. While many other groups have expressed β-glucuronidase, none have developed a system where the enzyme has been expressed at high enough levels for commercial production. Secondly, recombinant protein levels need to be stable over generations. Our results demonstrate stability of avidin and GUS expression over at least seven generations. Third, the recombinant and native proteins must be functionally equivalent. We have clearly demonstrated that avidin and β-glucuronidase purified from maize seed are functionally the same as the native proteins. We have illustrated that downstream activities in the production of these recombinant proteins, such as stabilizing the germplasm and processing for purification, were accomplished without any major obstacles. This system provides both a more economical and easier method for the production of these proteins compared to methods currently in use. These examples represent the first cases in which plants have been used to produce recombinant proteins for commercial sale. These products are currently being marketed by Sigma Chemical Company (catalog numbers G2035 and A8706).

ACKNOWLEDGMENTS

Figures 2A, 3A and 3B, 4, 5, 6 and 7: and Tables 2, 3 and 5 are from Molecular Breeding 3, 1997, pp. 291–306, Hood et al., and are reprinted with kind permission from Kluwer Academic Publishers.

REFERENCES

An, G., Mitra, A., Choi, H.K., Costa, M.A., An, K., Thornburg, R.W., Ryan, C.A. Functional analysis of the 3' control region of the potato wound-inducible proteinase inhibitor II gene. *Plant Cell* **1989**, *1*, 115–122.

Austin, S., Bingham, E.T., Koegel, R.G., Mathews, D.E., Shahan, M.N., Straub, R.J., Burgess, R.R. An overview of a feasibility study for the production of industrial enzymes in trangenic alfalfa. *Ann. NY Acad. Sci.* **1994,** *721*, 234–244.

Benfey, P.N., Chua, N-H. Regulated genes in transgenic plants. *Science* **1989,** *244,* 174–181.

Brot, F.E., Bell, C.E., Sly, W.S. Purification and properties of β-glucuronidase from human placenta. *Biochemistry* **1978,** *17*, 385–391.

Christensen, A.M., Sharrock, R.A., Quail, P.H. Maize polyubiquitin genes: structure, thermal perturbation of expression and transcript splicing, and promoter activity following transfer to protoplasts by electroporation. *Plant Mol. Biol.* **1992,** *18,* 675–689.

Clements, J.M., O'Connel, L.I., Tsunasawa, S., Sherman, F. Expression and activity of a gene encoding rat cytochrome c in the yeast Saccharomyces cerevisiae. *Gene* **1989,** *83*, 1–14.

Cornejo, M., Luth, D., Blankenship, K., Anderson, O., Blechl, A. Activity of a maize ubiquitin promoter in transgenic rice. *Plant Mol. Biol.* **1993,** *23*, 567–581.

DeLange, R.J., Huang, T.S. Egg White Avidin III. Sequence of the 75-residue middle cyanogen bromide peptide. Complete amino acid sequence of the protein subunit. *J. Biol.Chem.* **1971,** *246*, 698–709.

Fisk, H.J., Dandekar, A.M. The introduction and expression of transgenes in plants. *Scientia Hort .* **1993,** *55*, 5–36.

Gehrmann, M.C., Opper, M., Sedlacek, H.H., Bosslet, K., Czech, J. Biochemical properties of recombinant human β-glucuronidase synthesized in baby hamster kidney cells. *Biochem J.* **1994,** *301*, 821–828.

Gope, M.L., Keinanen, R.A., Kristo, P.A., Conneely, O.M., Beattie, W.G., Zarucki-Schulz, T., O'Malley, B.W., Kulomaa, M.S.: Molecular cloning of the chicken avidin cDNA. *Nuc. Acids Res.* **1987,** *15*, 3595–3606.

Heney, G., Orr, G.A. The purification of avidin and its derivatives on 2-iminobiotin-6-aminohexyl-sepharose 4B. *Anal. Biochem.* **1981,** *114*, 92–96.

Hiller, Y., Gershoni, J.M., Bayer, E.A., Wilchek, M. Biotin binding to avidin. *Biochem J.* **1987,** *248*, 167–171.

Hirel, P-H., Schmitter, J-M., Dessen, P., Fayat, G., Blanquet, S. Extent of N-terminal methionine excision from *Escherichia coli* proteins is governed by the side-chain length of the penultimat amino acid. *Proc Natl Acad Sci USA,* **1989,** *86,* 8247–8251.

Ho, K-J. A large scale purification of β-glucuronidase from human liver by immunoaffinity chromatography. *Biotech and Appl Biochem* **1991,** *14*, 296–305.

Hood, E.E., Witcher, D.R., Maddock, S., Meyer, T., Baszczynski, C., Bailey, M., Flynn, P., Register, J., Marshall, L., Bond, D., Kulisek, E., Kusnadi, A., Evangelista, R., Nikolov, Z., Wooge, C., Mehigh, R.J., Hernan, R., Kappel, W.K., Ritland, D., Li, C.P., Howard, J.A. Commercial production of avidin from transgenic maize: Characterization of transformant, production, processing, extraction and purification. *Mol. Breeding* **1997**, *3*:291–306.

Holzman, D. Agracetus grows monoclonals in soybeans and corn plants. *Genet. Eng. News* **1994,** *14 (16)*, 1, 34.

Jefferson, R.A., Burgess, S.M., Hirsch, D. β-Glucuronidase from Escherichia coli as a gene-fusion marker. *Proc Natl Acad Sci* **1986,** *83,* 8447–8451.

Jefferson, R.A. Assaying chimeric genes in plants: The GUS gene fusion system. *Plant Molec Biol Rep 5,* **1987,** 387–405.

Jefferson, R.A., Kavanagh, T.A., Bevan, M.W. GUS fusions: β-glucuronidase as a sensitive and versatile gene fusion marker in higher plants. *EMBO J* , **1987,** *6*, 3901–3907.

Jhingan, A.K. A novel technology for DNA isolation. *Methods Mol Cell Biol,* **1992,** *3,* 15–22.

Keinanen, R.A., Laukkanen, M-L., Kulomaa, M.S. Molecular cloning of three structurally related genes for chicken avidin. *J. Steroid Biochem.* **1988,** *30,* 17–21.

Keinanen, R.A., Wallen, M.J., Kristo, P.A., Laukkanen, M.O., Toimela, T.A., Helenius, M.A., Kulomaa, M.S. Molecular cloning and nucleotide sequence of chicken avidin-related genes 1–5. *Eur. J. Biochem.* **1994,** *220,* 615–21.

Krebbers, E., Bosch, D., Vandekerckhove, J. Prospects and progress in the production of foreign proteins and peptides in plants. In *Plant Protein Engineering* Shewry, P.R., Gutteridge, S., Eds.; Cambridge University Press, pp.315–325 Cambridge, MA **1993**.

Kusnadi, A.R., Hood, E.E., Witcher, D.R., Howard, J.A., Nikolov, Z.L. Production and purification of two recombinant proteins from transgenic corn. *Biotechnology Progress.* **1998**, *14*, 147–155.

Matsudaira, P. Sequence from picomole quantities of proteins electroblotted onto polyvinylidene difluoride membrnes. *J. Biol. Chem.* **1987,** *262,* 10035–10038.

Nagy, F., Odell, J.T., Morelli, G., Chua, N.H. Properties of expression of the 35S promoter from CaMV in transgenic tobacco plants. In *Biotechnology in plant science: relevance to agriculture in the eighties.* Zaitlin, M., Day, P., Hollaender, A. Eds.; Academic Press, p. 227–235, Orlando, FL **1985**.

Pen, J., Verwoerd, T.C., van Paridon, P.A., Beudeker, R.F., van den Elzen, P.J.M., Geerse, K., van der Klis, J.D., Versteegh, H.A.J., van Ooyen, A.J.J., Hoekema, A. Phytase-containing transgenic seeds as a novel feed additive for improved phosphorus utilization. *Bio/Technol.,* **1993A,** *11*, 811–814.

Pen, J., Sijmons, P.C., van Ooijen, A.J.J., Hoekema, A. Protein production in transgenic crops: Analysis of plant molecular farming. In *Industrial Crops Production.* Elsevier, Amsterdam. pp. 241–250 **1993B.**

Rogers, J.C. Two barley alpha-amylase gene families are regulated differently in aleurone cells. *J. Biol. Chem.* **1985,** *260*, 3731–3738.

Rethinaswamy, A., Yang, C-H., Srivastava, P.N. Purification and characterization of β-glucuronidase from bull seminal plasma and its role in fertilization. *Mol Reprod and Dev.* **1994**, *38* , 404–409.

Schmidt, J., Herfurth, E., Subramanian, A.R. Purification and characterization of seven chloroplast ribosomal proteins: Evidence that organelle ribosomal protein genes are functional and that NH_2-terminal processing occurs via multiple pathways in chloroplasts. *Plant Mol Biol.* **1992,** *20,* 459–465.

Termignoni, C., Freitas, J.O., Guimaraes, J.A. Methionyl aminopeptidase from rat liver: distribution of the membrane-bound subcellular enzyme. *Mol Cell Biochem.,* **1991,** *102,* 101–113.

White J., Chang, S-Y.P., Bibb, M.J., Bibb, J.M. A cassette containing the bar gene of Streptomyces hygroscopicus: a selectable marker for plant transformation. *Nuc. Acids Res.* **1990,** *18,* 1062.

Whitelam, G.C., Cockburn, B., Gandecha, A.R., Owen, M.R.L. Heterologous protein production in transgenic plants. *Biotechnol. Genetic Eng. Rev.* **1993,** *11,* 1–29.

Wilson, C.M. Proteins of the kernel. In *Corn: chemistry and technology.* Watson, S.A., Ramstad, P.E. Eds., American Association of Cereal Chemists, Inc. St. Paul, MN **1987,** pp. 273–305.

Witcher, D.R., Hood, E.E., Peterson, D., Bailey, M., Bond, D., Kusnadi, A., Evangelista, R., Nikolov, Z., Wooge, C., Mehigh, R., Kappel, W., Register, J., and Howard, J.A. Commercial production of β-glucuronidase (GUS): a model system for the production of proteins in plants. Molecular Breeding **1998,** *4*, 301–312.

12

IMPROVEMENTS IN HUMAN HEALTH THROUGH PRODUCTION OF HUMAN MILK PROTEINS IN TRANSGENIC FOOD PLANTS

Takeshi Arakawa,[1,3] Daniel K. X. Chong,[1] Charles W. Slattery,[2,4] and William H. R. Langridge[1,2,3,*]

[1]Center for Molecular Biology and Gene Therapy
[2]Department of Biochemistry
[3]Department of Microbiology and Molecular Genetics
[4]Department of Pediatrics
Loma Linda University
Loma Linda, California, 92350

ABSTRACT

Plants are particularly suitable bioreactors for the production of proteins, as their eukaryotic nature frequently directs the appropriate post-translational modifications of recombinant proteins to retain native biological activity. The autotrophic growth of plants makes this *in vivo* biosynthesis system economically competitive for supplementation or replacement of conventional production systems in the future. For the production of biologically active proteins, food plants provide the advantage of direct delivery via consumption of transformed plant tissues. Here we describe the production of recombinant human milk proteins in food plants for improvements in human nutrition and health, with emphasis on enhanced nutrition for non-breast fed infants as well as children and adults. Nutritional improvements in edible plants generated through advancements in recombinant DNA technology are rapidly repositioning the world for enjoyment of a more healthful diet for humans in all age groups.

[*] Center for Molecular Biology and Gene Therapy, School of Medicine, Loma Linda University, Loma Linda, California 92350, Telephone: (909) 824-4300, Ext. 81362, Fax: (909) 478-4177, E-mail: bLangridge@ccmail.llu.edu

Chemicals via Higher Plant Bioengineering, edited by Shahidi *et al.*
Kluwer Academic / Plenum Publishers, New York, 1999.

INTRODUCTION

Transgenic Plants as Bioreactors for Nutritional and Therapeutic Proteins

More than a decade has passed since transgenic farm animals were first considered as biological factories for the production of high value proteins in body fluids such as blood, urine and milk. Molecular farming has become an important aspect of contemporary biotechnology. Animal blood is readily available for purification of recombinant proteins. For example, transgenic mice producing a human growth hormone in their urine have been generated by introducing the human growth hormone gene under control of a bladder-specific promoter for the uroplakin gene II (Kerr *et al.*, 1998). The human growth hormone is secreted into the urine from the bladder epithelium, resulting in more than 100 ng/ml of recombinant protein. The milk of large farm animals (e.g., cattle and goats) is a particularly useful source of recombinant proteins. Novel proteins can be synthesized at high levels and may often be readily extracted, or even consumed directly. With the aid of recombinant DNA methods and the development of embryo manipulation techniques, increasing numbers of foreign genes have been introduced and expressed in the milk of transgenic livestock. Examples include human protein C in pigs (van Cott *et al.*, 1997), human tissue-type plasminogen activator in goats (Ebert *et al.*, 1991), and bovine β-casein in mice (Hitchin *et al.*, 1996).

The use of mammary glands as bioreactors for production of biologically active proteins with pharmaceutical and therapeutic value has several advantages over conventional bacterial production systems. The production of antimicrobial peptides in bacterial systems can result in self-toxicity, transgene protein insolubility, and lack of requisite post-translational modifications (e.g. glycosylation and phosphorylation). Due to their eukaryotic origin, animal cell culture production systems often modify and fold proteins. However, the requirement for an expensive culture medium can make animal cell systems economically prohibitive for large-scale recombinant protein production. In contrast, mammary gland-specific expression of foreign genes and secretion of their products in high amounts in the milk of large animals (e.g., cow, goat, pig, and sheep), can be reliably achieved using mammary gland-specific promoters and transcriptoinal regulatory elements. However, economic drawbacks of using large animals such as cows are their long time to reach sexual maturity, 8–9 month gestation period, and time to reach peak lactation, making it at least a 3 year period before product availability.

Plant-based molecular farming has attracted the pharmaceutical industry for several reasons: (1) potential economic advantages over animal-based production systems, (2) lack of contamination with animal microbial pathogens and other infectious agents like prions, and (3) the reduction in ethical issues linked with the creation of transgenic animals. Plants can be grown inexpensively due to their photoautotrophic nature; they require only water, CO_2, a small amount of soil minerals, nitrogen, and sunlight for growth. Perhaps of equal importance, plant cells, like mammalian cells, carry out post-translational modifications essential to retain the biological activity of the recombinant eukaryotic proteins.

Stable integration of foreign genes into the plant chromosome and subsequent clonal expansion of the transgenic plant e.g., potato, can pass on the integrated foreign genes from generation to generation in a Mendelian fashion and if desired, homozygotes with increased levels of foreign gene expression can be selected. Transgenic plants possess many

Table 1. Biological effects of some human milk proteins

Protein	Function and characteristics
β-Casein	Diverse biological effeccts including: binding to divalent cations (e.g., Ca, Mg, and P), chemotactic for leukocytes generates β-casomorphins (opioid agonists)
α-Lactalbumin fraction	Involved in lactose synthesis, a major protein of the albumin fraction
Secretory IgA, IgG	Anti-adhesion, opsonization, complement fixation, (anti-microbial)
Complement system	Complement fixation, opsonization
Lactoferrin, transferrin	Iron-binding (especially anti-gram negative bacteria)
Lactoperoxidase	Antibacterial (Streptococci, Psudomonas, Salmonella, *E. coli*)
Lysozyme	Bacteriolysis (gram-negative and gram-positive)
Interferons	Antiviral, anti-tumor

human health applications, including production of active immunization antigens, immunotolerizing autoantigens for autoimmune diseases (Haq *et al.*, 1995; Mason *et al.*, 1992, 1996; Arakawa *et al.*, 1997, 1998; Ma *et al.*, 1997), and antibody production for passive immunization against microbial pathogens (Ma *et al.*, 1995). Despite relatively low production levels achieved through stable transformation of food plants, ranging from 0.01% to 0.4% of total plant protein, stable transformation, in comparison with plant-virus based transient production systems which require extraction of recombinant protein products or modified viral particles, provide a distinct advantage in situations where extraction processes are impractical due to economic or technological limitations or where the presence of viral proteins are undesirable.

Human Milk Protein Composition

Human milk has long been recognized as a well-balanced diet for infants, supplying a high nutritional value in casein and whey proteins as well as providing a source of antimicrobial proteins such as lactoferrin and lysozyme (Table 1). The composition of human milk changes to meet the nutritional needs of the growing infant (Lawrence, 1989). Casein protein (3–3.5 g/l) of human milk makes up about 30–35% of total milk protein (Miller *et al.*, 1990). There are two major subclasses of human casein: β-casein the most abundant form (70%) and κ-casein which accounts for about 27% of total casein in human milk (Kunz and Lonnerdal, 1990). The β-casein is a 25-kDa globular protein that exists in several isoforms depending on its phosphorylation state. Up to 5 phosphate groups can be attached to serine and threonine residues located within the first 10 amino acids of the polypeptide (Greenberg *et al.*, 1984; Hansson *et al.*, 1993). The amphipathic structure of the casein molecule is due to uneven distribution of hydrophilic and hydrophobic amino acid residues as well as phosphorylation of the molecule. The binding of calcium, magnesium, and phosphate ions facilitates formation of large casein aggregates (micelles) giving milk its white color (Dickson and Perkins, 1971; Blakeborough *et al.*, 1983; Hansson *et al.*, 1993). In addition to its emulsification properties, human β-casein has diverse biological effects such as enhancement of absorption of calcium and other divalent cations (Naito and Suzuki, 1974; Mykkanen and Wasserman, 1980), inhibition of angiotensin I-converting enzymes (Kohmura *et al.*, 1989), opiate agonist effects (β-casomorphins) (Brantl, 1984), immunostimulating and modulating effects (Migliori-Samour and Jolles, 1988), and antibacterial functions (Mitra and Zhang, 1994; Lahov and Regelson, 1996)

Table 2. Differences in human and bovine milk composition

Characteristic	Human milk	Bovine milk	Reference
Protein content 1990	0.9 g/dl	3.5 g/dl	Miller *et al.*, 1990
Casein to whey ratio 1990	2 : 3	4 : 1	Miller *et al.*, 1990
Predominant casein 1990	β-casein	α-casein	Miller *et al.*, 1990
Cys to Met ratio	High	Low	Packard, 1982
Phe and Tyr levels	Low	High	Packard, 1982
Ca to P ratio	2.3 : 1	1.2 : 1	Packard, 1982
Taurine level	Substantial amount	Essentially absent	Packard, 1982

The Nutritional Value of Human Milk versus Bovine Milk

Milk is a biochemically complex mixture of proteins, lipids, carbohydrates, vitamins and minerals, designed to nurture mammalian newborns. Based on acid treatment, bovine milk as well as human milk is divided into the casein phosphoproteins and the whey proteins. There are three types of casein phosphoprotein subclasses: α-, β-, and κ-caseins. The whey proteins include lactoferrin, α-lactalbumin, β-lactoglobulin, and serum albumin (Miller *et al.*, 1990). Milk protein content and composition varies dramatically between mammalian species (Table 2). Cow's milk contains 3.5 g/dl of total protein in comparison with only 0.9 g/dl total protein in human milk. The predominant casein subclass of bovine milk, α-casein, is found only in trace amounts in human milk (Miller *et al.*, 1990). The casein-to-whey ratio in bovine milk is 80/20 in comparison with 40/60 in human milk.

Significant amino acid differences exist between human and cow's milk (Packard, 1982), and can present physiological problems in feeding cow's milk to infants. Human milk has a higher cysteine to methionine ratio than cow's milk and also contains a substantial amount of taurine, a nonprotein nitrogen constituent essentially absent from cow's milk. The liver and brain of the human fetus and pre-term infant are completely lacking cystathionase, an enzyme which converts methionine to cysteine, an essential amino acid for central nervous system development. Taurine, which is derived from cysteine through the action of cysteinesulfonic acid decarboxylase, is essential for infant brain and retinal development as well as for the function and conjugation of bile salts. Optimum levels of cysteine and taurine required for adequate infant nutrition cannot be obtained from cow's milk. The higher levels of phenylalanine and tyrosine in cow's milk versus mother's milk may present an additional problem. The lower levels of phenylalanine and tyrosine in human milk match the infant's limited ability to metabolize these amino acids, which at higher levels in the diet, can accumulate to cause phenylketonuria (PKU) and subsequent brain damage.

Milk caseins are responsible for binding essential minerals (especially calcium, phosphate, iron and zinc), to maintain them in an easily ingestible suspension for aiding absorption. The Ca to P ratio is higher in human milk than in cow's milk so that formulas must be adjusted to the appropriate Ca to P ratio. The fact that minerals are more efficiently absorbed from human milk due to structural and functional differences between bovine and human casein proteins must be taken into consideration for determination of the appropriate mineral ratio (Lonnerdal, 1985). These marked differences between cow's milk and human milk are not surprising as the nutritional requirements for growth of bovine and human newborns differ significantly.

Food Plants as Production and Oral Delivery Vehicles for Human Milk Proteins

It is generally accepted that human milk is superior to all other milk substitutes for the growing infant. However, there remains a large market for milk substitutes since not all mothers are able to nurture their infants by breast milk feeding due to adverse physical, psychological or socio-economic conditions. Infant formulas based on bovine milk or soybean have been, until recently, traditional substitutes for mother's milk. However, bovine milk has been designed by nature to facilitate the rapid growth of a 100 lb calf, which explains why bovine milk is substantially different from human milk in both protein content and composition. Although, cow's milk has been recognized as an excellent source of proteins, vitamins and minerals, its consumption is often related to excessive weight gain and cow's milk protein allergy and intolerance (CMPA/CMPI), which may affect gastrointestinal tract, respiratory tract, skin and blood (Johansson *et al.*, 1994; Wilson and Hamburger, 1988). The CMPA is a disease of infancy and usually appears in the first few months of life. Prenatal or early neonatal exposure to cow's milk protein increases the risk, not only of adverse reactions to bovine milk proteins but also of development of allergies to other foods, especially soy and egg (Host, 1991, 1995). Over 25 protein fractions in bovine milk have been demonstrated to induce an allergic response in humans, and many are capable of inducing IgE-mediated (type I) reactions. Most clinical cases of infant food allergy, induced by consumption of cow's milk or its products, are believed to be related to bovine casein (Perkin, 1990; Karjalainen *et al.*, 1992). Soy-based infant formulas have been recommended as hypoallergic alternatives for nonbreast-fed infants. However, many infants with food allergy to bovine milk are also allergic to soy proteins. Therefore, soy formulas can no longer be considered suitable substitutes for cow's milk formulas under these conditions. Although soybean allergenic proteins have not been fully identified, the globulin components (2S, 7S and 11S) and hemagglutinin (soybean trypsin inhibitor) have been implicated as allergens (Perkin, 1990).

There are a number of reports indicating a possible link between consumption of cow's milk and development of insulin-dependent diabetes mellitus (IDDM) (Daneman *et al.*, 1987; Savilahti *et al.*, 1988; Karjalainen *et al.*, 1992). Animal experiments have indicated that development of IDDM can be prevented in diabetes-prone rodents if animals are provided a bovine milk-free diet in early life (Daneman *et al.*, 1987). Shorter duration of breast-feeding and early introduction of dairy products was found to be associated with an increased risk of IDDM in Finnish and Brazilian children (Virtanen *et al.*, 1991, 1994; Gimeno and de Souza, 1997). It has also been reported that patients with IDDM have elevated levels of serum IgG and IgA anti-BSA (bovine serum albumin) antibodies which cross react with p69, one of the target β-cell antigens in IDDM patients (Karjalainen *et al.*, 1992).

Infant formulas, containing essential protein components of human milk derived from food plants, could provide alternative nutrition for nonbreast-fed infants as well as assisting in the prevention of immunological problems associated with the consumption of bovine milk or soy-based synthetic infant formulas. Moreover, food plants are not only a feasible production system for multiple protein components of human milk but can also be considered as an oral delivery system for people in all age groups. Cultivation and harvest of transformed plants producing human milk proteins require only conventionally-available agricultural technologies, after which the plant materials can be extracted for formula constituents or simply delivered by the plant in the diet. Furthermore, production of several human milk proteins in vegetables and fruits could provide a novel source of improved nutrition for children and malnourished human populations in economically emerging countries. The generation of individual transgenic food plants producing several milk proteins in addition to vaccine protein antigens can contribute to a cost-effective and nutritionally superior vegetable-based diet for people in need of more complete nutrition and protection against diseases.

Production of Human Milk β-Casein in Transgenic Plants

Agrobacterium-mediated stable transformation methods rapidly generate a large number of transgenic plants through vegetative methods and provide an opportunity to introduce several genes into an individual edible plant. The synthesis of a major component of human milk protein, β-casein, in potato plants (*Solanum tubersosum* cv. Bintje) was recently accomplished in our laboratory (Chong *et al.*, 1997). This is the first published report of stable genomic integration and expression of a human milk protein gene in food plants. Human β-casein cDNA was cloned into a multiple cloning site under control of the *Agrobacterium tumefaciens* mannopine synthase bi-directional promoter (*mas* P2) in the plant transformation vector, pPCV701 (Chong *et al.*, 1997). The bacterial luciferase fusion gene (Escher *et al.*, 1989), a detectable marker gene, was placed under control of the *mas* P1 promoter. The β-casein cDNA was introduced into tobacco and potato plants by *A. tumefaciens*-mediated transformation methods (Horsch *et al.*, 1985). Potato leaf cell transformation resulted in the stable integration of human β-casein cDNA into the plant genome. Full length β-casein mRNA was detected by RT-PCR, and the protein product identified in transformed potato tissue homogenates by immunoblot analysis. An initial β-casein expression level of 0.01% of total soluble protein was detected in leaf and tuber tissues. The plant-synthesized β-casein appears to be a single peptide of approximately 24 kDa, approximately 1–1.5 kDa smaller than the human β-casein isoform phosphorylated at two sites. The reason for the apparent reduction in molecular size between plant and animal cells is not clearly understood. Immunoaffinity column purification experiments are in progress to obtain amounts of protein sufficient for N-terminal amino acid sequence and phosphorylation pattern analyses. Recently, human β-casein cDNA was introduced into a tomato (*Lycopersicon esculentum* cv. Tiny Tim). The presence of β-casein cDNA, mRNA, and protein were detected (Sovyanhadi *et al.*, unpublished data).

Production of Human Lactoferrin and α-Lactalbumin in Transgenic Plants

Lactoferrin, one of the major whey proteins in human milk, is an iron-binding glycoprotein with an approximate molecular weight of 80 kDa. High levels of lactoferrin have been identified in neutrophils (Mason *et al.*, 1969) and in lactating mammary glands (Green and Pastewka, 1978). Although, little is known about the function of milk lactoferrin *in vivo*, functions ascribed to lactoferrin *in vitro* include promotion of cell growth, regulation of hematopoiesis, antimicrobial activity and immune-modulating properties. Antibacterial as well as antiviral activities of lactoferrin were reported (Arnold *et al.*, 1977; Lu *et al.*, 1987). Mitra and Zhang (1994) first reported expression of the human lactoferrin gene under control of the CaMV 35S promoter in tobacco cells. Transgenic tobacco cells produced a truncated form of lactoferrin with a molecular mass of 48 kDa. The extent of protein glycosylation in tobacco was undetermined. The antibacterial properties of transgenic callus extracts were tested against four different phytopathogenic bacterial strains: *Xanthomonas campestris* pv. phaseoli, *Pseudomonas syringae* pv. phaseolicola,

Table 3. Amount of human milk proteins supplied by transgenic potatoes

		% Protein[b]		Potato #[e]	
Milk protein	Protein[a] (mg)	P^{mas}[c]	$P^{patatin}$[d]	P^{mas}	$P^{patatin}$
β-casein	225	9.2[f]	62	11	1.6
κ-casein	90	22.5	155	4.3	0.6
Lactoferrin	360	6	39	16.7	2.6
α-lactalbumin	540	4	26	25	3.8
Lysozyme	100	21	140	4.8	0.7

Amount of milk protein in a cup (250 ml) of human milk (Miller *et al.*, 1990).
Percent protein per cup of human milk in an average potato (300g).
(c,d) Calculations based on the *mas*P2 or patatin promoters which show 0.3% or 2% foreign gene expression levels respectively in potato tubers.

Pseudomonas syringae pv. syringae, and *Clavibacter flaccumfaciens* pv. flaccumfaciens. In colony-forming unit reduction assays, transgenic calli containing recombinant lactoferrin exhibited substantially higher antibacterial activity than native lactoferrin.

Recently we have introduced human lactoferrin cDNA in a stable form into potato and tomato plants and found the recombinant lactoferrin to be expressed at a level of 0.05% of soluble proteins (Chong *et al.*, unpublished results). Potato and tomato plants produced lactoferrin protein with close to the expected molecular weight of 80 kDa. The potato and tomato-synthesized full-length human lactoferrin is currently being tested in colony forming unit reduction assay for its antimicrobial activity.

A cDNA encoding human α-lactalbumin has recently been introduced into tobacco plants to produce biologically functional human α-lactalbumin exhibiting an apparent molecular weight identical to human milk-derived α-lactalbumin (Takase and Hagiwara, 1998). When combined with galactosyltransferase, the tobacco-derived α-lactalbumin was found to be fully active in the synthesis of lactose.

Yield of Human Milk Protein from Transgenic Potato

The amount of a human milk protein, β-casein, in transgenic potato tuber tissue was determined to be 0.01% of total soluble tuber protein (Chong *et al.*, 1997), but may possibly be increased several-fold through addition of a microsomal retention signal to the protein (Arakawa *et al.*, 1997). Under control of the class I tuber-specific patatin promoter the yield of β-casein may be potentially increased to as much as 2% of total soluble tuber protein (Spychalla *et al.*, 1994). Since potato contains approximately 2.3% soluble protein (Pennington, 1993), an average sized potato (300 g) contains approximately 7 g of total soluble protein. Therefore, based on *mas* or patatin promoter gene expression, from 21 mg (0.3%) to 140 mg (2%) of recombinant human milk protein could be produced. Based on the amount of protein in human milk (Miller *et al.*, 1990), and from the 0.3–2% gene expression levels available through *mas* or patatin promoter expression, the percent of major human milk proteins in a cup of milk (250 ml) which could be provided by a 300 g of potato can be calculated (Table 3). Significantly less potato tissue is required to supplement dietary casein as compared with the whey proteins. For example, there is approximately 4 times less κ-casein (0.36 mg/ml) than lactoferrin (1.44 mg/ml) protein in human milk (Table 3).

Table 4. Relation between human milk and transgenic potato

Milk protein	Milk (liters)[a]	Acres/100 infants[b]
β-casein	7,700[c]	1.3[d]
κ-casein	19,000	0.5
Lactoferrin	4,900	2.0
α-lactalbumin	3,200	3.0
Lysozyme	17,500	0.6

[a]Volume (liters) of human milk equivalent in 1 acre of potatoes. Calculation is based on patatin promoter expression levels (2% of total soluble protein).
[b]Number of acres of potatoes equivalent to provide each milk protein to one hundred infants a year, assuming that a human produces approximately 100 liters of milk per lactation (1 year). Calculation is based on patatin promoter expression levels (2% of total soluble protein).
[c]Calculation is as follows: 15,000 kg potatoes/acre (Department of Commerce, 1995); 23 g total soluble protein/kg potatoes (2.3%) (Pennington, 1993); 20 mg β-casein/g total soluble protein (2%) (Spychalla *et al.*, 1994); 1 L human milk/900 mg b-casein (Miller *et al.*, 1990) = 7666.7 L human milk/acre
[d]Calculation is as follows: 1 acre/7666.7 L human milk;100 L human milk/lactation; 100 lactation/100 infants = 1.3 acres/100 infants

The volume of milk protein equivalent to 1 acre of transgenic potatoes can be calculated (Table 4). Thus, 1 acre of potatoes can produce the same amount of β-casein found in 7,700 liters of human milk. Further, based on an average volume of human milk (100 liters) produced per human lactation (in 1 year), the number of acres of potatoes required to produce the amount of protein synthesized during one lactation can also be calculated (Table 4). Thus, the amount of β-casein required to feed 100 infants for one year can be provided in 1.3 acres of transgenic potatoes.

Reconstitution of Human Milk Proteins in Food Plants

Several recombinant milk proteins have been synthesized in the milk of different transgenic animal species (Lee *et al.*, 1988; Vilotte *et al.*, 1989; Wall *et al.*, 1991; Platenburg *et al.*, 1994; Hitchin *et al.*, 1996; Gutierrez *et al.*, 1996). However, only a few human milk proteins such as human β-casein (Chong *et al.*, 1997), α-lactalbumin (Takase and Hagiwra, 1998), and lactoferrin (Mitra and Zhang, 1994; Chong *et al.*, unpublished data) have so far been synthesized in plants. The relatively inexpensive costs for plant cultivation and harvest coupled with development of gene transfer methods for edible plant species including potato, tomato, alfalfa and banana can in the near future make construction of transgenic plant-based bioreactors a more economically attractive alternative to animal-based production systems.

Production of milk proteins in vegetable and fruit plants capable of growth with little investment in economic resources in diverse climatic regions of the world must be developed to fully explore the opportunities for increased human nutrition. Transgenic food plants with improved nutritional quality will enhance the food value of crop plants. This may be especially important for economically emerging countries of the world where a small improvement in nutritional quality of food can be linked with antimicrobial proteins in food plants to return the highest yield in protection of crop and human infant and aging population against starvation and disease. Since human milk protein genes are available, reconstitution of human milk proteins including the caseins (β- and κ-) and whey proteins (a-lactalbumin, serum albumin, lactoferrin, lysozyme, and immunoglobulins) can now be achieved in transgenic food plants. In addition to the antimicrobial activities of lactoferrin and lysozyme, production of small antimicrobial peptides such as casecidin, isracidin, and lingual antimicrobial peptide in food plants may increase protection against enteropathogenic infections in human populations.

Future for Enhancement of the Biological Effects of Recombinant Human Milk Proteins in Plants

Each plant-synthesized recombinant milk protein must be evaluated *in vivo* as well as *in vitro*, before claims can be asserted that proteins produced in transgenic food plants may be used as milk substitutes. Once the nutritional and protective or therapeutic benefits of plant-delivered proteins have been determined, production and delivery systems must be optimized for highest yield and biological activity. The physiological and pharmacological activity of human milk proteins in transgenic plants can be further enhanced by improvements in the following areas. 1) Increased production levels; genetic manipulation of genes encoding milk protein may include the addition of transcriptional and translational enhancers, tissue-specific regulatory elements, leader peptides and/or endoplasmic reticulum and vacuolar retention signals. Optimization of human milk protein codon usage for plants as well as insertion of 5' and 3' translational enhancers may also increase milk protein production levels. Systems may be constructed in which multiple proteins can be synthesized in one plant, based on dual promoters, combination of two promoters in one plant expression vector, and internal ribosome entry site (IRES) elements inserted between adjacent milk protein genes. In addition to the *mas* dual promoters, strong constitutive promoters (such as the enhanced CaMV 35S promoter), and tissue-specific promoters (such as potato tuber specific patatin promoters) are currently under evaluation in our laboratory to achieve increased levels of milk protein gene expression. 2) Evaluation of correct post-translational modifications to the proteins such as phosphorylation and glycosylation, and their effects on biological and physiological activities *in vivo* using, for example, microbial inhibition assays for lactoferrin, casecidin and isracidin. Accurate measurement of the physiological effects of casein cleavage products such as casomorphins and antimicrobial peptides in cell cultures, animals and humans including infants may be required. 3) Selection of appropriate plant species; in the case of heat-labile proteins, production of the protein can be performed in plants which may be consumed in the raw state e.g., tomato and banana. When proteins require frequent consumption, plants must be selected which are a substantial part of the normal diet and which can be readily grown and harvested using the available agricultural technologies. 4) Modification of milk protein properties; modern genetic engineering methods now provide the opportunity to further improve digestibility, to increase protection against microbial pathogens by production of antimicrobial peptides e.g., lactoferrin, isracidin, casecidin, and to enhance the feeling of well being during periods of stress by producing human α-casein derived neurotropic peptides (casomorphins).

The ultimate objective will be to reconstitute a panel of essential human milk proteins such as b-casein, a-lactalbumin, lysozyme, lactoferrin, and b-lactoglobulin with the goal of enhancing the beneficial properties of milk proteins used for both infant formulas and in dairy food products for simultaneous improvement in human nutrition and the prevention of autoimmune and infectious diseases from the cradle to the grave, so that we can just say with certainty that "we will never outgrow our need for milk".

REFERENCES

Arakawa, T.; Chong, D. K. X.; Merritt, J. L.; Langridge, W. H. R. Expression of cholera toxin B subunit oligomers in transgenic potato plants. *Transgenic Res.* **1997**, *6*, 403–413.

Arakawa, T.; Chong, D. K. X.; Langridge, W. H. R. Efficacy of a food plant-based oral cholera toxin B subunit vaccine. *Nat. Biotechnol.* **1998**, *16*, 292–297.

Arnold, R. R.; Cole, M. F.; McGhee, J. R. A bactericidal effect for human lactoferrin. *Science* **1977**, *197*, 263–265.

Blakeborough, P.; Salter, D. N.; Gurr, M. I. Zinc binding in cow's milk and human milk. *Biochem. J.* **1983**, *209*, 505–512.

Brantl, V. Novel opioid peptide derived from human β-casein. *Eur. J. Pharmacol.* **1984**, *106*, 213–214.

Chong, D. K. X.; Roberts, W.; Arakawa, T.; Illes, K.; Bagi, G.; Slattery, C. W.; Langridge, W. H. R. Expression of the human milk protein, β-casein in transgenic potato plants. *Transgenic Res.* **1997**, *6*, 289–296.

van Cott, K. E.; Lubon, H.; Russel, C. G.; Butler, S. P.; Gwazdauskas, F. C.; Knight, J.; Drohan, W. N.; Velander, W. H. Phenotypic and genotypic stability of multiple lines of transgenic pigs expressing recombinant human protein C. *Transgenic Res.* **1997**, *6*, 203–212.

Daneman, D.; Fishman, L.; Clarson, C.; Martin, J. M. Dietary triggers of insulin-dependent diabetes in the BB rat. *Diabetes Res.* **1987**, *5*, 93–97.

Dickson, I. R.; Perkins, D. J. Studies on the interactions between purified bovine caseins and alkaline-earth-metal ions. *Biochem. J.* **1971**, *124*, 235–240.

Ebert, K. M.; Selgrath, J. P.; DiTullio, P.; Denman, J.; Smith, T. E.; Memon, M. A.; Schindler, J. E.; Monastersky, G. M.; Vitale, J. A.; Gordon, K. Transgenic production of a variant of human tissue-type plasminogen activator in goat milk: generation of transgenic goats and analysis of expression. *Biotechnology (NY)* **1991**, *9*, 835–838.

Escher, A.; O'Kane, D. J.; Lee, J.; Szalay, A. A. Bacterial luciferase alpha-beta fusion protein is fully active as a monomer and highly sensitive *in vivo* to elevated temperature. *Proc. Natl. Acad. Sci. USA* **1989**, *86*, 6528–6532.

Gimeno, S. G.; de Souza, J. M. IDDM and milk consumption. A case-control study in Sao Paulo, Brazil. *Diabetes Care* **1997**, *20*, 1256–1260.

Green, M. R.; Pastewka, J. V. High levels of lactoferrin expression in lactating mammary glands. *Endocrinology* **1978**, *103*, 1510–1513.

Greenberg, R.; Groves, M. L.; Dower, H. J. Human β-casein. Amino acid sequence and identification of phosphorylation sites. *J. Biol. Chem.* **1984**, *259*, 5132–5138.

Gutierrez, A.; Meade, H. M.; Ditullio, P.; Pollock, D.; Harvey, M.; Jimenez-Flores, R.; Anderson, G. B.; Murray, J. D.; Medrano, J. F. Expression of a bovine kappa-CN cDNA in the mammary gland of transgenic mice utilizing a genomic milk protein gene as an expression cassette. *Transgenic Res.* **1996**, *5*, 271–279.

Hansson, L.; Bergstrom, S.; Hernell, O.; Lonnerdal, B.; Nilsson, A. K.; Stromquist, M. Expression of human milk β-casein in *Escherichia coli*: comparison of recombinant protein with native isoforms. *Protein Expr. Purif.* **1993**, *4*, 373–381.

Haq, T. A.; Mason, H. S.; Clements, J. D.; Arntzen, C. J. Oral immunization with a recombinant bacterial antigen produced in transgenic plants. *Science* **1995**, *268*, 714–716.

Hitchin, E.; Stevenson, E. M.; Clark, A. J.; McClenaghan, M.; Leaver, J. Bovine beta-casein expressed in transgenic mouse milk is phosphorylated and incorporated into micelles. *Protein Expr. Purif.* **1996**, *7*, 247–252.

Host, A. Importance of the first meal on the development of cow's milk allergy and intolerance. *Allergy Proc.* **1991**, *12*, 227–232.

Host, A.; Jacobsen, H. P.; Halken, S.; Holmenlund, D. The natural history of cow's milk protein allergy/intolerance. *Eur. J. Clin. Nutr.* **1995**, *49 Suppl 1*, S13-S18.

Horsch, R. B.; Fry, J. E.; Hoffmann, N. L.; Eichholtz, D.; Rogers, S. G.; Fraley, R. T. A simple and general method for transferring genes into plants. *Science* **1985**, *227*, 1229–1231.

Johansson, C.; Samuelsson, U.; Ludvigsson, J. A high weight gain early in life is associated with an increased risk of type 1 (insulin-dependent) diabetes mellitus. *Diabetologia* **1994**, *37*, 91–94.

Karjalainen, J.; Martin, J. M.; Knip, M.; Ilonen, J.; Robinson, B. H.; Savilahti, E.; Åkerblom, H. K.; Dosch, H. - M. A bovine albumin peptide as a possible trigger of insulin-dependent diabetes mellitus. *N. Engl. J. Med.* **1992**, *327*, 302–307.

Kerr, D. E.; Liang, F.; Bondioli, K. R.; Zhao, H.; Kreibich, G.; Wall, R. J.; Sun, T. - T. The bladder as a bioreactor: urothelium production and secretion of growth hormone into urine. *Nat. Biotechnol.* **1998**, *16*, 75–79.

Kohmura, M.; Nio, N.; Kubo, K.; Minoshima, Y.; Munrkana, E.; Ariyoshi, Y. Inhibition of angiotensin-converting enzymes by synthetic peptides of human β casein. *Agric. Biol. Chem.* **1989**, *53*, 2107–2114.

Kunz, C.; Lonnerdal, B. Casein and casein subunits in preterm milk, colostrum and mature milk. *J. Pediatr. Gastroenterol. Nutr.* **1990**, *10*, 456–461.

Lahov, E.; Regelson, W. Antibacterial and immunostimulating casein-derived substances from milk: casecidin, isracidin peptides. *Food Chem. Toxicol.* **1996**, *34*, 131–145.

Lawrence, R. A. Breastfeeding. *A guide for the medical profession.* 3rd ed.; The C. V. Mosby Co., 1989.

Lee, K. F.; DeMayo, F. J.; Atiee, S. H.; Rosen, J. M. Tissue-specific expression of the rat beta-casein gene in transgenic mice. *Nucleic Acids Res.* **1988**, *16*, 1027–1041.

Lonnerdal, B. Methods for studying trace-element binding ligands in human milk. In *Human Lactation. Milk Components and Methodologies*; Jensen, R. G.; Neville, M. C., Eds.; Plenum Press: New York, 1985.

Lu, L.; Hangoc, G.; Oliff, A.; Chen, L. T.; Shen, R. - N.; Broxmeyer, H. E. Antivirus activity of cloned lactoferrin protein. *Cancer Res.* **1987**, *47*, 4184–4188.

Ma, J. K. - C.; Hiatt, A.; Hein, M.; Vine, N. D.; Wang, F.; Stabila, P.; van Dolleweerd, C.; Mostov, K.; Lehner, T. Generation and assembly of secretory antibodies in plants. *Science* **1995**, *268*, 716–719.

Ma, S. - W.; Zhao, D. - L.; Yin, Z. - Q.; Mukherjee, R.; Singh, B.; Qin, H. - Y.; Stiller, C. R.; Jevnikar, A. M. Transgenic plants expressing autoantigens fed to mice to induce oral immune tolerance. *Nat. Med.* **1997**, *3*, 793–796.

Mason, P. L.; Heremans, J. F.; Schonne, E. Lactoferrin, an iron-binding protein in neutrophilic leukocytes. *J. Exp. Med.* **1969**, *130*, 643.

Mason, H. S.; Ball, J. M.; Shi, J. - J.; Jiang, X.; Estes, M. K.; Arntzen, C. J. Expression of Norwalk virus capsid protein in transgenic tobacco and potato and its oral immunogenicity in mice. *Proc. Natl. Acad. Sci. USA* **1996**, *93*, 5335–5340.

Mason, H. S.; Lam, D. M. - K.; Arntzen, C. J. Expression of hepatitis B surface antigen in transgenic plants. *Proc. Natl. Acad. Sci. USA* **1992**, *89*, 11745–11749.

Migliori-Samour, D.; Jolles, P. Casein, a prohormone with an immunomodulating role for the newborn. *Experientia* **1988**, *44*, 188–193.

Miller, M. J. S.; Witherly, S. A.; Clark, D. A. Casein: a milk protein with diverse biologic consequences. *Proc. Soc. Exp. Biol. Med.* **1990**, *195*, 143–159.

Mitra, A.; Zhang, Z. Expression of a human lactoferrin cDNA in tobacco cells produces antibacterial protein(s). *Plant Physiol.* **1994**, *106*, 977–981.

Mykkanen, H. W.; Wasserman, R. H. Enhanced absorption of calcium by β-casein phosphopeptides in rachitic and normal chicks. *J. Nutr.* **1980**, *10*, 2141–2148.

Naito, H.; Suzuki, H. Further evidence for the formation in vivo of phospholipids in the intestinal lumen from dietary β-casein. *Agric. Biol. Chem.* **1974**, *38*, 1543–1545.

Packard, V. S. Macronutrients and Energy. In *Human Milk and Infant Formula*; Stwart, G. F., Schweigert, B. S.; Hawthorn, J., Eds.; Academic Press: New York, 1982.

Pennington, J. A. T. Vegetables, Vegetable Products & Vegetable Salads. In *Bowes and Church's food values of proteins commonly used* 16th edition. J.B. Lippincott Company, Philadelphia, Pennsylvania, 1993.

Perkin, J. E. In *Food allergies and adverse reactions*; Aspen Publishers, Inc., 1990.

Platenburg, G. J.; Kootwijk, E. P.; Kooiman, P. M.; Woloshuk, S. L.; Nuijens, J. H.; Krimpenfort, P. J.; Pieper, F. R.; de Boer, H. A.; Strijker, R. Expression of human lactoferrin in milk of transgenic mice. *Transgenic Res.* **1994**, *3*, 99–108.

Savilahti, E.; Åkerblom, H. K.; Tainio, V. - M.; Koskimies, S. Children newly diagnosed insulin dependent diabetes mellitus have increased levels of cow's milk antibodies. *Diabetes Res.* **1988**, *7*, 137–140.

Spychalla, J. P.; Scheffler, B. E.; Sowoknos, J. R.; Bevan, M. W. Cloning, antisense RNA inhibition, and the coordinated expression of UDP-glucose pyrophosphorylase with starch biosynthetic genes in potato tubers. *J. Plant Physiol.* **1994**, *144*, 444–453.

Takase, K.; Hagiwara, K. Expression of human alpha-lactalbumin in transgenic tobacco. *J. Biochem.* **1998**, *123*, 440–444.

Vilotte, J. L.; Soulier, S.; Stinnakre, M. G.; Massoud, M.; Mercier, J. C. Efficient tissue-specific expression of bovine alpha-lactalbumin in transgenic mice. *Eur. J. Biochem.* **1989**, *186*, 43–48.

Virtanen, S. M.; Rasanen, L.; Aro, A.; Lindstrom, J.; Sippola, H.; Lounamaa, R.; Toivanen, L.; Tuomilehto, J.; Akerblom H. K. Infant feeding in Finnish children less than 7 yr of age with newly diagnosed IDDM. Childhood diabetes in Finland study group. *Diabetes Care* **1991**, *14*, 415–417.

Virtanen, S. M.; Saukkonen, T.; Savilahti, E.; Ylonen, K.; Rasanen, L.; Aro, A.; Knip, M.; Tuomilehto, J.; Akerblom, H. K. Diet, cow's milk protein antibodies and the risk of IDDM in Finnish children. Childhood diabetes in Finland study group. *Diabetologia* **1994**, *37*, 381–387.

Wall, R. J.; Pursel, V. G.; Shamay, A.; McKnight, R. A.; Pittius, C. W.; Hennighausen, L. High-level synthesis of a heterologous milk protein in the mammary glands of transgenic swine. *Proc. Natl. Acad. Sci. USA* **1991**, *88*, 1696–1700.

Wilson, N. W.; Hamburger, R. N. Allergy to cow's milk in the first year of life and its prevention. *Ann. Allergy* **1988**, *61*, 323–327.

13

FOOD PLANT-DELIVERED CHOLERA TOXIN B SUBUNIT FOR VACCINATION AND IMMUNOTOLERIZATION

Takeshi Arakawa,[1,2] Jie Yu,[1,2] and William H. R. Langridge[1,2,3,*]

[1]Center for Molecular Biology and Gene Therapy
[2]Department of Microbiology and Molecular Genetics
[3]Department of Biochemistry
School of Medicine, Loma Linda University
Loma Linda, California 92350

ABSTRACT

Developments in recombinant DNA technology have enabled molecular biologists to introduce a variety of novel genes into plant species for specific purposes. From crop improvement to vaccine antigen and antibody production, plants are attractive bioreactors for production of recombinant proteins, as their eukaryotic nature often permits appropriate post-translational modification of recombinant proteins to retain native biological activity. The autotrophic growth of plants requires only soil minerals, water, nitrogen, sunlight energy and carbon dioxide for the synthesis of constituent proteins. Furthermore, production of biologically active proteins in food plants provides the advantage of direct delivery through consumption of edible transformed plant tissues. The production of cholera toxin B subunit in potato plants and applications for prevention of infectious and autoimmune disease are explained in this contribution.

[*] Biology and Gene Therapy, School of Medicine, Loma Linda University, Loma Linda, California 92350, Telephone: (909) 824-4300, Ext. 81362, Fax: (909) 478-4177, E-mail: blangridge@ccmail.llu.edu

Chemicals via Higher Plant Bioengineering, edited by Shahidi *et al.*
Kluwer Academic / Plenum Publishers, New York, 1999.

INTRODUCTION

Genetic Engineering of Crop Plants

The practice of agriculture is a unique human activity which began over 3,000 years ago and has gradually enabled man to improve the nutritional and therapeutic value of plants, eventually permitting humans to escape their nomadic existence for the development of civilization. Until recently, the methods of crossing and inbreeding of plant species was the only tool for improvement in the quality of plant products. However, during the past two decades the advent of modern recombinant DNA methods has enabled us to clone a large number of plant genes, to identify their genetic regulatory signals, and improve and confer valuable traits to plant species through the transfer of genes from other organisms from bacteria to man. Examples include the introduction of: a yeast halotolerance gene for increased resistance of melons to salinity (Bordas *et al.*, 1997); a gene encoding flounder antifreeze protein into tobacco for improved frost resistance (Kenward *et al.*, 1993); genes from the resurrection plant *Craterostigma* into tobacco for improved drought resistance (Iturriaga *et al.*, 1996); plant genes of the phenylpropanoid pathway for increased resistance to plant pathogens (Logemann *et al.*, 1995); the *Bacillus thuringiensis* delta endotoxin gene into cotton for greater resistance to insect pests (Sutton *et al.,* 1992); the 5-enolpyruvylshikimate-3-phosphate synthase gene and mutant photosystem B genes for improved resistance of crop plants to the herbicides glyphosate and atrazine (Suh *et al.*, 1993; Smeda *et al.*, 1993).

More recently, improvements have been made in the protein quality of food plants for human nutrition e.g., the introduction of lysine and methionine biosynthetic genes from bacteria into grain proteins to improve wheat protein quality for bread-making (Keeler *et al.*, 1997). The introduction of cDNAs encoding polygalacturonase and the DNA sequence for production of antisense RNA to down-regulate the ethylene biosynthetic pathway have delayed tomato fruit ripening for the improvement of tomato and melon storage and flavor quality (Flavr Savr) (Meyer, 1995). "Flavr Savr" is a transgenic tomato in which a gene encoding an antisense RNA to the ethylene forming enzyme has been introduced which results in slower ripening permitting shipping in a hard but more ripe stage of development. Genes encoding omega-3 fatty acids have been introduced into plants for the prevention of cancer and arterial plaque formation in order to reduce the possibility of atherosclerosis and stroke (Arondel *et al.*, 1992). Recently, the gene encoding a human milk protein, β-casein, has been introduced into potato and tomato to supplement the mother's milk protein for improving the nutritional quality of infant formula and baby foods (Chong *et al.*, 1997). Lactoferrin, a human milk protein with antimicrobial activity, has been synthesized in tobacco callus (Mitra and Zhang, 1994) and recently in potato and tomato plants (Chong *et al.*, unpublished results). The production of additional milk proteins such as α-lactalbumin and β-lactoglobulin in food plants may increase nutritional quality and prevent enteric infectious diseases. Substitution of major bovine milk proteins with plant-synthesized human milk proteins in baby formulas can contribute to the prevention of cow's milk protein allergy and intolerance (Host *et al.*, 1995), and the suppression of autoimmune diabetes (Virtanen *et al.*, 1991, 1994) for non-breast fed infants, without total sacrifice of nutritionally important milk proteins.

The goals of plant genetic engineering have been expanded to include the addition of medicinal functions, such as genes encoding immunoglobulins (Hiatt *et al.*, 1989; de Neve *et al.*, 1993; Ma *et al.*, 1995, 1998), and vaccine antigens for protection against bacterial enterotoxins (Haq *et al.*, 1995; Arakawa *et al.*, 1997, 1998) and viral infections (Mason *et al.*, 1992, 1996; Thanavala *et al.*, 1995).

Antigen Delivery by Transgenic Plants

Plant-synthesized recombinant proteins with vaccine and nutritional values can be delivered to the body through ingestion of edible plant tissues. Proteins are generally degraded with significant loss of biological activity in the gastrointestinal tract, however, degradation is often partial with intact antigen directly absorbed via the intestinal epithelial cells. Partial digestion of certain peptides may be essential for maintenance of biological activity e.g., induction of oral tolerance. Recombinant proteins, synthesized in transgenic plants, can be fed to humans for induction of immunological responses against infectious diseases, to confer antimicrobial responses, to induce immunotolerance for autoimmune diseases, and to provide novel nutritional peptides. The efficacy of recombinant proteins can be determined by measuring the extent of protection against pathogenic agents such as *Eschericha coli* heat-labile enterotoxin (Haq *et al.*, 1995), hepatitis B virus (Thanavala *et al.*, 1995), cholera toxin (Arakawa *et al.*, 1998), Norwalk virus (Mason *et al.*, 1996), as well as protection against autoimmune diseases demonstrated later in this chapter.

Effective prevention or treatment of infectious and autoimmune diseases can be achieved by ingestion of food plants containing therapeutic proteins. Several reports indicate that infectious diseases caused by viruses and bacteria may be prevented by the use of plant-based vaccines. Neutralizing peptide antigens of human hepatitis B virus surface antigen (HBsAg) and Norwalk virus capsid protein (NVCP) were produced in plants and found to be immunogenic in mice (Mason *et al.*, 1992, 1996; Thanavala *et al.*, 1995). Diarrhea-causing bacterial enterotoxins were produced in plants and found to be immunogenic in mice (Haq *et al.*, 1995; Arakawa *et al.*, 1998). Transgenic plants carrying these protein antigens conferred partial protection against both virus and toxin challenge. Generation of transgenic plants synthesizing rotavirus capsid proteins VP2, VP6, and the viral enterotoxin NSP4 are currently underway to evaluate the effectiveness of plants for production of multi-layered virus-like particles and the induction of immunity against rotavirus neutralizing antibodies in mice.

Transgenic plants prevent autoimmune diabetes in nonobese diabetic (NOD) mice. Mouse glutamic acid decarboxylase (GAD67) produced in potato and tobacco plants was used successfully to prevent insulitis and diabetic symptoms (Ma *et al.*, 1997; Arakawa *et al.*, unpublished results). These experiments are the first examples of the use of plants as production and oral delivery systems for the prevention of autoimmune diseases. Prevention of a variety of autoimmune diseases may also be possible through oral administration of edible plant tissues synthesizing disease-specific autoantigens. Thus, the plant-based oral tolerization system may be useful for prevention of autoimmune diseases such as encephalomyelitis (Khoury *et al.*, 1990; Miller *et al.*, 1992; Sun *et al.*, 1996), type I diabetes (Zhang *et al.*, 1991), collagen-induced arthritis (Trentham *et al.*, 1993), and multiple sclerosis (Weiner *et al.*, 1993).

PRODUCTION OF THE CHOLERA TOXIN B SUBUNIT IN FOOD PLANTS

Oral administration of *Vibrio cholerae* enterotoxin (CT) and its nontoxic B subunit (CTB) induce both systemic and mucosal antibody production in animals and human (Fujita and Finkelstein, 1972; Fuhrman and Cebra, 1981; Lycke *et al.*, 1983; Elson and Ealding, 1984; Lycke *et al.*, 1985). The CT molecule acts as a strong immunological adjuvant for co-administered antigens (Lycke and Holmgren, 1986; Jackson *et al.*, 1993; Holmgren *et al.*,

1993). The strong adjuvant effect of CT is due to its ability to interact with many vertebrate cell types by elevating intracellular cyclic adenosine mono phosphate (cAMP) levels through activation of adenylate cyclase (Gill and Meren, 1978). Although, cholera toxin B subunit protein (CTB) is not generally considered to be a strong adjuvant for co-administered antigens, it has been demonstrated to be an effective carrier molecule for induction of mucosal immunity to polypeptides to which it is chemically or genetically conjugated (McKenzie and Halsey, 1984; Czerkinsky *et al.*, 1989; Dertzbaugh and Elson, 1993a,b). In addition, CTB has been shown to function as a carrier of conjugated peptides for induction of immunological tolerance (Sun *et al.*, 1994, 1996). The characteristic of CTB as a immunomodulatory carrier molecule may be largely due to its ability to avidly bind its natural receptor G_{M1}-ganglioside on the surface of mammalian intestinal epithelial cells, including M cells of the gut-associated lymphoid tissues (GALT) (Svennerholm, 1976).

Construction of Transgenic Potato Plants Synthesizing the CTB Protein Pentamer

A gene encoding the cholera toxin B subunit protein (CTB), fused to a microsomal retention signal Ser Glu Lys Asp Glu Leu (SEKDEL) was inserted adjacent to the bi-directional mannopine synthase P2 promoter in a plant expression vector (Figure 1). The CTB:SEKDEL fusion gene was inserted into the plant expression vector pPCV701luxF resulting in pPCV701luxF-CTB:SEKDEL. Plant expression vector pPCV701luxF harbors a 430 bp DNA fragment containing the *A. tumefaciens* bidirectional mannopine synthase (*mas* P1, P2) promoters (Koncz *et al.*, 1987). The P1 promoter is fused to the bacterial luciferase reporter gene (*luxF*) (Escher *et al.*, 1989), and the P2 promoter is linked to the CTB:SEKDEL fusion gene. We changed the oligonucleotide sequence surrounding the translation initiation codon of the CTB gene to a preferred nucleotide context for translation in eukaryotic cells (Kozak, 1981). The 21-amino acid leader peptide of the CTB protein, absent from secreted *V. cholerae* cholera toxin molecules, is thought to function as a leader peptide in eukaryotic cells for translocation of nascent CTB polypeptides into the lumen of the plant endoplasmic reticulum (ER). The ER retention signal (SEKDEL) fused to the C-terminus of the plant CTB may sequester the protein within the plant ER and may increase stability of the protein.

Potato leaf explants were transformed by *Agrobacterium tumefaciens* carrying the vector and kanamycin-resistant plants were regenerated. The CTB-SEKDEL fusion gene was identified in the genomic DNA. Immunoblot analysis indicated that plant-derived CTB was antigenically indistinguishable from bacterial CTB, and that pentameric CTB molecules (50 kDa) were the dominant molecular species isolated from transgenic potato leaf and tuber tissues. Similar to bacterial CTB, plant-synthesized CTB dissociated into monomers (15 kDa) during heat or acid treatment. Heat-induced dissociation of CTB pentamers into monomers is a characteristic of bacterial enterotoxins (Gill, 1976; Hirst and Holmgren, 1987; Hardy *et al.*, 1988). Acid-induced dissociation of plant-derived CTB protein, similar to that seen for bacterial enterotoxin was also detected. However, when potato tissues were directly heated by boiling, we did not observe as extensive pentamer dissociation as seen when plant homogenates were boiled. This result suggests that vaccine antigens may be more stable in the plant tissue as compared with the extracted form. Vaccine antigens accumulated in the intracellular compartment of plant cells may provided the natural protective shield against antigen degradation. The plant endoplasmic reticulum functions analogous to the periplasmic space of Gram-negative bacteria (Hirst and Holmgren, 1987), thus providing a concentrated intracellular environment in which

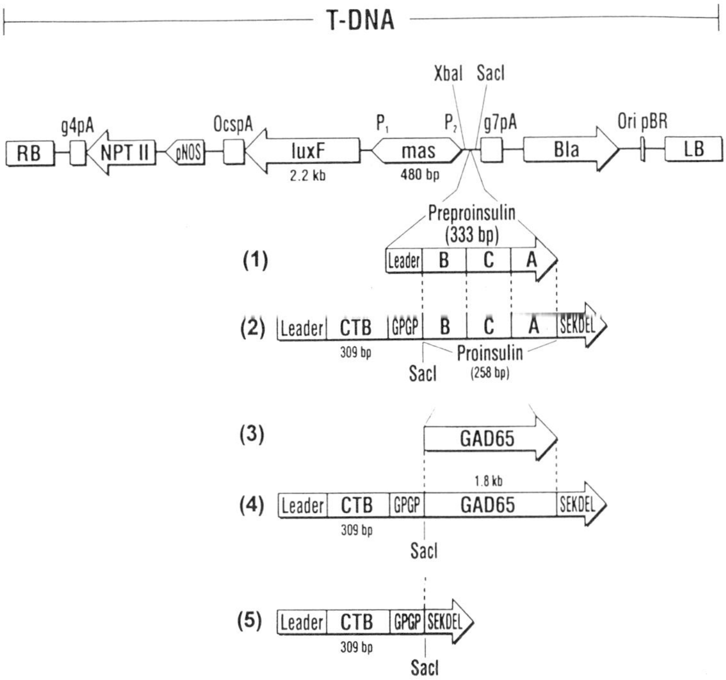

Figure 1. Plant expression vector pPCV701luxF The T-DNA sequence flanked by the right and left borders (RB and LB) for insertion into the plant genome contains the bacterial luciferase (*luxF*)/IDDM autoantigen gene expression cassette transcriptionally regulated by the bi-directional mannopine synthase (*mas*) P1 and P2 promoter respectively. The *luxF* is a detectable marker gene for transformed agrobacteria and plants. The β-lactamase (Bla) expression cassette is included to confer for ampicillin resistance in *Escherichia coli* and carbenicillin resistance in *Agrobacterium tumefaciens*. The NPT II expression cassette linked to the nopaline synthase promoter (pNOS) generates kanamycin resistance in plants. The g7pA, g4pA and OcspA polyadenylation signals are from the *A. tumefaciens* TL-DNA, gene 7 and gene 4, the octopine synthase gene, respectively. Ori pBR is the origin of replication from plasmid pBR322. **(1)** human preproinsulin cDNA, **(2)** CTB-INS fusion gene, **(3)** human GAD65 cDNA, **(4)** CTB-GAD fusion gene, and **(5)** the CTB gene were inserted between the *Xba*I and *Sac*I cloning sites downstream from the *mas* P2 promoter. The CTB and the CTB fusion genes are flanked at the 5' end by the bacterial CTB leader sequence and at the 3' end by an 18-bp DNA sequence encoding the hexapeptide SEKDEL microsomal retention signal. A DNA sequence encoding a flexible hinge tetrapeptide GPGP with less frequently used codons in plants, was inserted between the CTB and the proinsulin or the GAD65 gene.

monomeric CTB subunits may accumulate and assemble into pentamers. Presence of the SEKDEL hexapeptide at the C-terminus has been shown to significantly increase protein accumulation within plant tissues e.g., potato leaves and tubers, tobacco and alfalfa leaves as well as in COS cells (Munro and Pelham, 1987; Wandelt *et al.*, 1992; Haq *et al.*, 1995; Arakawa *et al.*, 1997), thereby facilitating protein subunit pentamerization.

Both pentameric and monomeric forms of plant-derived CTB were of slightly higher molecular weight than bacterial CTB (50 kDa versus 45 kDa for the pentamer and 15 kDa versus12 kDa for the monomer, respectively). This result may be due to presence of the hexapeptide ER retention signal, or possibly to retention of the CTB leader peptide. Molecular weight differences of approximately 3 kDa between the monomers of bacteria and plant-derived CTB suggest that the presence of both ER retention signal and extra N-terminal peptide contribute to the molecular weight increase of the plant-synthesized CTB molecule. Based on gel electrophoresis data, the apparent molecular weight of bacterial

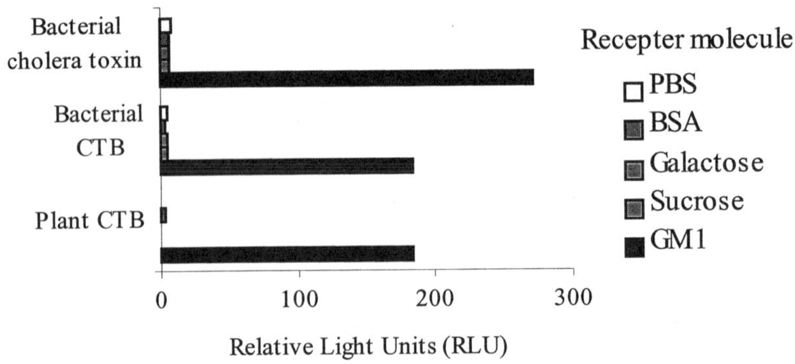

Figure 2. GM1 binding ability of plant-derived CTB, bacterial CTB, and bacterial CTX The G_{M1}-ELISA assay was performed by coating microtiter plates with either G_{M1}-monosialoganglioside, sucrose, galactose, or BSA (300 ng/well). For binding bacterial CTB and cholera toxin, 30 ng of each molecule was mixed with untransformed plant homogenate containing 10 μg of total soluble protein per well. For binding the plant-derived CTB, a transformed plant tissue homogenate containing approximately 30 ng of recombinant CTB in 10 μg of total soluble protein was used per well. Relative binding affinity of plant CTB, bacterial CTB, and cholera toxin for receptor molecules was expressed as relative light units (RLU).

and plant CTB pentamers is slightly smaller than the molecular weight of their monomers multiplied fivefold, which suggests a compact oligomeric configuration.

The maximum amount of CTB detected in auxin-induced transgenic potato leaf and tuber tissues by G_{M1}-ELISA methods was approximately 0.3% of total soluble plant protein. Based on the results of the chemiluminescent G_{M1}-ELISA and immunoblot assays, 1 g fresh weight of potato tissue contained approximately 30 μg of recombinant plant CTB protein. Plant-synthesized CTB protein bound specifically to G_{M1}-ganglioside, the natural membrane receptor of cholera toxin (Figure 2). The ability of plant-derived CTB to bind G_{M1}-ganglioside indicates that the specific protein-ganglioside binding interactions between amino acid residues forming the G_{M1} binding sites and the oligosaccharide moiety of G_{M1}-ganglioside are conserved. The strong binding efficiencies of plant CTB for G_{M1} indicate that subunit binding to G_{M1} is cooperative (Schön and Freire, 1989; Merritt *et al.*, 1994). The apparent cooperativity exhibited by plant CTB–G_{M1}-ganglioside binding reinforces the probability that monomeric B subunits accumulate within the lumen of the plant cell ER where self-assembly into pentameric, G_{M1} binding forms occurs.

Immunogenicity in Mice of Plant-Synthesized CTB Protein

Three groups of mice were orally immunized (i.e., 30 μg bacterial CTB, 1 g potato, and 3 g potato) a total of four times at weekly intervals with a final booster dose. Systemic and intestial CTB-specific antibody titers were determined in both serum and feces collected from the immunized mice by the class-specific chemiluminescent ELISA method (Arakawa *et al.*, 1998).

The titers of the three serum anti-CTB antibody isotypes (IgG, IgA and IgM) were determined on days 35 and 70. Anti-CTB titers for IgG and IgA increased slightly from day 35 to day 70, while the IgM titer decreased. Both intestinal IgA and IgG titers reached the highest level around day 28, four days after the 4th feeding, and gradually decreased over the next 40 days until the booster feeding on day 65. In all three groups of CTB immunized mice, fecal IgA and IgG titers increased after the booster dose. For mice fed 3 g

Table 1. Oral immunization with bacterial or plant CTB protected the small intestine against CT-induced diarrhea[a]

Immunization	vol (μl)/cm ± s.e.m.	Protection (%)
Unimmunized	157.1 ± 12.5	
Bacterial CTB	71.0 ± 7.7	55 ($P < 0.05$)
Potato (1 g)	91.0 ± 8.6	42 ($P < 0.05$)
Potato (3 g)	59.9 ± 10.0	62 ($P < 0.05$)

[a]The intestinal loops were challenged with cholera toxin (125 ng/loop). The intestinal fluid accumulation per unit length of ligated ileal loops from unimmunized and immunized mice was determined [intestinal fluid volume (μl)/intestinal loop length (cm) ± standard error]. The reduction in fluid secretion into the ileal loop in the immunized mice in comparison with unimmunized mice is expressed as % protection. A significant difference in reduction of fluid accumulation in mice orally immunized with plant or bacterial CTB was observed in comparison with mice fed untransformed potato tissues ($P < 0.05$)

of transformed potato, a single booster feeding increased the IgA titer to its highest level on day 28. Our animal immunization experiments demonstrated that antibody titers can be boosted simply by oral administration of additional vaccine food plant tissues when the protective titer declines.

Antiserum was analyzed for neutralization capability against cholera holotoxin using toxin-sensitive mammalian Vero cell culture. Mice orally immunized four times followed by a final booster of 30 μg bacterial CTB showed a titer of 1:32 for protection, in comparison with a titer of 1:8 for mice fed on 3 g of transformed potato tissues and a titer of 1:2 for mice fed on 1 g of transformed potato tissues.

Incubation of cholera toxin with antiserum derived from potato-immunized mice resulted in approximately one third less toxin binding to the G_{M1}-ganglioside in comparison with the same amount of cholera toxin incubated with pooled antiserum derived from unimmunized mice, which indicates that protection against cholera toxin is based on antibody-mediated toxin binding interference to the cell surface.

Protection of Mice against Cholera Toxin-Induced Diarrhea

To evaluate the protective efficacy of oral immunization, the volume of fluid accumulated in ileal loops from immunized and non-immunized animals was measured and expressed as the ratio of volume to loop length [vol (μl)/cm] (Table 1). Mice orally immunized with bacterial or plant-derived CTB showed significantly less diarrheal fluid accumulation in comparison with mice fed untransformed potato tissues. Mean values of volume to loop length for each group of mice were: 157.1 ± 12.5 standard error of the mean (SEM) for untransformed potato-immunized mice; 71.0 ± 7.7 SEM for 30 μg bacterial CTB-immunized mice ($p < 0.05$); 91.0 ± 8.6 SEM for 1 g transgenic potato-fed mice ($p < 0.05$); and 59.9 ± 10.0 SEM for 3 g transgenic potato-fed mice ($p < 0.05$). Considering the reduction in the intestinal-fluid accumulation as a measure of protective efficacy of oral immunization, approximately 55% protection was observed for mice immunized with 30 μg of bacterial CTB, 42% protection for mice immunized with 1 g of transformed potato tissues, and 62% protection for mice immunized with 3 g of transgenic potato tissues.

Infectious agents which invade through mucosal surfaces can often be controlled more effectively by mucosal immunization than parenteral immunization. Oral immunization of CD-1 mice with plant-synthesized pentameric CTB induced both mucosal and systemic immune response at levels sufficient to generate protection against the cytotoxic effects of cholera toxin. The extent of cholera toxin neutralization in both Vero cell and ileal loop experiments suggests that anti-CTB antibodies prevent toxin binding to cellular G_{M1}-ganglioside. This result is consistent with *in vivo* experiments in which anti-CT antibodies prevented cholera toxin from binding to the cell surface (Apter *et al.*, 1993). A reduction in diarrheal fluid secretion in plant-CTB immunized mice indicates that plant-based oral immunization is effective in inducing protective immunity against the effects of cholera toxin.

The increased protective efficacy of a killed-whole cell vaccine strain mixed with CTB (BS-WC) in comparison with the killed-cell vaccine strain alone (WC) in a large-scale field trial in Bangladesh clearly demonstrated the significant protective immunogenicity of the CTB subunit (Clemens *et al.*, 1991, 1992). Induction of antitoxic immunity alone does not provide protection against pathological effects caused by the live *V. cholerae*. The antibacterial (vibriocidal; i.e., anti-lipopolysaccharide and anti-outer membrane protein antigens) immune mechanism is critical for prevention of *V. cholerae* infection (Levine *et al.*, 1979). In addition to enterotoxin antigens, it may be desirable to produce edible vaccine plants containing neutralizing epitopes of structural antigens such as pilus-associated adhesins for induction of antibacterial immunity.

Since the CTB can function as an effective carrier for induction of a mucosal immune response to polypeptides to which it is conjugated, plant-synthesized pentameric CTB may be useful for providing safe and cost-effective mucosal immunization against various diseases which can be effectively controlled by recombinant subunit vaccines. Increased immunogen concentration at the sites of the mucosal lymphoid tissues may reduce the requirement for high levels of antigen biosynthesis in food plants. In addition, the CTB subunit has been shown to function as a mucosal carrier for chemically conjugated polypeptides for induction of systemic immunological tolerance.

PRODUCTION OF CTB-AUTOANTIGEN FUSION PROTEINS IN FOOD PLANTS

Oral administration of disease-specific autoantigens can prevent or delay the onset of autoimmune diseases. We have generated transgenic potato plants synthesizing two major autoantigens of insulin-dependent diabetes mellitus (IDDM); insulin and glutamic acid decarboxylase (GAD65). We were able to obtain transgenic potato plants which produced the autoantigens at levels of up to 0.05% of total soluble protein. Human insulin or GAD was linked to the C-terminus of the CTB subunit to direct delivery of the autoantigens to the GALT. Transgenic potato tubers produced the pentameric CTB-INS or CTB-GAD fusion peptide which retained G_{M1}-ganglioside binding affinity and displayed native antigenicity of both CTB, insulin or GAD.

Nonobese diabetic (NOD) mice fed transformed potato tuber tissues containing microgram levels of the CTB-INS or CTB-GAD fusion peptide showed a substantial reduction in pancreatic islet inflammation (insulitis), and a delay in the progression of clinical diabetes, indicating that food plants are feasible production and convenient CTB-autoantigen conjugate delivery system for immunotolerization against a T cell-mediated autoimmune disease.

Insulin-Dependent Diabetes Mellitus (IDDM) and Oral Tolerance for Immunotherapy

IDDM is an autoimmune disease in which immunological events lead to the destruction of insulin-producing pancreatic β cells (Eisenbarth, 1986; Tisch and McDevitt, 1996). Both humoral and cellular mechanisms of immunity are involved in disease pathogenesis, although destruction of pancreatic β cells appears to be mediated predominantly by the activity of CD4+ and CD8+ T-cells. There are several autoantigens involved in IDDM, however, insulin and GAD appear to play a major role in the diabetogenic response and it has been demonstrated that T-cells infiltrating the islet respond to the autoantigens in the nonobese diabetic (NOD) mouse (Eisenbarth, 1986).

Oral administration of pancreatic tissue-specific autoantigens provides a convenient and safe clinical approach for prevention of spontaneous autoimmune diabetes (Zhang *et al.*, 1991; Trentham *et al.*, 1993; Weiner *et al.*, 1993, 1994). However, the therapeutic potential of this approach is limited by the requirement for repeated administration of large amounts of autoantigen, and tolerization is usually less efficient in previously sensitized hosts (Zhang *et al.*, 1991). The CTB molecule was employed to overcome such limitations by serving as a mucosal carrier for conjugated autoantigens required for induction of oral tolerance. Fusion with the cholera toxin B subunit reduces the required amount of antigen for oral tolerization induction by 100–1000 fold (Sun *et al.*, 1994, 1996; Bergerot *et al.*, 1997). Conjugation with CTB may greatly facilitate antigen delivery and presentation to the GALT due to CTB's affinity for the cell surface receptor G_{M1}-ganglioside located on cell membranes of the GALT including the membranous (M) cells as well as enterocytes for increased uptake and immunological recognition.

Construction of Transgenic Potato Plants Producing CTB-IDDM Autoantigen Conjugates

Potato plants transformed with the plant expression vector pPCV701luxF containing the human insulin, GAD, CTB-INS, or CTB-GAD fusion genes were generated by an *Agrobacterium tumefaciens*-mediated leaf explant transformation method. Regenerated transformed plants were analyzed for the presence of the transgenes, their mRNAs and the protein products.

The amount of insulin protein synthesized in transformed tuber tissues, estimated by chemiluminescent ELISA, was approximately 0.05% of the total soluble tuber protein. Immunoblot analysis of transgenic potato plants transformed with the CTB-INS fusion gene revealed a pentameric CTB-INS fusion protein with a higher molecular weight than either bacterial CTB or the plant-synthesized CTB pentamer. Production level was calculated to be 0.1% of potato soluble protein by chemiluminescent G_{M1}-ELISA method (Figure 3). The monomeric fusion peptide extracted from transgenic plants showed an approximate molecular weight of 30 kDa. The CTB-INS pentamer retained specific affinity for G_{M1}-ganglioside and both CTB and insulin antigenicity. The CTB-INS fusion protein obtained from potato leaf tissues exhibited identical biochemical and antigenic properties to the tuber-derived fusion peptide.

Potato plants produced up to 0.01% of potato soluble protein as GAD peptide with an approximate molecular weight of 60 kDa. Transgenic potato plants transformed with the CTB-GAD fusion gene produced the fusion peptide with an approximate molecular weight of 75 kDa, closely corresponding to the sum of the plant-synthesized CTB (15

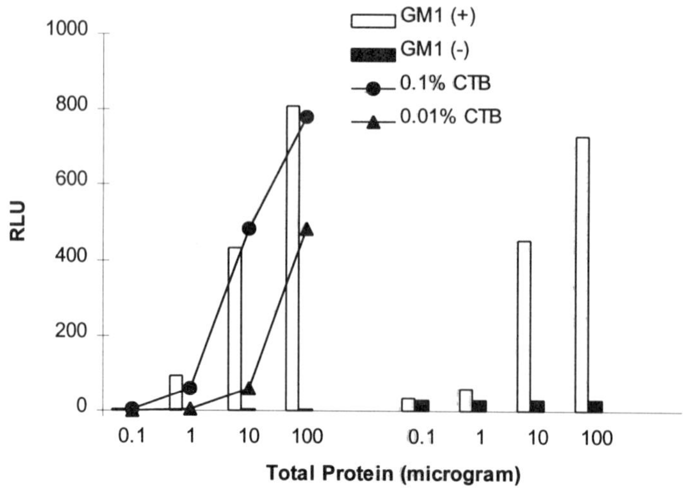

Figure 3. Quantitative analysis of the pentameric CTB-INS fusion protein by chemiluminescent GM1 ELISA The CTB-INS fusion peptide exhibiting specific binding affinity for G_{M1}-ganglioside made up approximately 0.1% of total soluble tuber protein. The levels of the CTB-GAD fusion peptide was similarly measured and was found to be approximately 0.001% of total soluble protein.

kDa) and GAD (60 kDa). Potato tuber produced approximately 0.001% of soluble protein as the CTB-GAD fusion protein. A 100 fold less CTB-GAD protein production in comparison with the CTB-INS levels (0.1%) is probably due to increased steric hindrance caused by large GAD polypeptide fusion to the CTB subunit. Similar to the CTB-INS fusion peptide, the CTB-GAD fusion protein retained specific affinity for G_{M1}-ganglioside and both CTB and GAD65 antigenicity.

The proinsulin or GAD65 peptide was conjugated with CTB via the flexible hinge peptide containing two glycine and two proline amino acids, which may reduce steric hindrance between the CTB and the fusion peptide moieties facilitating CTB subunit assembly in plant cells (Fig. 1). The application of less frequently used codons in plants within the hinge peptide may promote translational arrest during the peptide elongation process, facilitating CTB subunit folding prior to translation of the insulin message. The microsomal retention signal has been found to permit sequestration of recombinant proteins in the plant endoplasmic reticulum, increasing protein concentration for more effective subunit interactions for the pentamerization process (Haq *et al.*, 1995). The fusion peptides retain G_{M1}-ganglioside binding affinity which indicates that they exist predominantly as a pentamer since only the pentameric CTB configuration can bind to the ganglioside receptor. The presence of the pentameric chimeras as the predominant protein species in plant tissues suggests efficient pentamerization within plant cells.

Oral Immunogenicity of the Plant-Produced CTB-Autoantigen Protein in NOD Mice

We have previously demonstrated that feeding plant tissues synthesizing CTB induced both mucosal and serum antibody responses. The feeding of CTB-INS or CTB-GAD potato tuber tissues to NOD mice induced both serum and intestinal anti-CTB antibodies. In addition, serum anti-insulin or anti-GAD IgG level was elevated in animals fed CTB-INS or CTB-GAD potato tissues in comparison to animals fed INS or GAD only potato or untransformed potato tissues.

Induction of both systemic and intestinal anti-CTB antibodies in mice fed transformed potato tubers synthesizing CTB-fusion protein suggests that the fusion peptide was effectively delivered to the GALT. This finding was reinforced by the observation of slightly increased serum anti-insulin and anti-GAD IgG antibody titers in mice fed the transformed potato tissues. The humoral response mounted against both CTB and the autoantigen peptides may be a collateral effect occurring in response to stimulation of Th2 cells by oral tolerance (Challacombe and Tomasi, 1980; Husby *et al.*, 1994). Since there is a decreasing gradient of sensitivity to induction of oral tolerance with the Th1 cells being the most susceptible to tolerization followed sequentially by Th2 cells and B cells, it is possible that induction of systemic T cell tolerance may occur simultaneously with systemic and mucosal B cell priming (Elson and Zivny, 1996; Fujihashi *et al.*, 1996).

Suppression of Pancreatic Inflammation and Diabetic Symptoms

To determine the effects of feeding transgenic potato tissues producing autoantigens and their CTB conjugate proteins on the reduction of insulitis, female NOD mice were fed transgenic potato tuber tissues containing INS, GAD, CTB-INS, or CTB-GAD once per week for five weeks starting at 5 weeks of age. The mice were sacrificed at 10 weeks of age for histopathological analysis of pancreatic tissues. A substantial reduction in insulitis was detected in mice fed CTB-INS potato tissues (1.9 ± 0.5 SEM, $P = 0.001$, Student's *t* test) or CTB-GAD potato tissues (2.44 ± 0.28 SEM, $P = 0.003$) in comparison with those fed untransformed potato tissues (3.9 ± 0.4 SEM) (Figure 4). Although there was a detectable reduction in insulitis severity in NOD mice fed transformed potato tissues containing 30 μg of insulin (3.8 ± 0.4 SEM) or 3 μg of GAD (3.67 ± 0.38 SEM) in comparison with the unimmunized animals, the difference was statistically insignificant (Fig. 4). To evaluate whether simultaneous oral administration of both IDDM autoantigens conjugated with CTB have synergistic effect on the pancreatic anti-inflammatory response, mice were fed a mixture of CTB-GAD and CTB-INS potato tissues. A substantial reduction in isulitis was detected (1.65 ± 0.33 SEM, $P < 0.0001$), suggesting that enhanced tolerization is achieved through multiple antigen delivery (Fig. 4).

The onset of diabetes, as determined by hyperglycemia, was suppressed in mice fed the CTB-INS or CTB-GAD potato tissues in comparison with animals fed transgenic potato tissues synthesizing CTB alone (Figure 5). For the period between 16 and 26 weeks of age, the incidence of diabetes was significantly lower in the CTB-INS or CTB-GAD-fed animal group in comparison to the CTB-fed group ($P < 0.01$).

The T cell-mediated destruction of pancreatic β cells is mainly responsible for diabetogenesis in NOD mice, thus, a reduction in pancreatic lymphocyte infiltration and subsequent suppression of diabetic symptoms due to such mechanisms as bystander suppression, peripheral deletion, and induction of clonal anergy is direct histological and clinical evidence of oral tolerance induction. Although, from our experimental results, it is difficult to confirm the sole mechanism of oral tolerance responsible for the observed suppression of insulitis and diabetes in the NOD mice, recent reports from several laboratories indicate that a subset of protective T cells generated by feeding autoantigens migrates to the pancreatic lymph nodes where they interfere with initiation of the inflammatory response from migratory diabetogenic T lymphocytes. Further reports indicate that feeding autoantigens which stimulate the gut associated lymphoid tissues (GALT) preferentially generate a Th2-lymphocyte response including secretion of lymphokines such as interleukin (IL)-4, IL-10, and transforming growth factor (TGF)-β (Weiner *et al.*, 1994; Han-

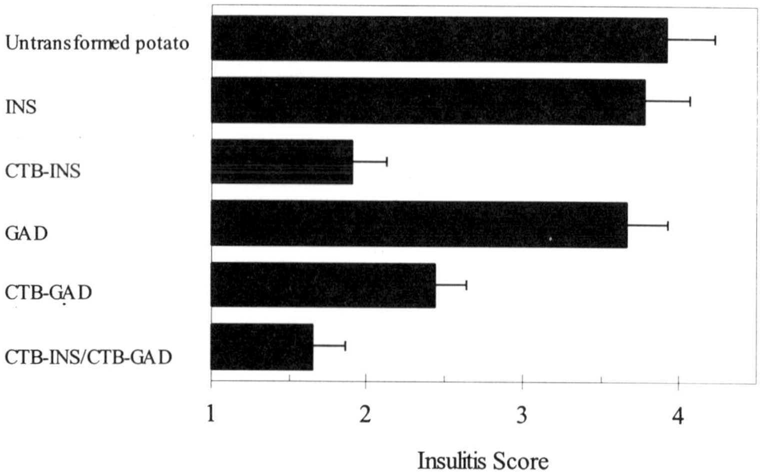

Figure 4. Reduction of insulitis in NOD mice. Five-week-old female NOD mice were fed 3 g of potato tubers once per week until they reached 9 weeks of age (a total of five feedings). Each feeding of transgenic potato tuber tissues was found to deliver approximately 30 μg of insulin, 20 μg of insulin as the CTB-INS fusion protein, 3 μg of GAD, or 2 μg of GAD as the CTB-GAD fusion protein. The animals were sacrificed at 10 weeks of age for histopathological examination of pancreatic islets. Approximately 10 islets from each ahimal were scored with semiquantitative scales. The degree of insulitis was scored based on a 7-level semiquantitative scale ranging from 0 to 6, where score 0 is normal islet morphology with no sign of T-cell infiltration and scores 1–3 and 4–6 indicate increasing levels of peri-insulitis and intra-insulitis, respectively. Data are expressed as the mean score of each group ± s.e.m. ($P = 0.001$ for mice fed CTB-INS potato tubers, $P = 0.003$ for mice fed CTB-GAD potato tissues, and $P < 0.0001$ for mice fed a mixture of CTB-INS and CTB-GAD in comparison with a group fed untransformed potato tubers). No statistically significant difference was observed between groups fed INS or GAD, and untransformed potato tuber tissues. Scores were compared by Student's *t* test for unpaired samples.

cock *et al.*, 1995). The recombinant CTB subunit without the CTA subunit was recently shown to induce mucosal antibody production mediated through TGF-β, a major mediator of oral tolerance (Weiner *et al.*, 1994; Kim *et al.*, 1998). Therefore, it is possible that plant-synthesized CTB-INS fusion peptide delivered to the GALT stimulates lymphoid cells to secrete TGF-β, which in turn, mediates induction of systemic immunotolerance i.e., suppression of insulitis and subsequent diabetic symptoms, and concurrent induction of both mucosal and systemic humoral responses against CTB and autoantigens.

Oral immunotolerization against autoimmune diseases traditionally requires repeated administration of substantial amounts (milligram levels) of autoantigen proteins. Although, stable transformation of food plants may provide an opportunity for cost-effective, safe and convenient methods for production and delivery of recombinant proteins, the use of edible plants for this purpose has been limited primarily due to low recombinant protein production levels in transgenic plants. Cholera toxin B subunit has been used to specifically target the autoantigens to the GALT for enhanced immunological tolerance. The pentameric structure of the CTB fusion protein complex not only facilitates site-specific delivery and presentation of conjugated polypeptides to the GALT, it also increases the molar concentration of the antigen per molecule of CTB pentamer. This increase in antigen concentration at the GALT may reduce the requirement for antigen biosynthesis in food plants.

This study demonstrates that feeding microgram amounts of food plant-produced autoantigens conjugated with CTB subunit can effectively suppress development of pancreatic inflammation and subsequent diabetes in non obese diabetic mice (NOD) mice. The plant-synthesized CTB-autoantigen conjugate protein is effective at doses at least 100 fold

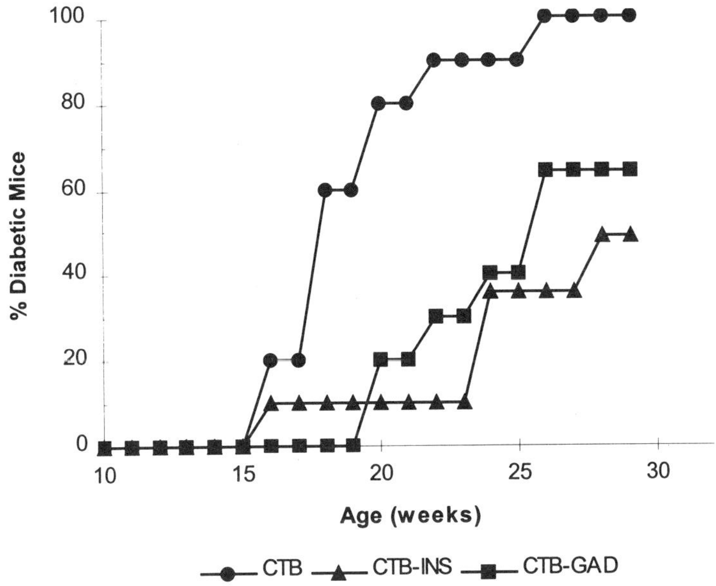

Figure 5. Suppression of diabetes in NOD mice by feeding transgenic potato tubers Five-week old female NOD mice (10 animals per group) were fed with either transgenic potato tuber tissue producing CTB, CTB-INS, or CTB-GAD once per week for a total of five weeks until animals reached 10 weeks of age. The mice were monitored weekly for glycosuria starting at 10 weeks of age. Glycosuric mice were bled from the tail vein and the blood assayed for glycemia. Clinical diabetes was confirmed by detection of hyperglycemia (> 13.8 mM blood glucose) for two consecutive weeks.

less than generally reported for unconjugated autoantigens. Feeding comparable amounts of insulin or GAD protein alone did not effectively tolerize the animals. This finding reinforces the importance of antigen conjugation with CTB for effective oral tolerance induction.

Based on fusion of the autoantigen with CTB subunit, transgenic food plants may now be feasible production and mucosal delivery systems for disease-specific CTB-autoantigen fusion proteins for suppression of autoimmune diseases. Transgenic food plants may provide a cost-effective and convenient production systems especially useful in economically emerging nations, as cultivation of antigen-producing food plants requires only sunlight energy, water, and conventional or indigenous agricultural technology. This novel CTB-autoantigen conjugate method for induction of oral tolerization makes available food plant-based prevention for a variety of T cell-mediated autoimmune diseases.

Multiple Antigen Delivery for Improved Efficacy of Vaccine Plants

For enhanced preventive efficacy of recombinant proteins against infectious as well as autoimmune diseases, it is desirable to deliver multiple vaccine antigens to sensitize animals against broad range of neutralizing epitopes, or to deliver multiple antimicrobial peptides for production of synergistic effects. The following improvements can be considered for this purpose:

1. *Consumption of plant tissue mixture*: To deliver multiple antigens a mixture of individual transgenic plant tissues containing recombinant proteins is orally admin-

istered to mice by feeding. For example, a mixture of two transformed potato plants each producing CTB-INS or CTB-GAD will tolerize animals against both insulin and glutamic acid decarboxylase. However, a disadvantage of this strategy is that since an animal can consume only limited amounts of edible tissues, the amount of each recombinant protein delivered to an animal will be limited.

2. *Sexual crossing between multiple transgenic plants*: One method to achieve multiple gene expression within the same plant tissue is by generating individual transgenic plants and subsequently to make sexual crosses between them to generate a single plant producing multiple recombinant proteins. Successive sexual crosses between four different transgenic tobacco plants expressing four individual components of secretory immunoglobulin (a light chain, a heavy chain, a joining chain, and a secretory component) resulted in a plant that expressed all four chains simultaneously, and assembled all the recombinant proteins into a functional, high molecular weight secretory IgA molecule that specifically recognized its target, the native streptococcal antigen I/II cell surface adhesion molecule (Ma *et al.*, 1995). This result indicates the possibility that sexual crosses between multiple plants is a feasible method to generate a plant simultaneously producing multiple recombinant proteins within the same plant. A possible disadvantage of this method is that sexual crosses are not always easy to achieve for certain plant species.
3. *Dual promoter systems*: Multiple gene expression systems may provide a distinct advantage when multiple proteins need to be produced in one plant. For example, multiple gene expression systems are required when two recombinant protein antigens are desired for construction of a vaccine against multiple neutralizing epitopes of a single or multiple pathogens. More than one protein can be synthesized simultaneously in one plant using the bidirectional *mas* P1, P2 promoters (Langridge *et al.*, 1989). The *mas* P1 and P2 promoters can be induced with auxin to produce foreign proteins at levels greater than those obtained by known strong constitutive promoters such as the cauliflower mosaic virus (CaMV) 35S promoter (Mason *et al.*, 1992, 1996; Haq *et al.*, 1995). Without auxin induction, expression levels are so low that no phytotoxicity is detected, which is a problem frequently encountered with the use of constitutive promoter like the 35S promoter (Dr. C. Arntzen, personal communication). The results of these experiments indicate that the *mas* dual promoter system can serve as an excellent model system for simultaneous expression of two desired gene products in food plant tissues. Further, use of auxin-upregulated promoters such as *mas* promoters may provide a certain advantage in the gene expression in climacteric food plants like tomato and banana. Auxin levels are usually elevated during fruit ripening stage for climacteric plants, which switches on the gene expression from auxin-inducible promoters, possibly achieving highest protein accumulation within the fruit tissues.
4. *Promoter combination*: A number of different promoters can be inserted within a single plant expression vector. For example, cauliflower mosaic virus 35S promoter, the *mas* P1, P2 promoter, and potato constitutive tuber-specific patatin promoter or stem-specific gene 5 promoter can be located on the same plant expression vector.
5. *Cholera toxin subunit CTA2/CTB double fusion vector*: Several CTB fusion proteins have been constructed to carry polypeptides fused either at the N- or C-terminus of the CTB protein. Short oligopeptides were fused at C-terminus without

disrupting the pentamerization of CTB chimeric monomers. However, the N-terminus of the CTB molecule seems to be critical for the pentamerization process of the protein as addition of amino acids with increasing length gradually decreased the affinity of the chimera for G_{M1}-ganglioside (Dertzbaugh and Elson, 1993b). In our experience plants at present can produce up to 0.1% of total plant soluble protein as the pentameric CTB chimera with polypeptide of more than 100 amino acids length at the C-terminus without disrupting the affinity for G_{M1}-ganglioside. Addition of a flexible hinge region (Gly-Pro-Gly-Pro) between the CTB and fused oligopeptide may reduce the steric hindrance created by the fused oligopeptide, facilitating pentamerization of the CTB-antigen chimera. Although CTB is an effective carrier molecule to target and present conjugated polypeptides to the gut-associated lymphoid tissues (GALT), it may be limited in the length of oligopeptide it can carry. In contrast to the CTB subunit, the toxic 22-kDa CTA_1 moiety of the dimeric CTA subunit is dispensable and can be replaced with a polypeptide with a molecular weight of up to 45 kDa (Hajishengallis *et al.*, 1995). This chimeric CTA_2-conjugated polypeptide was demonstrated to assemble into heterohexamers with the CTB pentamer molecule. Since the CTA_1 moiety of CTA peptide is exposed at the surface of the CTB pentamer structure, the peptide molecule which replaces the CTA_1 moiety will presumably be exposed as well. The exposed protein antigen may be effectively presented to the GALT by oral administration of the transgenic food plant tissues producing the chimera. The oligonucleotide encoding the leader peptide of the CTA (18 amino acids) was retained to target the fusion CTA_2 peptide to the endoplasmic reticulum. Since the original CTA_2 molecule has the eukaryotic ER sequestration signal (KDEL) at the C-terminus, the CTA_2 chimera can be expected to accumulate in the ER of plant cells. The ER sequestration signal sequence was added to the C-terminus of CTB chimeric proteins. We can, therefore, colocalize the CTB chimera and CTA_2 chimera in the plant ER microenvironment which mimics periplasmic space of Gram-negative bacteria to facilitate formation of heterohexameric molecules carrying a short oligopeptide attached to the CTB moiety and a larger peptide fused to the CTA_2 subunit. This double fusion CTA_2/CTB chimeric peptide could be used for multivalent vaccine delivered by food plants.

CONCLUSION

Plants not only have the ability to express, fold, and assemble complex protein molecules which retain their native biological activity, they can also deliver active or passive immunization to animals and humans. The plant-based production of immunotherapeutic proteins such as antibodies and vaccine antigens prepares the way for construction of large-scale, cost-effective production systems which may in near future become sufficiently economically competitive to supplement or even replace more conventional vaccine production systems. The synthesis in plants of recombinant proteins with demonstrated therapeutic value will contribute to improved public health especially in developing countries of the world. Palatable food plant varieties native to developing and industrialized countries must be evaluated for their ability to serve as recombinant protein production and delivery systems. Due to their palatability and common availability they would be particularly useful as an oral delivery source for children. These research efforts, enable us to visualize the application of plant molecular biology to preventive medicine for continued enhancement of our well-being, while reminding us of our intimate dependence upon the useful nature of the plant world surrounding us.

REFERENCES

Arakawa, T.; Chong, D. K. X.; Merritt, J. L.; Langridge, W. H. R. Expression of cholera toxin B subunit oligomers in transgenic potato plants. *Transgenic Res.* **1997**, *6*, 403–413.

Arakawa, T.; Chong, D. K. X.; Langridge, W. H. R. Efficacy of a food plant-based oral cholera toxin B subunit vaccine. *Nat. Biotechnol.* **1998**, *16*, 292–297.

Arondel, V.; Lemieux, B.; Hwang, I.; Gibson, S.; Goodman, H. M.; Somerville, C. R. Map-based cloning of a gene controlling omega-3 fatty acid desaturation in Arabidopsis. *Science* **1992**, *258*, 1353–1355.

Bergerot, I.; Ploix, C.; Petersen, J.; Moulin, V.; Rask, C.; Fabien, N.; Lindblad, M.; Mayer, A.; Czerkinsky, C.; Holmgren, J.; Thivolet, C. A cholera toxoid-insulin conjugate as an oral vaccine against spontaneous autoimmune diabetes. *Proc. Natl. Acad. Sci. USA* **1997**, *94*, 4610–4614.

Bordas, M.; Montesinos, C.; Dabauza, M.; Salvador, A.; Roig, L. A.; Serrano, R.; Moreno, V. Transfer of the yeast salt tolerance gene HAL1 to Cucumis melo L. cultivars and in vitro evaluation of salt tolerance. *Transgenic Res.* **1997**, *6*, 41–50.

Challacombe, S.J.; Tomasi, T. B. Systemic tolerance and secretory immunity after oral immunization. J. Exp. Med. **1980**, *152*, 1459–1472.

Chong, D. K. X.; Roberts, W.; Arakawa, T.; Illes, K.; Bagi, G.; Slattery, C. W.; Langridge, W. H. R. Expression of the human milk protein, β-casein in transgenic potato plants. *Transgenic Res.* **1997**, *6*, 289–296.

Clemens J. D.; Sack, D. A.; Rao, M. R.; Chakraborty, J.; Khan, M. R.; Kay, B.; Ahmed, F.; Banik, A. K.; van Loon, F. P. L.; Yunus, M.; Harris, J. R. Evidence that inactivated oral cholera vaccines both prevent and mitigate *Vibrio cholerae* O1 infections in a cholera-endemic area. *J. Infect. Dis.* **1992**, *166*, 1029–1034.

Clemens J. D., van Loon, F.; Sack, D. A.; Chakraborty, J.; Rao, M. R.; Ahmed, F.; Harris, J. R.; Khan, M. R.; Yunus, M.; Huda, S.; Kay, B. A.; Svennerholm, A. - M.; Holmgren, J. Field trial of oral cholera vaccines in Bangladesh: serum vibriocidal and antitoxic antibodies as markers of the risk of cholera. *J. Infect. Dis.* **1991**, *163*, 1235–1242.

Czerkinsky, C.; Russell, M. W.; Lycke, N.; Lindblad, M.; Holmgren, J. Oral administration of a streptococcal antigen coupled to cholera toxin B subunit evokes strong antibody responses in salivary glands and extramucosal tissues. *Infect. Immun.* **1989**, *57*, 1072–1077.

de Neve, M.; de Loose, M.; Jacobs, A.; van Houdt, H.; Kaluza, B.; Weidle, U.; van Montagu, M.; Depicker, A. Assembly of an antibody and its derived antibody fragment in *Nicotiana* and *Arabidopsis*. *Transgenic Res.* **1993**, *2*, 227–237.

Dertzbaugh, M.T.; Elson, C. O. Comparative effectiveness of the cholera toxin B subunit and alkaline phosphatase as carriers for oral vaccines. *Infect. Immun.* **1993a**, *61*, 48–55.

Dertzbaugh, M.T.; Elson, C. O. Reduction in oral immunogenicity of cholera toxin B subunit by N-terminal peptide addition. *Infect. Immun.* **1993b**, *61*, 384–390.

Eisenbarth, G. Type I diabetes mellitus: a chronic autoimmune disease. *N. Engl. J. Med.* **1986**, *314*, 1360–1368.

Elson, C. O.; Ealding, W. Generalized systemic and mucosal immunity in mice after mucosal stimulation with cholera toxin. *J. Immunol.* **1984**, *132*, 2736–2741.

Elson, C. O.; Zivny, J. Oral tolerance: a commentary. In *Essentials of Mucosal Immunology*; Kagnoff, M. F., Kiyono, H., Eds.; Academic Press: San Diego, California, 1996.

Escher, A.; O'Kane, D. J.; Lee, J.; Szalay, A. A. Bacterial luciferase alpha-beta fusion protein is fully active as a monomer and highly sensitive *in vivo* to elevated temperature. *Proc. Natl. Acad. Sci. USA* **1989**, *86*, 6528–6532.

Fuhrman, J. A.; Cebra, J. J. Special features of the priming process for a secretory IgA response: B-cell priming with cholera toxin. *J. Exp. Med.* **1981**, *153*, 534–544.

Fujihashi, K.; McGhee, J. R.; Yamamoto, M.; Hiroi, T.; Kiyono, H. Role of gamma delta T cells in the regulation of mucosal IgA response and oral tolerance. *Ann. N. Y. Acad. Sci.* **1996**, *778*, 55–63.

Fujita, K.; Finkelstein, R. A. Antitoxic immunity in experimental cholera: comparison of immunity induced perorally and parenterally in mice. *J. Infect. Dis.* **1972**, *125*, 647–655.

Gill, D.M. The arrangement of subunits in cholera toxin. *Biochemistry* **1976**, *15*, 1242–1248.

Gill, D.M.; Meren, R. ADP-ribosylation of membrane proteins catalyzed by cholera toxin: Basis of the activation of adenylate cyclase. *Proc. Natl. Acad. Sci. USA* **1978**, *75*, 3050–3054.

Hajishengallis, G.; Hollingshead, S. K.; Koga, T.; Russell, M. W. Mucosal immunization with a bacterial protein antigen genetically coupled to cholera toxin A2/B subunits. *J. Immunol.* **1995**, *154*, 4322–4332.

Hancock, W. W.; Polanski, M.; Zhang, J.; Blogg, N.; Weiner, H. L. Suppression of insulitis in non-obese diabetic (NOD) mice by oral insulin administration is associated with selective expression of interleukin-4 and -10, transforming growth factor-β, and prostaglandin-E. *Am. J. Pathol.* **1995**, *147*, 1193–1199.

Haq, T. A.; Mason, H. S.; Clements, J. D.; Arntzen, C. J. Oral immunization with a recombinant bacterial antigen produced in transgenic plants. *Science* **1995**, *268*, 714–716.

Hardy, S. J. S.; Holmgren, J.; Johansson, S. Sanchez, J.; Hirst, T. R. Coordinated assembly of multisubunit proteins: Oligomerization of bacterial enterotoxins *in vivo* and *in vitro*. *Proc. Natl. Acad. Sci. USA* **1988**, *85*, 7109–7113.

Hiatt, A.; Cafferkey, R.; Bowdish, K. Production of antibodies in transgenic plants. *Nature* **1989**, *342*, 76–78.

Hirst, T. R.; Holmgren, J. Conformation of protein secreted across bacterial outer membranes: a study of enterotoxin translocation from *Vibrio cholerae*. *Proc. Natl. Acad. Sci. USA* **1987**, *84*, 7418–7422.

Holmgren, J.; Lycke, N.; Czerkinsky, C. Cholera toxin and cholera B subunit as oral-mucosal adjuvant and antigen vector systems. *Vaccine* **1993**, *11*, 1179–1184.

Host, A.; Jacobsen, H. P.; Halken, S.; Holmenlund, D. The natural history of cow's milk protein allergy/intolerance. *Eur. J. Clin. Nutr.* **1995**, *49 Suppl 1*, S13–S18.

Husby, S.; Mestecky, J.; Moldoveanu, Z.; Holland, S.; Elson, C. O. Oral tolerance in humans. T cell but not B cell tolerance after antigen feeding. *J. Immunol.* **1994**, *152*, 4663–4670.

Iturriaga, G.; Leyns, L.; Villegas, A.; Gharaibeh, R.; Salamini, F.; Bartels, D. A family of novel myb-related genes from the resurrection plant Craterostigma plantagineum are specifically expressed in callus and roots in response to ABA or desiccation. *Plant Mol. Biol.* **1996**, *32*, 707–716.

Jackson, R. J.; Fujihashi, K.; Xu-Amano, J.; Kiyono, H.; Elson, C. O.; McGhee, J. R. Optimizing oral vaccines: induction of systemic and mucosal B-cell and antibody responses to tetanus toxoid by use of cholera toxin as an adjuvant. *Infect. Immun.* **1993**, *61*, 4272–4279.

Keeler, S. J.; Maloney, C. L.; Webber, P. Y.; Patterson, C.; Hirata, L. T.; Falco, S. C.; Rice, J. A. Expression of de novo high-lysine alpha-helical coiled-coil proteins may significantly increase the accumulated levels of lysine in mature seeds of transgenic tobacco plants. *Plant Mol. Biol.* **1997**, *34*, 15–29.

Kenward, K. D.; Altschuler, M.; Hildebrand, D.; Davies, P. L. Accumulation of type I fish antifreeze protein in transgenic tobacco is cold-specific. *Plant Mol. Biol.* **1993**, *23*, 377–385.

Khoury, S. J.; Lider, O.; Al-Sabbagh, A.; Weiner, H. L. Suppression of experimental autoimmune encephalomyelitis by oral administration of myelin basic protein. *Cell. Immunol.* **1990**, *131*, 302–310.

Kim, P.-H.; Eckmann, L.; Lee, W. - J.; Han, W.; Kagnoff, M. F. Cholera toxin and cholera toxin B subunit induce IgA switching through the action of TGF-β1. *J. Immunol.* **1998**, *160*, 1198–1203.

Koncz, C.; Olsson, O.; Langridge, W. H. R.; Schell, J.; Szalay, A. A. Expression and assembly of functional bacterial luciferase in plants. *Proc. Natl. Acad. Sci. USA* **1987**, *84*, 131–135.

Kozak, M. Possible role of flanking nucleotides in recognition of the AUG initiator codon by eukaryotic ribosomes. *Nucleic Acid Res.* **1981**, *9*, 5233–5252.

Langridge, W. H. R.; Fitzgerald, K. L.; Koncz, C.; Schell, J.; Szalay, A. A. Dual promoter of *Agrobacterium tumefaciens* mannopine synthase gene is regulated by plant growth hormones. *Proc. Natl. Acad. Sci. USA* **1989**, *86*, 3219–3223.

Levine, M. M.; Nalin, D. R.; Craig, J. P.; Hoover, D.; Bergquist, E. J.; Waterman, D.; Holley, H. P.; Hornick, R. B.; Pierce, N. P.; Libonati, J. P. Immunity of cholera in man: relative role of antibacterial versus antitoxic immunity. *Trans. R. Soc. Trop. Med. Hyg.* **1979**, *73*, 3–9.

Logemann, E.; Parniske, M.; Hahlbrock, K. Modes of expression and common structural features of the complete phenylalanine ammonia-lyase gene family in parsley. *Proc. Natl. Acad. Sci. USA* **1995**, *92*, 5905–5909.

Lycke, N.; Holmgren, J. Strong adjuvant properties of cholera toxin on gut mucosal immune responses to orally presented antigens. *Immunology* **1986**, *59*, 301–308.

Lycke, N.; Lindholm, L.; Holmgren, J. IgA isotype restriction in the mucosal but not in the extramucosal immune response after oral immunizations with cholera toxin or cholera B subunit. *Int. Arch. Allergy Appl. Immunol.* **1983**, *72*, 119–127.

Lycke, N.; Lindholm, L.; Holmgren, J. Cholera antibody production *in vitro* by peripheral blood lymphocytes following oral immunization of humans and mice. *Clin. Exp. Immunol.* **1985**, *62*, 39–47.

Ma, S. - W.; Zhao, D. - L.; Yin, Z. - Q.; Mukherjee, R.; Singh, B.; Qin, H. - Y.; Stiller, C. R.; Jevnikar, A. M. Transgenic plants expressing autoantigens fed to mice to induce oral tolerance. *Nat. Med.* **1997**, *3*, 793–796.

Ma, J. K.-C.; Hiatt, A.; Hein, M.; Vine, N. D.; Wang, F.; Stabila, P.; van Dolleweerd, C.; Mostov, K.; Lehner, T. Generation and assembly of secretory antibodies in plants. *Science* **1995**, *268*, 716–719.

Mason, H. S.; Ball, J. M.; Shi, J.-J.; Jiang, X.; Estes, M. K.; Arntzen, C. J. Expression of Norwalk virus capsid protein in transgenic tobacco and potato and its oral immunogenicity in mice. *Proc. Natl. Acad. Sci. USA* **1996**, *93*, 5335–5340.

Mason, H. S.; Lam, D. M.-K.; Arntzen, C. J. Expression of hepatitis B surface antigen in transgenic plants. *Proc. Natl. Acad. Sci. USA* **1992**, *89*, 11745–11749.

McKenzie, S. J.; Halsey, J. F. Cholera toxin B subunit as a carrier protein to stimulate a mucosal immune response. *J. Immunol.* **1984**, *133*, 1818–1824.

Merritt, E. A.; Sarfaty, S.; Akker, F. V.-D.; L'Hoir, C.; Martial, J. A.; Hol, W. G. J. Crystal structure of cholera toxin B-pentamer bound to receptor G_{M1} pentasaccharide. *Protein Sci.* **1994**, *3*, 166–175.

Meyer, R. Detection of genetically engineered plants by polymerase chain reaction (PCR) using the FLAVR SAVR tomato as an example. *Zeitschriff Fur Lebensmittel-Untersuchung und-Forschung* **1995**, *201*, 583–586.

Miller, A.; Lider, O.; Roberts, A. B.; Sporn, M. B.; Weiner, H. L. Suppressor T cells generated by oral tolerization to myelin basic protein suppress both *in vitro* and *in vivo* immune responses by the release of transforming growth factor β after antigen-specific triggering. *Proc. Natl. Acad. Sci. USA* **1992**, *89*, 421–425.

Mitra, A.; Zhang, Z. Expression of a human lactoferrin cDNA in tobacco cells produces antibacterial protein(s). *Plant Physiol.* **1994**, *106*, 977–981.

Miyazaki, J.; Ishii, M.; Tashiro, F. Current studies on the identification of susceptibility genes for IDDM in NOD mice. *Nippon Rinsho* **1994**, *52*, 2772–2777.

Miyazaki, A.; Hanafusa, T.; Yamada, K.; Miyagawa, J.; Fujino-Kurihara, H.; Nakajima, H.; Nonaka, K.; Tarui, S. Predominance of T lymphocytes in pancreatic islets and spleen of pre-diabetic non-obese diabetic (NOD) mice: a longitudinal study. *Clin. Exp. Immunol.* **1985**, ***60***, 622–630.

Munro, S.; Pelham, H. R. B. A C-terminal signal prevents secretion of luminal ER proteins. *Cell* **1987**, *48*, 899–907.

Schön, A.; Freire, E. Thermodynamics of intersubunit interactions in cholera toxin upon binding to the oligosaccharide portion of its cell surface receptor, ganglioside G_{M1}. *Biochemistry* **1989**, *28*, 5019–5024.

Smeda, R. J.; Hasegawa, P. M.; Goldsbrough, P. B.; Singh, N. K.; Weller, S. C. A serine-to-threonine substitution in the triazine herbicide-binding protein in potato cells results in atrazine resistance without impairing productivity. *Plant Physiol.* **1993**, *103*, 911–917.

Suh, H.; Hepburn, A. G.; Kriz, A. L.; Widholm, J. M. Structure of the amplified 5-enolpyruvylshikimate-3-phosphate synthase gene in glyphosate-resistant carrot cells. *Plant Mol. Biol.* **1993**, *22*, 195–205.

Sun, J.-B.; Holmgren, J.; Czerkinsky, C. Cholera toxin B subunit: an efficient transmucosal carrier-delivery system for induction of peripheral immunological tolerance. *Proc. Natl. Acad. Sci. USA* **1994**, *91*, 10795–10799.

Sun, J.-B.; Rask, C.; Olsson, T.; Holmgren, J.; Czerkinsky, C. Treatment of experimental autoimmune encephalomyelitis by feeding myelin basic protein conjugated to cholera toxin B subunit. *Proc. Nat.l Acad. Sci. USA* **1996**, *93*, 7196–7201.

Sutton, D.W.; Havstad, P. K.; Kemp, J. D. Synthetic cryIIIA gene from *Bacillus thuringiensis* improved for high expression in plants. *Transgenic Res.* **1992**, *1*, 228–236.

Svennerholm, L. Interaction of cholera toxin and ganglioside G(M1). *Adv. Exp. Med. Biol.* **1976**, *71*, 191–204.

Thanavala, Y.; Yang, Y. - F.; Lyons, P.; Mason, H. S.; Arntzen, C. J. Immunogenicity of transgenic plant-derived hepatitis B surface antigen. *Proc. Natl. Acad. Sci. USA* **1995**, *92*, 3358–3361.

Tisch, R.; McDevitt, H. Insulin-dependent diabetes mellitus. *Cell* **1996**, *85*, 291–297.

Trentham, D. E.; Dynesius-Trentham, R. A.; Orav, E. J.; Combitchi, D.; Lorenzo, C.; Sewell, K. L.; Hafler, D. A.; Weiner, H. L. Effects of oral administration of type II collagen on rheumatoid arthritis. *Science* **1993**, *261*, 1727–1730.

Virtanen, S. M.; Rasanen, L.; Aro, A.; Lindstrom, J.; Sippola, H.; Lounamaa, R.; Toivanen, L.; Tuomilehto, J.; Akerblom, H. K. Infant feeding in Finnish children less than 7 yr of age with newly diagnosed IDDM. Childhood diabetes in Finland study group. *Diabetes Care* **1991**, *14*, 415–417.

Virtanen, S. M.; Saukkonen, T.; Savilahti, E.; Ylonen, K.; Rasanen, L.; Aro, A.; Knip, M.; Tuomilehto, J.; Akerblom, H. K. Diet, cow's milk protein antibodies and the risk of IDDM in Finnish children. Childhood diabetes in Finland study group. *Diabetologia* **1994**, *37*, 381–387.

Wandelt, C. I.; Khan, M. R. I.; Craig, S.; Schroeder, H. E.; Spencer, D.; Higgins, T. J. V. Vicilin with carboxy-terminal KDEL is retained in the endoplasmic reticulum and accumulates to high levels in the leaves of transgenic plants. *Plant J.* **1992**, *2*, 181–192.

Weiner, H. L.; Friedman, A.; Miller, A.; Khoury, S. J.; al-Sabbagh, A.; Santos, L.; Sayegh, M.; Nussenblatt, R. B.; Trentham, D. E.; Hafler, D. A. Oral tolerance: immunologic mechanisms and treatment of animal and human organ-specific autoimmune diseases by oral administration of autoantigens. *Annu. Rev. Immunol.* **1994**, ***12***, 809–837.

Weiner, H. L.; Mackin, G. A.; Matsui, M.; Orav, E. J.; Khoury, S. J.; Dawson, D. M.; Hafler, D. A. Double-blind pilot trial of oral tolerization with myelin antigens in multiple sclerosis. *Science* **1993**, *259*, 1321–1324.

Zhang, Z. J.; Davidson, L.; Eisenbarth, G.; Weiner, H. L. Suppression of diabetes in nonobese diabetic mice by oral administration of porcine insulin. *Proc. Natl. Acad. Sci. USA* **1991**, *88*, 10252–10256.

14

AUTOANTIGENS PRODUCED IN PLANTS FOR ORAL TOLERANCE THERAPY OF AUTOIMMUNE DISEASES

Shengwu Ma and A. M. Jevnikar

John P. Robarts Research Institute and Siebens-Drake Research Institute
University of Western Ontario; and London Health Sciences Centre
London, Ontario, N6G 2V4 Canada

ABSTRACT

Oral administration of protein antigens can induce antigen-specific immune hyporesponsiveness and may be useful in treating autoimmune diseases or preventing transplant rejection. However, the therapeutic value of oral tolerance may be limited when candidate autoantigens cannot be produced by conventional system in quantities sufficient for clinical studies. Plants may be ideally suited for this purpose, as they can produce hugh quantities of functional mammalian proteins at extremely competitive cost. Furthermore, transgenic food plants could provide a simple and direct method of autoantigen delivery for oral tolerance. Here we show that the diabetes-associated autoantigen glutamic acid decarboxylase (GAD) is efficiently expressed in both tobacco and potato plants, and that mice, when fed with fresh transgenic potato tubers, are fully protected from diabetes.

INTRODUCTION

Oral tolerance, defined as a state of specific immunological hyporesponsiveness resulting from orally administered protein antigens, has been regarded increasingly as an innovative and effective strategy for the treatment of autoimmune disease and transplant rejection. Currently available non-selective immunosuppressive drug therapies are inadequate because they exhibit limited efficacy, low target selectivity, considerable toxicity and many side effects including the development of opportunistic infections and late malignancies. Therefore, it is highly desirable to develop immunologically-specific therapies that would selectively target autoreactive immune processes that are associated with the

Chemicals via Higher Plant Bioengineering, edited by Shahidi *et al.*
Kluwer Academic / Plenum Publishers, New York, 1999.

disease without affecting the remainder of the immune system. Induction of immunological tolerance by administration of autoantigens is one potent approach to achieve this goal. Although there are several routes of antigen administration, the oral route offers several attractive advantages, and two of the most notable ones are the simplicity and low risk of sensitization to the antigen administered. Parenteral administration (e.g. intravenous, intrathymical and subcutaneous) of antigens to induce immunological tolerance for treatment of disease poses a risk of inducing antigen-sensitization which could lead to further health complications. Current data suggests that immunological tolerance induced by oral antigen administration carries minimal sensitization. Oral tolerance therapy has been used successfully to treat several autoimmune disorders in experimental animals and its utility is now being tested in the treatment of human autoimmune disease (reviewed by Weiner 1997).

The clinical value of oral tolerance as an effective immunotherapy remains limited by practical problems, in that large quantities of antigens (e.g., mg to kg) are required to induce tolerance in experimental animals as well as humans by the oral route. To circumvent this practical disadvantage, several conventional expression systems including microorganisms and mammalian cells, have been tested for possible large-scale production of candidate proteins, but all with limited success. The major problem with bacterial expression systems is their inability to affect the post-translational modifications such as glycosylation, which is critical for the function of many mammalian proteins. Use of mammalian cells improves the probability of obtaining recombinant foreign proteins with defined functionality, but their relatively low yields and expensive culture media and equipment make it unrealistic for production of proteins in bulk at low cost. In contrast, transgenic plants have the potential to avoid these problems.

Plants as an expression system offer a number of unique advantages. Plants are higher eukaryotes and thus are able to properly fold, glycosylate and assemble proteins in a fashion similar to a mammalian system. This may be particularly important in producing plant recombinant protein antigens that have *in vivo* activities identical to their animal counterparts. Furthermore, unlike fermentation-based bacterial and mammalian cell systems, protein production in plants is not restricted by equipment and the volume of the cells cultured on bench. Using agriculture scales, sufficient amounts of antigenic proteins that are realistically required for extensive clinical studies can be generated at an extremely competitive cost. In addition, safety issues regarding the potential transmission of pathogens from animal sources can be avoided using plant systems. The heat required to sterilize animal-born pathogens simultaneously can cause extensive denaturation of the intended autoantigen which eliminates its capacity to induce tolerance. More importantly, the ability to produce autoantigen proteins in edible portions of plants raises the possibility of generating food plants that could serve directly as oral tolerogens in oral tolerance study, thus avoiding the need for expensive down-stream purification processes and simplifying the delivery of candidate autoantigens.

Using the diabetes-associated autoantigen glutamic acid decarboxylase (GAD67) as a model protein, we are exploring the potential of genetically engineered plants as a low-cost production system for GAD, and the possibility of edible transgenic plants as a simple and direct method of GAD delivery for oral tolerance. As described below, our results suggest that transgenic plants express a high level of GAD with retained authentic immunogenicity, and that a GAD-expressing transgenic plant, given as a dietary supplement, inhibits the development of diabetes in the non-obese diabetic (NOD) mouse. In this chapter, we will review some of the basic immunological natures of diabetes, the basics of oral tolerance and oral tolerance therapy, and our results in the production of murine

GAD67 in transgenic tobacco and potato plants, as well as the effect of oral GAD plant feeding on disease prevention. Lastly, we will also briefly review our results on the production of murine major histocompatibility complex (MHC) class II antigens in transgenic tobacco.

INSULIN-DEPENDENT DIABETES MELLITUS (IDDM): A T LYMPHOCYTE-MEDIATED DISEASE

Insulin-dependent diabetes mellitus (IDDM), also known as type I diabetes, is one of the most common and chronic autoimmune diseases of children and young adults (Atkinson and Maclaren, 1994). In this disorder, cells producing insulin (called β cells) within the pancreatic islets are destroyed (Bosi and Botazzo, 1995). Whereas IDDM can be thought of as an insulin deficiency disease with the majority of afflicted people showing evidence for an immunologically mediated pathogenesis, non-insulin dependent diabetes mellitus (NIDDM or type II diabetes mellitus) represents a collection of genetically determined diseases resulting from both insulin deficiency and insulin resistance, but often having no immune involvement (Maclaren and Atkinson, 1997). In the United States the lifetime risk for developing IDDM is approximately 1 in 300 individuals, with an average yearly incidence of approximately 15 per 100,000. In Canada, about 150,000 or 10% of all diabetics have IDDM, while a greater number of patients have adult-onset NIDDM. The worldwide incidence of IDDM varies widely, ranging from a low of 1–2 per 100,000 in Japan to a high of more than 40 per 100,000 in parts of Finland (Maclaren and Atkinson, 1997). Clinical onset of IDDM, once established, even with control of blood sugar levels by insulin administration, is associated with serious complications (Castano and Eisenbarth, 1990). IDDM is the leading cause of non-congenital blindness in adults, and is responsible for 25% of cardiac surgery and 40% of endstage kidney failure. It is also a major secondary cause of death from other diseases in both the United States and Canada; therefore, IDDM complications have staggering socioeconomic impact.

IDDM is an exemplary autoimmune disease resulting from the interaction of multiple susceptibility genes and the environment. Although the development of diabetes is associated with the major histocompatibility complex (MHC), especially MHC class II alleles, the development of disease is clearly polygenic with susceptibility related to both MHC and non-MHC genes (Wicker et al., 1995). However, the concordance for IDDM among identical twins is only 30–50%. This finding is consistent with the existence of genetic predisposition but at the same time implicates environmental modifiers that influence the full expression of the disease (Maclaren and Atkinson, 1997).

The immunological mechanisms underlying human IDDM are still not fully understood, partly due to the difficulty inherent in studying IDDM in humans, notably including inaccessibility of the diseased organ and complexity of the genetic make-up. To compensate for this, many studies have been performed in NOD mice, which is a serendipitously discovered spontaneous model of autoimmune diabetes and a widely accepted animal model for human IDDM (Castano and Eisenbart, 1990). Research evidence from studies of affected humans and NOD mice suggests that the development of diabetes is associated with a destructive process in the pancreas mediated by T lymphocytes that carry out critical immune actions. In NOD mice, it was shown that T lymphocytes from diabetic donor mice could transfer disease into immunocompatible recipients that normally do not become diabetic (Wicker et al., 1986).

Studies of this animal model have suggested important roles for several proinflamatory and immunoregulatory cytokines, including interferon γ (IFN-γ), interleukin 2 (IL-2), IL-4, IL-10 and tumor necrosis factor (reviewed by Rabinovich, 1994). In both NOD mice and humans, autoimmune responses that are directed against endogenous islet antigens (autoantigens) such as insulin, glutamic acid decarboxylase (GAD), islet cell antigen (ICA), heat shock protein (hsp60), carboxypeptides H, and cytoplasmic domains of transmembrane proteins (IA-2) are observed (Atkinson and Maclaren, 1994). While it is unclear how self-tolerance is lost to these naturally-occurring pancreatic proteins, it has been suggested that microorganisms such as bacteria, viruses and parasites may trigger and/or promote anti-self immune responses. Microbial antigens having structural homologies to self-antigens can serve as immunogenic stimuli that elicit cross-reactivity with self-tissue antigens. This might lead to a bypassing of normal mechanisms that maintain self-antigen tolerance, and thus induce and/or promote the development of a self-perpetuating autoimmune response. Such an autoimmune response is mediated by activated lymphocytes that cross-react with homologous self-antigens that are normally non-immunogenic. Thus, a protective immune response against an invading microorganism could inadvertantly set off an autoimmune response (Porcelli, 1993; Wucherpfening and Strominger, 1995). One example is the recent findings that an 18 amino acid peptide from Coxsackie virus P2-C protein, which shows striking sequence homology with human GAD, can induce T-cell responses in NOD mice immunized with either Coxsackie virus P2-C protein or the P2-C peptide containing the homologous region to GAD (Szopa et al., 1993). This may be relevant to IDDM onset since humans can be a host for Coxsackie B virus.

OVERVIEW OF ORAL TOLERANCE

Oral administration of protein antigens, in addition to eliciting a local secretory immunoglobulin A (S-IgA) antibody response in mucosal tissues such as in the gastrointestinal tract, can induce a state of systemic tolerance. This phenomenon was first described in 1911 by Wells, who found that guinea pigs fed hen egg proteins did not develop anaphylactic responses to subsequent systemic challenge (e.g., by injection) with the same antigen. In 1946, Chase showed that guinea pigs fed dinitrochlorobenzene had diminished capacity to develop an immune response when re-exposed to the same antigen introduced by a systemic route. Subsequently, numerous investigators have shown that ingestion of a range of proteins can lead to poor systemic immune responsiveness to those particular antigens (Mowat, 1987). Oral tolerance, as this enteric-induced process has come to be known as, may represent an important natural physiological property of the immune system, whereby the host can avoid dangerous reactions such as delayed-type hypersensitivity (DTH) to many ingested food proteins and other antigens, which produces undue inflammation and tissue damage (reviewed by Weiner et al., 1994). The gut-associated lymphoid tissue (GALT) may represent the major site in which the events leading to oral tolerance may be initiated. A striking feature of oral tolerance is the degree of specificity to the antigen administered, and this applies to both the humoral (antibody) and cellular immune responses. For example, feeding animals ovalbumin tolerizes them to this protein but not to unrelated antigens such as human γ-globulin (HGG) and human serum albumin (Titus and Chiller, 1981). Although the reduction of the systemic immune responses of the IgG, IgA, and IgM isotypes (B cell tolerance) were demonstrated in animals given oral antigen, orally induced tolerance often has a more profound and sustained effect on T cell-mediated responses. In animals given oral antigen the proliferation of the antigen-stimulated T lymphocytes is characteristically reduced when they are cultured in vitro (Weiner et al., 1994). It should be noted that oral tolerance is primarily induced against antigens recognized by T cells and that the tolerance is maintained at the T cell level.

The primary mechanisms of oral tolerance governing decreased T-cell responsiveness differ according to the amount of antigen that is fed. The mechanisms are clonal anergy and/or deletion of antigen-specific T cells and the generation of active cellular suppression. High doses of antigen ingestion favour the development of clonal anergy and/or deletion; probably in this condition the entry of the fed antigen (intact or fragmented) into the systemic circulation directly induces unresponsiveness of the antigen-specific T cell function primarily via clonal anergy (Weiner et al, 1994; Weiner, 1997). On the other hand, low doses of fed antigen favor indirect active suppression, which is chiefly mediated through induction of regulatory T cells in GALT, such as Peyer's patches. These GALT-originated T cells migrate to systemic sites and then mediate suppression predomonantly by production of tumor growth factor-beta (TGF-β), a cytokine with potent activity for inhibiting T cell growth and differentiation (Weiner et al., 1994). Several reports have also suggested that the induction of oral tolerance can be explained by altering the interaction between T helper 1 (Th1) and T helper 2 (Th2) type cells. It has been shown that orally administered antigen preferentially induces the activation of Th2 cells and production of the Th2 cytokines interleukin-4 (IL-4) and IL-10, which may downregulate the disease-associated Th1 cell-mediated immune responses (Liblau et al., 1995; Powrie and Coffman, 1993). Accordingly, Th1 type cells are more sensitive to anergy induction following oral administration of protein antigens. These findings suggest the possibility that oral tolerance is associated with selective downregulation of Th1 cells by Th2 cells via specific cytokines in the systemic immune compartment. This possibility is consistent with the fact that oral tolerance has more profound suppressive effects on Th1-regulated immune responses than on Th2 cell-mediated humoral (antibody) responses.

Although mucosal administration of antigens favors oral tolerance, the risk of sensitization (priming) to the fed antigen can not be completely neglected, as this could potentially accelerate the condition of an existing autoimmune disease if the fed protein is an immunostimulatory autoantigen. There are several factors that could influence the induction as well as the efficacy of oral tolerance. These include the nature of the antigen, such as whether it is a particulate or soluble agent, susceptibility to degradation by gut proteases, persistence in the intestinal tract, dosage and absorption of the antigen administered and immune status of subjects (prior antigen exposure or not?) or genetic differences of the subjects. For example, an efficient induction of oral tolerance may require administration of large quantities of soluble protein antigens and likewise maintainence of the tolerance status may require repeated administration of the same antigen (Mowat, 1987).

Identification and characterization of specific autoantigens in several autoimmune diseases is a major step toward using mucosally induced tolerance as a means of preventing and treating autoimmune diseases. As summarized in Table 1, induction of oral tolerance to autoantigens has been successfully used in animal models, such as chronic experimental allergic encephalomyelitis (EAE), uveitis, myasthenia, collagen- and adjuvant-induced arthritis and NOD diabetes. Human trials have been initiated in rheumatoid arthritis, multiple sclerosis, uveitis, and diabetes. Although the available results of initial human phaseI/II trials are not conclusive and the final demonstration of clinical efficacy of oral antigen in these diseases awaits the results of large-scale phase III trails, so far there has been no observed toxicity or exacerbation of the disease, and the results have paralleled in many respects what has been observed in animals. Specifically, in uveitis, pa-

Table 1. Suppression of autoimmunity by oral tolerance

	Protein fed
(a) Animal models	
EAE	MBP, PLP
Arthritis (CIA,AIA)	CII, HSP
Uveitis	S-Ag, IRBP
Myasthenia gravis	AchR
Diabetes (NOD mouse)	Insulin, GAD65, GAD67
Transplantation	Alloantigen, MHC peptide
Thyroiditis	Thyroglobulin
Colitis	Haptenozed colonic proteins
(b) Human disease trials	
Multiple sclerosis	Bovine myelin
Rheumatoid arthritis	Chicken CII
Uveitis	Bovine S-Ag
Type I diabetes	Human insulin

Abbreviations: AIA, adjuvant- and antigen-induced arthritis: AchR, acetyecholine receptor; CII, type II collagen; CIA, collagen- induced arthritis; EAE, experimental allergic encephalomyelitis; GAD, glutamic acid decarboxylase; HSP, heat shock protein; IRBP,] inter-photoreceptor retinoid-binding protein; MBP, myelin basic protein; MHC, major histocompatibility complex; NOD, nonobese diabetic; PLP, proteolipid protein; S-Ag, S antigen.

tients given S protein antigen orally were more compliant to the tapering off of their medication as compared to control groups tested (Nussenblatt et al., 1997). In MS patients, myelin basic protein (MBP) and proteolipid protein (PLP)-specific TGF-secreting cells have been observed in the peripheral blood of myelin-treated but not untreated patients, suggesting induction of tolerizing mechanism, (Fukaura et al., 1996). As discussed above, these TGF-β lymphocytes have the potential to suppress the responses of autoreactive T lymphocytes. In rheumatoid arthritis, a recently completed 280-patient, double-blind, phase II dosing trial of collagen type II in doses ranging from 20 to 2500μg demonstrated statistically positive effects in the group fed the lowest dose (20 μg) (Weiner, 1997). These results support that it is possible to use oral tolerance to treat human autoimmune diseases.

GAD: AN KEY EARLY ISLET CELL AUTOANTIGEN IN IDDM

GAD is one of the most extensively characterized islet-cell antigens in IDDM. Antibodies to GAD precede overt diabetes both in patients (Scott et al., 1994) and in NOD mouse (Atkinson and Maclaren, 1988). GAD antibodies provide sensitive and consistent diagnostic markers for evidence of anti-islet immunity.

GAD catalyzes the conversion of glutamic acid to γ-aminobutyric acid (GABA), which is the major membrane-stabilizing neurotransmitter of the central nervous system. There are two forms of GAD, known as GAD67 and GAD65, based on their molecular weight, and these are encoded by separate genes (Bu et al., 1992). GAD65 is the predominant form within the human pancreatic islet cells, while GAD67 predominates in mouse islets. Both forms of GAD have been implicated in the pathogenesis of IDDM (Elliott et al., 1994).

In NOD mice, T cell responses specific for GAD become apparent from four weeks of age, before responses to other β-cell autoantigens including insulin, heat-shock proteins, peripherin and carboxypeptidase are detected (Kaufman et al., 1993; Tisch et al., 1993). The response to GAD by T cells appears to coincide with the onset of insulitis, an early stage of IDDM which involves infiltration of lymphocytes into the pancreatic β cells. The injection of large quantities of GAD into young NOD recepients can prevent diabetes (Kaufman et al., 1993, Tisch et al., 1993). These data suggest that GAD may represent a key early islet cell autoantigen in IDDM and can become an attractive target for antigen specific therapies, such as inducing oral tolerance to GAD. The role of oral GAD in the prevention of IDDM has not been tested clinically due to the inaccessibility of recombinant GAD in industrial-sized quantities. Poor GAD protein solubility in bacteria and inadequate production capacity from mammalian cells have precluded the use of these conventional expression systems for GAD production.

EXPRESSION OF THE MURINE AUTOANTIGEN GAD67 IN TOBACCO AND POTATO

The murine GAD67 is a cytoplasmic protein of 67 kDa. To achieve its expression in plants, the coding sequence of GAD67 cDNA was modified by PCR to introduce a NcoI site as part of the ATG start codon. The modified GAD67 was inserted into a plant expression and transformation vector (Ma et al., 1997). GAD expression is under control of a modified CaMV (cauliflower mosaic virus) 35S promoter with a double enhancer sequence linked to a 5' untranslated leader sequence derived from tobacco etch virus to maximize expression. The GAD construct was transferred into low alkaloid tobacco as well as potato plants by *Agrobacterium tumefaciens*-mediated transformation. Both transgenic tobacco and potato plants containing the GAD construct were produced. Characterization of selected plants at the DNA and RNA levels revealed stable genomic integration of GAD sequence and high levels of GAD transcripts.

To assess whether transformed plants containing transgene GAD expressed the GAD67 protein, protein extracts from transgenic tobacco leaves and from tubers of transgenic potato plants were assayed by western blot. As shown in Figure 1, when probed with anti-mouse GAD67 polyclonal antibodies, western blot showed a single protein band of the correct size. The level of GAD expression was estimated from blot densitometry and found to be approximately 0.4% of total soluble proteins. GAD67 expression was similar in tobacco leaves and potato tubers and was not detected in untransformed control plants or control plants transformed with expression vector alone (no GAD gene inserted).

EFFECT OF GAD PLANT FEEDING ON INCIDENCE OF DIABETES IN NOD MICE

The successful expression of murine GAD67 described above provided the impetus to assess whether the recombinant protein retained its intrinsic immunogenicity and whether it could be used as a tolerogen in oral tolerance. To test these, young NOD female mice were used in plant GAD feeding. NOD mice develop diabetes spontaneously with the incidence in females consistently higher than in males, typically around 75–85% at 6–7 months compared with <10% in males at the same age (Makino et al., 1981). The feeding was started at 5 weeks of age and lasted until they reached 8 months of age. Fresh potato tubers or freeze dried low-alkaloid tobacco leaves were added as a supplement to standard mouse chow. The amount of GAD consumed was calculated to be approximately 1mg per mouse daily. Con-

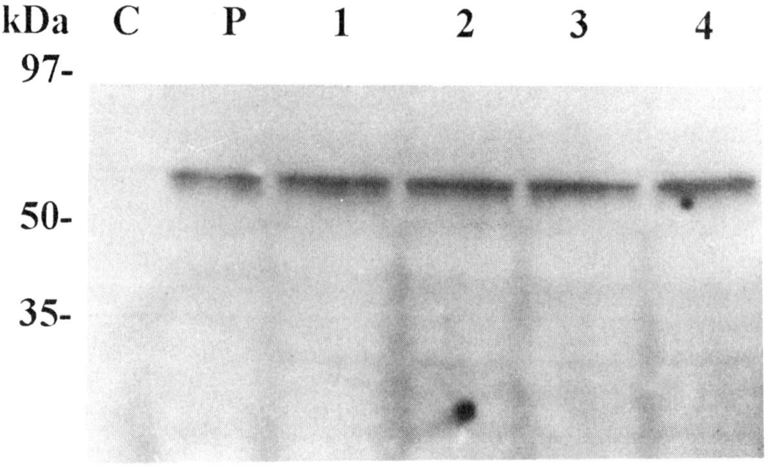

Figure 1. Western blot analysis of murine GAD67 expression in transgenic tobacco plants. Protein (30 μg) extracted from leaves were separated on a 12% SDS-polyacrylamide gel and GAD67 was identified by anti-GAD antibody. Lanes are as marked: C, extracts from a wild-type plant; P, purified mouse GAD67 (100 ng); lane 1 to 4: extracts from independent transgenic plants identified by northern blot analysis.

trol mice received an equivalent amount of corresponding tobacco or potato tissue from vector (without GAD insert) transformed plants. All mice were monitored weekly for urine glucose level with serum confirmation of positive mice (Glucoscan 2000 Strips, Lifescan, Milpitas, CA). Mice were scored as diabetic when blood glucose levels were greater than 16.7 mmol/l on 2 consecutive days. As illustrated in Figure 2, ten out of 12 NOD mice fed either tobacco (n=6) or potato GAD67 (n=6) remained free of disease (p=0.007 from controls) with equivalent protection using tobacco or potato. In contrast, 8 out of 12 control plant-fed mice (67%) equally divided between tobacco and potato supplementation, developed diabetes, which is approximately equivalent to the incidence of the stock NOD colony used. These data are consistent with previous data showing successful prevention of diabetes in NOD mice using different forms of GAD immunization (Elliott et al., 1994), and indicate that plant-derived GAD67 is an orally active tolerogen.

In a separate experiment, when NOD mice were given GAD67 plant material for only about two months (8 week duration) followed by standard chow, they showed a delay in disease-onset, but the incidence level of diabetes eventually returned to that of control groups by the study completion, suggesting that continuous GAD feeding is necessary in order to maintain the tolerance status. Mice tolerated potato and low-alkaloid tobacco leaf tissue and no differences were observed in the appearance or weight gain of mice between groups.

EFFECT OF ORAL PLANT GAD67 ON T LYMPHOCYTE FUNCTION IN NOD MICE

To determine whether the protection against diabetes in plant GAD-fed group was due to the specific suppression of GAD reactive T cells, we examined proliferative responses of GAD reactive T cells from plant GAD67 fed NOD mice. NOD mice were fed plant GAD tissue for about 4 weeks and then immunized with highly purified *Escherichia coli*-derived recombinant GAD67 (Ellitt et al., 1994), and 10 days later a splenic T cell

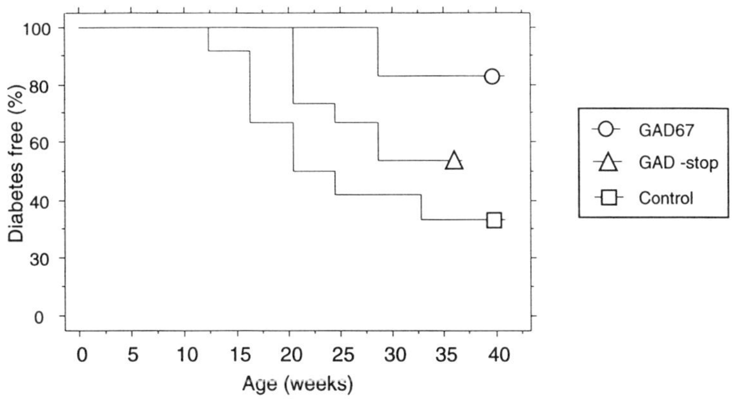

Figure 2. Oral administration of plant material expressing mouse GAD67 prevents NOD mice from developing spontaneous diabetes. Female NOD mice (n=12/group) were fed plant GAD67 or control plant material from 5 week to 8 months of age. All mice were followed for the onset of hyperglycemia (defined as blood glucose > 16.7 mmol/l). Ten out of 12 NOD mice fed either tobacco (n=6) or potato GAD67 (n=6) remained free of disease (p=0.007 from controls) with equivalent protection using tobacco or potato. In contrast, 8 out of 12 control plant-fed mice developed diabetes.

proliferation assay against GAD67 was conducted. As shown in Figure 3, mice fed plant GAD67 exhibited a significant reduction of T cell proliferation in comparison with control plant-fed groups. Such a reduction in T cell proliferation was independent of the amount of GAD67 used for *in vitro* stimulation. Insignificant proliferative responses occurred with the irrelevant protein ovalbuman (OVA) in cultures. To examine the effects of oral tolerance on Th1 and Th2 cells more directly, we examined the production of GAD-specific cytokines in vitro. IFN-γ were used as an indicator of Th1 activity, whereas IL-4 and IL-10 were used as markers of a Th2 response (Mossmann and Coffman, 1989). As shown in Figure 4, T cells from plant GAD-fed mice displayed reduced levels of IFN secretion in response to GAD67 challenge *in vitro*, as compared to that in control plant-fed NOD mice. Furthermore, while the level of secretion of IL-4 and IL-10 is generally low in both groups, an elevated level of IL-4 and IL-10 secretion was detected in plant GAD-fed mice, suggesting the activation of Th2 responses. These results are consistent with those reported previously (Tian et al., 1996). Autoimmune diabetes in NOD mice is related to a predominance of Th1 cell function, which chiefly supports cell-mediated immune responses involving IL-2 and IFN-γ-induced activation of cytotoxic macrophage and T-cell killing of islet β cells. In contrast, Th2 cells predominantly support humoral (antibody) immunity, and they are characterized by the secretion of IL-4, IL-5, IL-10 and IL-13, and they have a "protective" role against IDDM in NOD mice. Shifting the immunoregulatory balance from Th1 to Th2 immune responses can result in protection from the disease.

INDUCTION OF GAD67-SPECIFIC ANTIBODIES OF THE IgG1 ISOTYPE IN PLANT GAD-FED MICE

We next examined how GAD plant feeding affected T cell-dependent humoral (antibody) immune responses. Oral tolerance has been shown to bias T cells toward the development of Th2-type responses when they are re-exposed to antigen (Burstein and Abbas,

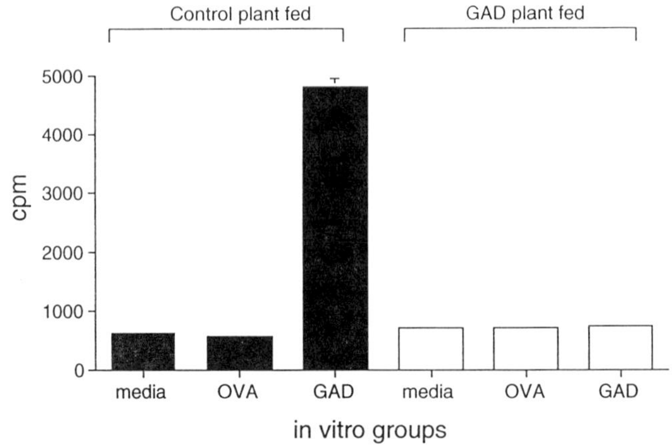

Figure 3. Loss of GAD specific T cell proliferation in plant GAD67 fed mice. NOD female NOD mice were fed either plant GAD67 or control plant and footpad challenged with 50 μg purified GAD67 after 4 weeks of feeding. Proliferative responses of splenocytes to GAD67 and OVA were determined after 10 days and results shown are representative of three experiments performed using 2 mice per group. GAD fed mice had responses equivalent to media background and a marked difference was seen between GAD and control plant (SI=7.6) fed mice ($p< 0.001$).

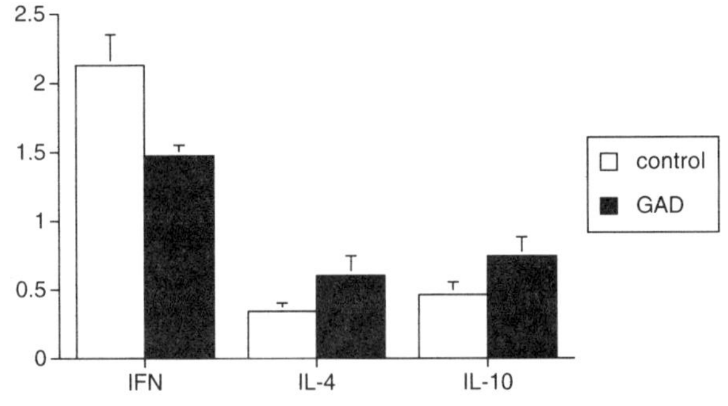

Figure 4. FN-γ expression is reduced ($p=0.01$) and IL-4 /IL-10 expression is increased in supernatants of splenocytes from plant GAD67 fed NOD mice. Prediabetic NOD mice were fed plant GAD67 tissue for 4 weeks and then immunized with 50 μg of highly purified *E. coli* derived recombinant GAD67, combined with incomplete Freund's adjuvant. Ten days later mice were killed, and a single-cell suspension of spleen cells were cultured for 24 or 48 hours in 24-well flat-bottomed plates with purified mouse GAD67 (10 μg/ml). Supernants were tested for IFN-γ, IL-4 and IL-10 by standard sandwich ELISA, and results are expressed as arbitrary absorbance units at 405 nm (A_{405}) for group comparisons.

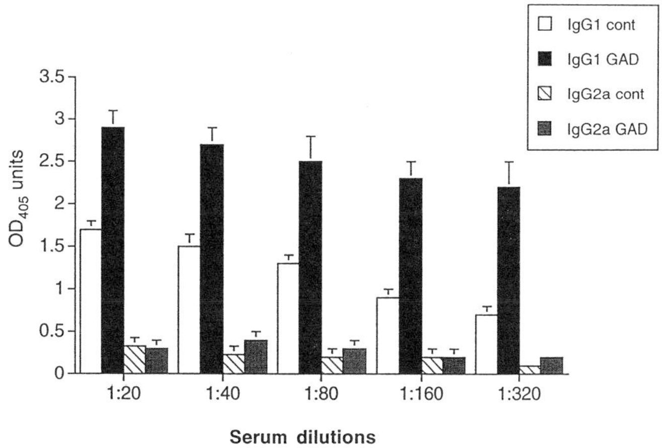

Figure 5. IgG1 anti GAD antibody response is augmented following plant GAD67 feeding. Mice fed GAD67 or control plants as in T cell proliferation studies were assessed for serum anti-GAD IgG antibody isotypes by ELISA. Mean ± SEM values of grouped samples (n=4) are presented as mean OD_{405} at various dilutions. IgG1 anti-GAD antibody levels were different between GAD67 and control plant fed mice at all dilutions ($p<0.002$). Results are representative of 3 experiments.

1993), which can result in predominantly IgG1 rather than IgG2a antibody responses (Finkelman et al., 1990). Both GAD67 and control mice were tested for humoral responses to GAD67. We found that the level of anti-GAD67 antibobies in plant GAD-fed mice were increased as compared to that in the control group (data not shown), indicating that the oral plant GAD feeding triggered a T-cell dependent anti-GAD67 B cell Ig response. To identify the type of T cell that provided help for GAD67 antibody production, we further characterized the isotype of anti-GAD67 antibodies. As can be seen in Figure 5, the increase of anti-GAD67 antibodies in GAD mice was primarily of the IgG1 subclass, which is indicative of a Th2 response (Finkelman et al., 1990). It is clear from previous studies (Tian et al., 1996) that the increase in the titre of IgG1 anti-GAD antibodies did not itself closely correlate with the extent of protection against diabetes in NOD mice, although all animals fed GAD67 did produce more IgG1 antibodies. In contrast, IgG2a GAD67 antibodies, which correlate with a Th1 response, were at very low levels and showed no changes in both GAD and control groups. These data indicate that feeding NOD mice of plant tissue containing GAD67 leads to activation of a GAD67-specific Th2 immune response.

EXPRESSION OF THE MURINE MAJOR HISTOCOMPATIBILITY COMPLEX (MHC) CLASS II MOLECULES IN TOBACCO

Major histocompatibility complex (MHC) class I and II molecules are highly polymorphic cell surface glycoproteins that play critical roles in self (endogenous) and nonself (foreign) antigen discrimination by the immune system (Klein, 1986) and in the presentation of antigenic peptides to T lymphocytes (Schwartz, 1985). T helper (Th) cells recognize foreign antigen in association with MHC class II molecules while cytotoxic T cells (having the ability to kill cells) recognize foreign antigen in association with MHC

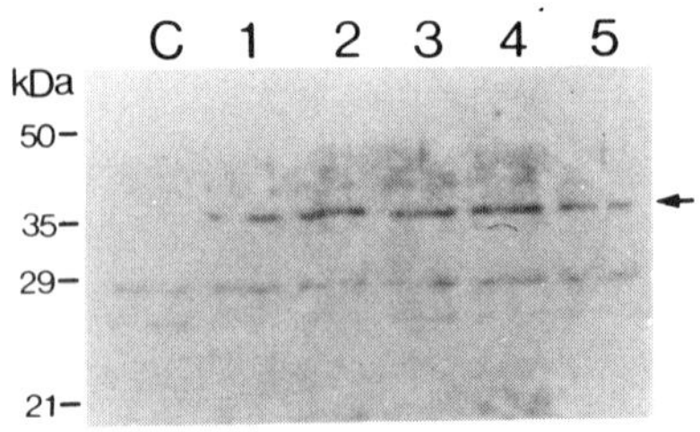

Figure 6. Western blot analysis of α chain expression in α gene tobacco plants. Protein (30 μg) extracted from tobacco leaves transformed with α gene construct were boiled, separated on a 12% SDS-polyacrylamide gel and α chain polypeptides were identified by immunoreaction with antibody specific for α chain protein. Lanes are as marked: C, extracts from untransformed control plants; lane 1 to 5: extracts from independent α gene transgenic plants identified by both southern and northern blot analysis. The arrow indicates the protein band with expected molecular weight.

class I molecules (Germain and Margulies, 1993). In addition, the principle targets of the immune response to allografts (a graft of tissue from non-self donor of the same species) are the MHC molecules, and T cell recognition of allo-MHC is the primary and central event that initiates allograft rejection (Krensky et al., 1990). Several groups have used orally administered polymorphic MHC class II peptides to suppress organ transplant rejection (Sayegh et al., 1992; Weiner, 1997). A practical limitation of this approach is the cost, as synthesis of many candidate peptides in quantities sufficient for long term studies is often prohibitively expensive. Therefore, we have utilized plants as an expression system for producing MHC antigen. There are two non-allelic forms of class II in the mouse, I-A and I-E. I-A is composed of a 34–36 kDα chain non-covalently linked to a 27–29 kD β chain subunit. Each chain is composed of two external globular domains and a connecting peptide that joins the external domains to the transmembrane and cytoplasmic domains. The cDNA containing the full length protein sequences of α or β chain subunit with an intact signal peptide were either separately or together cloned into the plant expression vector (Ma et al., 1997). Single chain and double chain transgenic tobacco plants were obtained by leaf disc cocultivation with *Agrobacterium* strains containing the respective plasmids. As shown in Figure 6, in α chain plants, antibody specific for I-Aα^k protein recognized a protein band with expected molecular weight of approximately 33 kDa, which was missing in untransformed control plants. The antibody we used showed some unspecified reactivities with plant proteins, as evidenced by several additional band in all plants. Likewise, as shown in Figure 7, in β chain plants, antibody specific for β protein recognized an unique band with the expected molecular weight. In both cases, the level of expression from single chain transgenic plants was rather low and the amount of protein produced was calculated to be less than 0.05% of total soluble protein. In double chain transgenic plants, when the same antibody specific for β protein was used, it showed a single strong protein band with expected size (Figure 8), although there was an additional smaller band detected with almost the same intensity in some transgenic lines. The level of MHC protein produced in double chain transgenic tobacco plants showed a marked increase, such that transgenic protein accounted for approximately 1–2% of total soluble protein. While further investigation is necessary, the occurrence of an unexpected smaller protein band may represent a premature form of the β polypeptide. This may suggest that it is feasible to use a plant production approach for the large-scale manufacturing of MHC

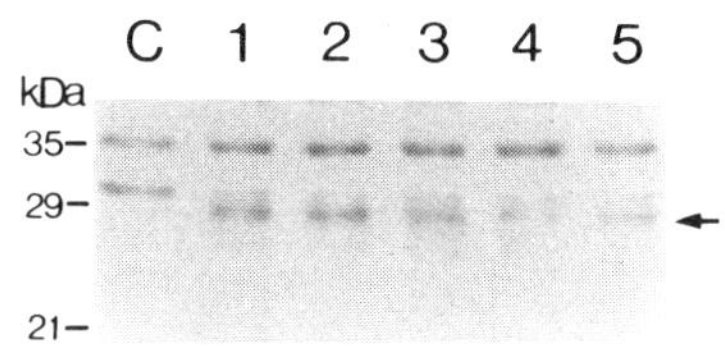

Figure 7. Western blot analysis of β chain expression in β gene tobacco plants. Protein (30 μg) extracted from tobacco leaves transformed with β gene construct were boiled, separated on a 12% SDS-polyacrylamide gel and β chain polypeptides were identified by immunoreaction with antibody specific for β chain protein. Lanes are as marked: C, extracts from untransformed control plants; lane 1 to 5: extracts from independent β gene transgenic plants. The arrow indicates the protein band with expected molecular weight.

antigen. Furthermore, expression of MHC class II antigen in edible transgenic plants may provide a simple method of inducing oral tolerance in organ transplantation.

FUTURE PROSPECTS

The demonstration that autoantigens can be produced in transgenic plants in their native, immunogenic form and that they can be further actively delivered to mucosal surfaces in a food source opens exciting possibilities for long term and cost-effective clinical studies of oral tolerance in treating autoimmune diseases. The most notable advantages of

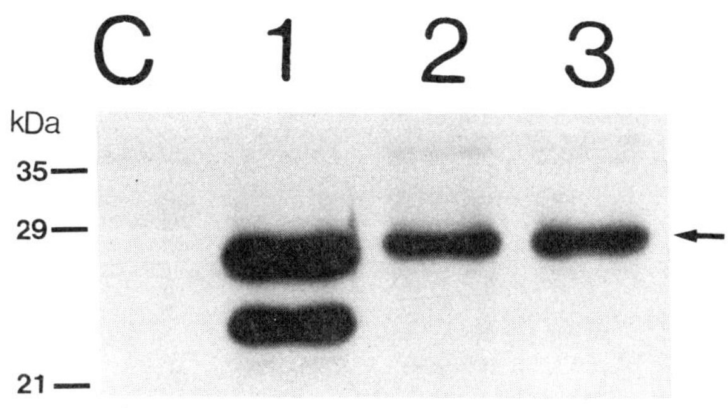

Figure 8. Western blot analysis of murine MHC class II expression in transgenic tobacco plants. Protein (30 μg) extracted from tobacco leaves transformed with both α and β gene constructs together were boiled, separated on a 12% SDS-polyacrylamide gel and MHC class II proteins were identified by immunoreaction with MHC class II antibody specific for β chain protein. Lanes are as marked: C, extracts from a wild-type plant; lane 1 to 3: extracts from independent transgenic plants identified by both southern and northern blot analysis. The arrow indicates the protein band with expected molecular weight.

plant expression systems are their capacity for high-volume production of proteins as well as their ability to secrete and accumulate proteins in certain parts of plants such as tubers, seeds and fruits. Since the tubers,seeds or fruits of many plants are edible, this make it possible to generate recombinant products that require little or no further processing, prior to oral administration. For proteins that intend to be used as autoantigens in oral tolerance or antigens in oral immunization, this provides the oppotunity to create "edible antigens", eliminating the need for expensive purification and processing of the foreign protein. In recent years, many investigators have taken these advantages of plant expression systems to develop oral vaccines against enteric bacterial and viral infections and made significant progress (Haq et al., 1995; MaGarvey et al., 1995; Ganz et al., 1996; Mason et al., 1996; Dalsgaard et al., 1997).

Unlike GAD, many mammalian proteins with the potentials for use in oral tolerance strategies have a complicated multi-unit structure and their maturation requires sophisticated post-transcriptional as well as post-translational modifications. For example, class II collagen, which has been used successfully as an oral tolerogen in the treatment of reumatoid arthristis, has three glycosylated polypeptide chains that are intertwined to form a triple-stranded helical rod. It is also recognized that expressing some antigenic proteins in transgenic plants will be challenging. It is evident that some recombinant proteins fail to accumulate in plants or have such a low expression level that hardly reaches the minimal dosing necessary for inducing immune responses. Therefore, selection of appropriate plant species, optimization of expression levels and stabilization of the foreign proteins in plant cells and/or tissues are some areas of ongoing efforts.

The prospects for plant application in clinical therapies are excellent. It is anticipated that the area of oral immunization by plants will attract considerable attention in future for the potential preventative treatment of multiple sclerosis, rheumatoid arthritis, lupus, uveitis, diabetes and perhaps transplantation rejection.

ACKNOWLEDGMENTS

The authors wish to thank D.-L. Zhao, Z.-Q. Yin for their technical assistance, Dr. Bruce Gill for a critical review and comments of the chapter; Dr. J. C. Carrington for providing the plasmid pRTL-2GUS. This work was supported by funds from the London Health Sciences Centre MOTS Research Fund and University Hospital Foundation.

REFERENCES

Atkinson, M. A.; Maclaren, N. K. Autoantibodies in nonobese diabetic mice immunoprecipitate 64,000-M_r islet antigen. *Diabetes* **1988**, *37*, 1587–1590.

Atkinson, M. A.; Maclaren, N.K. The pathogenesis *of* insulin-dependent diabetes millitus. *New. Eng. J. Med.* **1994**, *331*, 1428–1436

Atkinson, M. A.; Bowmam, M. A.; Campbell, L.; Darrow, B. L.; Kaufman, D. L.; Maclaren, N. K. Cellular immunity to a determinant common to glutamate decarboxylase and Coxsackie virus in insulin-dependent diabetes. *J. Clin. Invest.* **1994**, *94*, 2125–2129

Bosi, E.; Botazzo, G.F. Autoimmunity in insulin-dependent diabetes mellitus. 1995. *Clin. Immunother.* **1995**, *3*, 125–135

Bu, D. F.; Erlander, M.G. Two human glutamate decarboxylases 65-kDa GAD and 67- kDa GAD, are each encoded by a single gene. *Pro. Natl. Acad. Sci. USA.* **1992**, *89*, 2115–2119.

Burstein, H. J.; Abbas, A. K. In vivo role of interleukin 4 in T cell tolerance induced by aqueous protein antigen. *J. Exp. Med.* **1993**, *177*, 457–463.

Castano, L.; Eisenbarth, G. S. Type-I diabetes: a chronic autoimmune disease of human, mouse, and rat. *Annu. Rev. Immunol.* **1990**, *8*, 647–679.

Challacombe, S. J,; Tomasi, T. J. Systemic tolerance and secretory immunity after oral immunization. *J. Exp. Med.* **1980**, *152,* 1459–1472

Chase, M. Inhibition of experimental drug allergy by prior feeding of the sensitizing agent. *Proc. Soc. Exp. Bio. Med.* **1946**, *61*, 257–259.

Dalsgaard, K. Uttenthal, A.; Jones, T. D.; Xu, F.; Merryweather, A.; Hamilton, W. D. O.; Langeveld, J. P. M.; Boshuizen, R. S.; Kamstrup, S.; Lomonossoff, G. P.; Porta, C.; Vela, C.; Casal, J. I.; Meloen, R. H.; Rodgers, P. B. Plant-derived vaccine protects target animals against a viral disease. *Nature Biotechnol.* **1997**, *15*, 248–252.

Ellitt, J. F.; Qin, H.Y.; Bhatti, S.; Smith, D. K.; Singh, R. K.; Fraga, E.; Dillon, T.; Lauzon, J.; Singh, B. Immunization with the larger isoform of mouse glutamic acid decarboxylase (GAD67) prevents autoimmune diabetes in NOD mice. *Diabetes* **1994**, *43*, 1494–1499

Finkelman, F.D.; Holmes, J.; Katona, I.M.; Urban, Jr.; Beckmann, M.P.; Park, L.S.; Schooley, K.A.; Coffman, R.L.; Mosmann, T.R.; Paul, W.E. Lymphokine control of in vivo immunoglobulin isotype selection. *Annu Rev. Immunol.* **1990**, *8*, 303–333.

Fukaura, H.; Kent, S. C.; Pietrusewicz, M. J.; Khoury, S. J.; Weiner, H. L.; Hafler, D. A. Induction of circulating myelin basic protein and proteolipid protein specific TGF-β-1 secreting T cells by oral administration of myelin in multiple sclerosis patients. *J. Clin. Invest.* **1996**, *98*, 70–77.

Ganz, P. R.; Dudani, A. K.; Tackaberry, E. S.; Sardana, R.; Sauder,C.; Cheng, X. Y.; Altosaar, I. Expression of human blood proteins in transgenic plants: the cytokine GM-CSF as a model protein. In *transgenic plants: A production system for industrial and pharmaceutical proteins*. Owen, M. R. L., Pen, J., eds.; John Wiley & Sons, New York, 1996, pp.281–297.

Germain, R. N.; Margulies, D. M. The biochemistry and cell biology of antigen processing and presentation. *Annu. Rev. Immunol.* **1993**, *11*, 403–450.

Haq, T.A., Mason, H.S., Clements, J.D. and Arntzen, C.J. Oral immunization with a recombinant bacterial antigen produced in transgenic plants. *Science* **1995**, *268*, 714–716

Kaufman, D. L.; Clare-Salzler, M.; Tian, J.; Forsthuber, T.; Ting, G. S.; Robinson, P.; Atkinson, M. A.; Sercarz, E. E.; Tobin, A. J.; Lehmann, P. V. Spontaneous loss of T cell tolerance to glutamic acid decarboxylase in murine insulin dependent diabetes. *Nature* **1993**, *366,* 69–72.

Klein, J. In *Natural history of the major hiscocompatibility complex*. John Wiley and Sons, New York, **1986**.

Krensky, A. M.; Crabtree, W. G.; Davis, M. M.; Parham, P. T-lymphocyte-antigen interactions in transplant rejection. *New Engl. J. Med.* **1990**, *322,* 510–517.

Liblau, R. S.; Singer, S. M.; McDevitt, H. O. Th1 and Th2 CD4+ T cells in the pathogenesis of organ-specific autoimmune diseases. *Immunol. Today* **1995**, *16*, 34–38.

Ma, S. W.; Zhao, D. L.; Yin, Z. Q.; Mukherjee, R.; Singh, B.; Qin, H.Y.; Stiller, C.R.; Jevnikar, A.M. Transgenic plants expressing autoantigens fed to mice to induce oral immune tolerance. *Nature Medicine* **1997**, *3*, 793–796

Maclaren, N. K.; Atkinson, M. A. Insulin-dependent diabetes mellitus: the hypothesis of molecular mimicry between islet cell antigens and microorganisms. *Mol. Medi.Today* **1997**, *2*, 76–83.

McGarvey, P. B.; Hammond, J.; Dienelt, M. M.; Hooper, D. C.; Fu, Z. F.; Dietzschold, B.; Koprowski, H.; Michaels, F. H. Expression of the rabies virus glycoprotein in transgenic tomatoes. *Bio/Technology* **1995**, *13*, 1484–1487.

Makino, S.; Hunimoto, K.; Muraoka, Y.; Katagiri, K. Effects of castration on the appearance of diabetes in NOD mice. *Exp. Anim.* **1981**, *3*, 137–140.

Mason, H. S.; Ball, J. M.; Shi, J.-J.; Jiang, X.; Estes, M. K.; Arntzen, C. J. Expression of Norwalk virus capsid protein in transgenic tobacco and potato and its oral Immunogenicity in mice. *Proc. Natl. Acad. Sci. USA* **1996**, *93*, 5335–5340.

Mossmann, T.R.; Coffman, R.L. Th1 and Th2 cells: different patterns of lymphokine secretion lead to different functional properties. *Annu. Rev. Immunol.* **1989**, *7*, 145

Mowat, A.M. The regulation of immune responses to dietary protein antigens. *Immunol. Today* **1987**, *8*, 93–98.

Nussenblatt, R. B.; Gery, I.; Weiner, H. L.; Ferris, F. L.; Shiloach, J.; Remaley, N.; Perry, C.; Caspi, R. R.; Hafler, D. A.; Foster, C. S.; Whitcup, S. M. Traetment of uveitis by oral administration of retinal antigens: results of a phase I/II randomized masked trial. *Am. J. Ophthalmol.* **1997**, *123*, 583–592.

Porcelli, S. Molecular mimicry and the generation of autoimmune diseases. Rheumatol. Rev. **1993**, *2*, 41–50.

Powrie, F.; Coffman, R. L. Cytokine regulation of T-cell function: potential for therapeutic intervention. *Immunol. Today* **1993**, *14*, 270–274.

Rabinovich, A. Immunoregulatory and cytokine imbalances in the pathogenesis of IDDM. Therapeutic intervention by immunostimulation. *Diabetes* **1994**, *43*, 613–621.

Sayegh, M. H.; Khoury, S. J.; Hancock, W. W.; Weiner, H. L.; Carpenter, C. B. Induction of immunity and oral tolerance with polymorphic class II major histocompatibility complex allopeptides in the rat. *Proc. Natl. Acad. Sci. USA.* **1992**,*89*, 772–7766.

Schwartz, R. H. T lymphocyte recognition of antigen in association with gene products of the major histocompatibility complex. *Annu. Rev. Immunol.* **1985**, *3*, 237–261.

Scott, M.; Schatz, D.; Atkinson, M.; Krischer, J.; Mehta, H.; Vold, B.; Maclaren, N. GAD65 autoantibodies increase the predictability but not the sensitivity of islet cell and insulin antoantibodies for developing insulin dependent diabetes mellitus patients. *J. Autoimmun.* **1994**, *7*, 865–872.

Szopa, T. M.; Tichener, P. A.; Portwood, N. D.; Taylor, K. W. Diabetes mellitus due to viruses. Some recent developments. *Diabetologia* **1993**, *36*, 687–695.

Tian, J.; Atkinson, M. A.; Clare-Salzler, M.; Herschenfeld, A.; Forsthuber, T.; Lehmann, P.; Kaufman, D. L. Nasal administration of glutamic acid decarboxylase (GAD65) peptides induces Th2 responses and prevents murine insulin-dependent diabetes. *J. Exp. Med.* **1996**, *183*, 1561–1567.

Tisch, R.; Yang, X. D.; Singer, S. M.; Liblau, R. S.; Fugger, L.; McDevitt, H. O. Immune response to glutamic acid decarboxylase correlates with insulitis in nonobese diabetic mice. *Nature* **1993**, *366*, 72–75.

Titus, R. G.; Chiller, J. M. Oral induced tolerance. Definition at the cellular level. *Int. Arch. Allergy App. Immunol.* **1982**, *65*, 321–326.

Weiner, H. L. Oral tolerance for the treatment of autoimmune diseases. *Annu. Rev. Med.* **1997**, *48*, 341–351.

Weiner, H.L.; Friedman, A.; Miller, A.; Khoury, S.J.; Al-Sabbagh, A.; Santos, L.; Sayegh, M.; Nussenblatt. R.; Trentham, D.E.; Hafler, D.A. Oral tolerance: Immunologic mechanisms and treatment of animal and human organ-specific autoimmune diseases by oral administration of autoantigens. *Annu. Rev. Immunol.* **1994**, *12*, 809–837.

Wells, H. Studies on the chemistry of anaphylaxis. III. Experiments with isolated proteins, especially those of hen's egg. *J. Infect. Dis.* **1911**, *9*, 147–151.

Wicker, L. S.; Miller, B. J.; Mullen, Y. Transfer of autoimmune diabetes mellitus with splenocytes from non-obese diabetic (NOD) mice. *Diabetes* **1986**, *35*, 855–860.

Wicker, L. S., Todd, J. A.; Peterson, L. B. Genetic control of autoimmune diabetes in the NOD mouse. *Annu. Rev. Immunol.* **1995**, *13*, 179–200.

Wucherpfening, K. W.; Strominger, J. L. Molecular mimicry in T-cell mediated autoimmunity: viral peptides activate human T cell clones specific for myelin basic protein. *Cell* **1995**, *80*, 695–705.

15

PRODUCTION OF FOOD RELATED COLORANTS BY CULTURE OF PLANT CELLS

The Case of Betalains

A. Jiménez-Aparicio[1*] and G. Gutiérrez-López[2]

[1]CEPROBI-IPN
P.O. Box 24 Yautepec, C.P. 62731, Morelos (México)
[2]DGIA
ENCB-IPN Carpio and Plan de Ayala s/n México 11340 D.F. (México)

ABSTRACT

Betalains are vacuole pigments produced solely by Caryophyllales plants and few superior fungi. For many years, betalains have been used as food colorants. The aim of the present chapter is to present relevant aspects related to betalains characteristics and feasibility of their industrial production in bioreactors.

INTRODUCTION

The Food and Drug Administration (FDA) and the *Codex Alimentarius,* have restricted and banned the usage of an important number of synthetic food colorants, mainly red and yellow, due to potential risks involved when used as food additives. This situation has intensified interest in natural pigments, mainly in those considered as *non-conventional.* This is the case of pigments produced by microorganisms such as *Monascus anka* and *Aspergillus japonicus*, by flowers such as *Gardenia jasminoides*, insects such as *Laccifera lacca* and vegetable cells and tissues as *Beta vulgaris* (Ilke, 1987; Spears,1988; Francis, 1994; Havkin-Frenkel et al., 1997).

Natural pigments have some technological advantages over those from other sources (Spears, 1988). For example, they present good dispersability and solubility, which allows

* E-mail: AAPARICI@VMREDIPN.IPN.MX

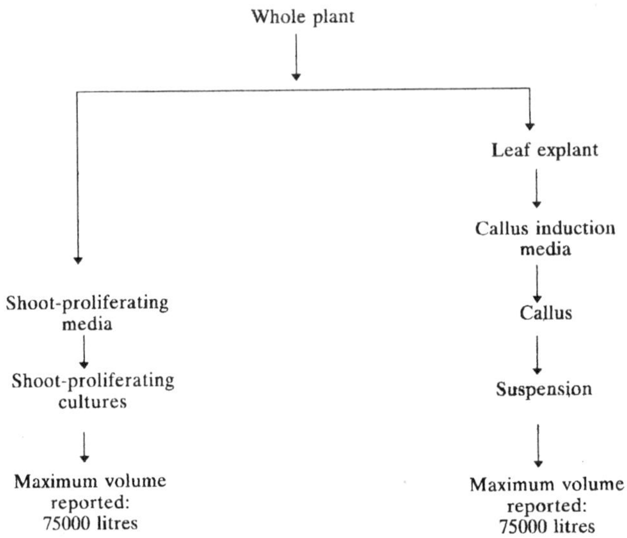

Figure 1. Simplified scheme describing the process to obtain a secondary metabolite using PCC.

the preparation of combined colors for application in a wide variety of food commodities such as sauces, snack-foods, beverages, soups, confectionery, meat and dairy products (Hayward, 1992). In general, natural colors are unstable when subjected to heat, some ions and pH change (Gullett, 1991).

Natural colorants may be classified as value-added products suitable for commercialization (Pras, 1992; Humphrey, 1994; Dörnenburg and Knorr, 1997). Vegetable cell culture, including root transformation, helps in the production of these pigments and there are reports on biosynthesis of antocyanins, carotenoids, antraquinones and betalains using different plants (Stafford, 1991). Examples of pigments obtained by cell and tissue culture are showed in table I.

Some of the strategies for developing a typical system for vegetable cell culture are shown in Figure 1. The following main stages are illustrated (Stafford, 1991; Havkin-Frenkel *et al.*, 1997): a) obtaining of aseptic culture from plants which produce the pigment in high quantities b) establishing of culture conditions to obtain non-differentiated tissue or "callus" by manipulating the chemical, physical and biological properties of the culture, c) selection of stable strains and their preservation, d) suspended cells culture, e) scaling up of the culture to a bioreactor level, f) application of strategies to allow the excretion of the pigments from the cells, and g) bioengineering studies to select the most suitable downstream process for the recovery of the pigment at the required level of purity (Dörnenburg and Knorr, 1995).

THE BETALAINS

Generalities

Betalains are natural pigments used as food colorants (Sobkowska et al., 1991). Their existence has been recognized for many years and used as adulterants in red wine produc-

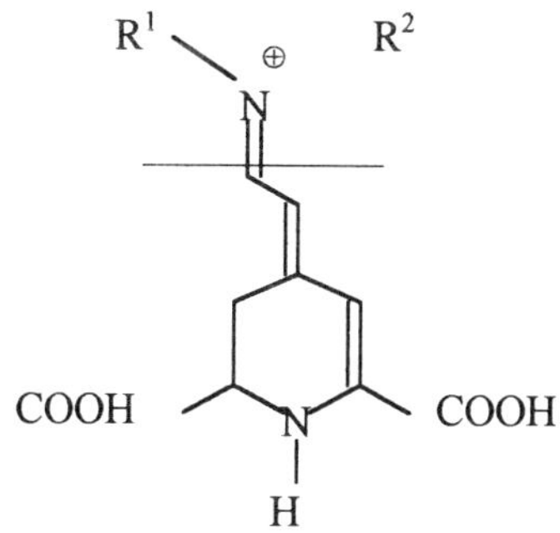

Figure 2. General betalain formula. (a) Betalamic acid moiety of all molecules. (b) R^1-N-R^2 / residue of cyclo-DOPA (higher plants) or cyclized stizolobic acid (*Amanita muscaria*) in betacyanins. (c) R^1-N-R^2 / residue of an amino acid or amine in betaxanthins.

tion. Apart from being an excellent colorant for wine, pokeberry juice (*Phytolacca americana*) also contains a laxative substance (von Elbe, 1977). Given their toxicity, this extract was banned by law in France as a food colorant since 1892. The best known pigment of this kind is the betanin from red-beet root, *Beta vulgaris* L. (Piatelli and Minale, 1964).

Betalains are used as natural colorants for use in foods, pharmaceutical products and cosmetics. These include: ice cream, dairy products, processed cold meat, canned fruits and vegetables and confectionery products (von Elbe, 1977; Spears, 1988). Nevertheless, the complexity of the betalanin molecule and difficulties to produce it by chemical synthesis as well as its sensibility to temperature, light and alkaline conditions, limits its use (Strack *et al.*, 1993).

Betalains are found in nine of the eleven Caryophyllales families (Rosendal-Jensen, 1989), as well as in certain superior fungi such as *Amanita, Hygrocybe* and *Hygrosporus* (Piatelli, 1981; Strack *et al.*, 1993). Betalains replace anthocyanins as colorants in different parts of the plant, *i.e.* roots of *Beta vulgaris*, stem of *Portulaca grandiflora*, fruits of *Opuntia ficus-indica* and seeds of *Amaranthus tricolor* (Piatelli, 1976; 1981). Betalains are generally found in solution within the vacuole, and can be located in the betalainoplasts, spherical organelles which are analogous structures as those described as antocyanoplasts (Strack *et al.*, 1993).

Betalains can be divided into two structural groups (Piatelli, 1981): betacyanins and betaxanthins. They have a pigmentation which goes from red to red-purple in betacyanins and orange to yellow in betaxanthins. On the basis of structural and biogenetic considerations, Mabry and Dreiding (1968) proposed the term "betalain" to differentiate them from anthocyanins.

The two groups have as a basic structure, the betalamic acid and differ mainly in the radicals bonded to the main structure (Figure 2). If the groups are amino acids or amino groups, the structure corresponds to betaxanthins (maximum absorbance at 480 nm), while in betacyanins the bonded group is 3,4-dihydroxyphenylalanine (DOPA) which may or may not be glycolized and having a maximum absorbance at 540 nm (Böhm and Rink, 1988; Strack *et al.*, 1993). These substances, have at least one heterocyclic carbon molecule and their color is dictated by the chemical resonance (Piatelli, 1981). Conjugation of a substituted aromatic moiety with 1,7-diazahepamethinium chromophore, in the betalamic acid, shifts the absorption peak from 480 to 540 nm.

There are reports that suggest at least 50 different structures for betalains, which are naturally produced. In most cases, betacyanins are found as isomeric C_{15} diasteroisomeric molecules (Piatelli, 1976). However, in some plants, the glycolysed betanidins which are primary products of metabolism are not found together with the isobetanidins which are formed by instantaneous epimerization, and not by enzymatic action (Minale *et al.*, 1966).

Biosynthesis

Biosynthesis of betalains has been reviewed since the 70's and is described in more detail elsewhere (Piatelli, 1976, 1981; Böhm and Rink, 1988; Steghis and Strack, 1991; Strack *et al.*, 1993), a summary of it is presented below.

The betalains are secondary metabolites, derived from shikimic acid and from the amino acid, tyrosine (Elliot, 1983). Their biosynthesis was deduced essentially from feeding experiments with labeled precursors (Piatelli, 1981) and *in vitro* cell cultures (Leathers et al., 1992). Biogenesis of betalains from tyrosine has not been thoroughly understood and only few enzymes involved in the biosynthetic route have been identified (Heuer and Strack, 1992; Müeller *et al.*, 1996).

Figure 3 shows the proposed pathway. Betalamic acid, a chromophore pigment, can be derived from DOPA by extradiol-fission of the aromatic ring and subsequent closure of the resultant intermediate to a heterocyclic system. The transition between C_4 and C_5 would produce one molecule of seco-DOPA and subsequent bonding of nitrogen to C_3.

The betalamic acid may condense with the imino group of cyclo-DOPA to produce the red-purple betacyanins (Mabry *et al.*, 1963) or with imino or amino group of amino acids to give the yellow betaxanthins (Trezzini and Zrÿd, 1991).

Complete glycosylation takes place with cyclo-DOPA, followed by condensation with betalamic acid. Free betanidin can be stored and this compound can be the main receptor of UDP-glucose from glucosyltransferase during biosynthesis of betanin (Heuer and Strack, 1992).

Subsequent acylation of glycosylated betanidins is due to the donation of acyl groups from 1-*O*-acyl glucosides. It is important to point out that this reaction (in which an acyl transferase is involved), seems to be a biosynthetic mechanism exclusive of betalanin producing plants in contrast with those reactions analogous in the biosynthesis of flavonoids, in which the acylated flavonoids have their origin in the route of the hydroxycinnamoyl coenzyme A (Bokern and Strack, 1988).

The presence of free betalamic acid in plants which produce betaxanthins or betacyanins and betaxanthins and its absence in plants that produce only betacyanins suggests a regulatory mechanism during its biosynthesis (Girod and Zrÿd, 1991; Strack *et al.*, 1993; Rodríguez *et al.*, 1996).

A co-ordinated condensation mechanism of betalamic acid with cyclo-DOPA or glycolized-cyclo-DOPA may exist in plants that produce betacyanins, inhibiting the accumulation of betalamic acid. This is opposed to events in plants producing betaxanthins in which accumulation of this compound takes place (Rink and Böhm, 1985; Böhm and Rink, 1988; Müeller *et al.*, 1997).

Little is known on the biosynthesis of betaxanthins (Piatelli, 1981; Trezzini and Zrÿd, 1991). However, it has been suggested that interchange of basic compounds might play a role in the main biosynthetic routes (Strack *et al.*, 1993). A spontaneous condensation between betalamic acid and an amino group inside the vacuole has been considered on the grounds of genetic and biochemical studies on clones of *P. grandiflora* (Trezzini and Zrÿd, 1990; Hempel and Böhm, 1997).

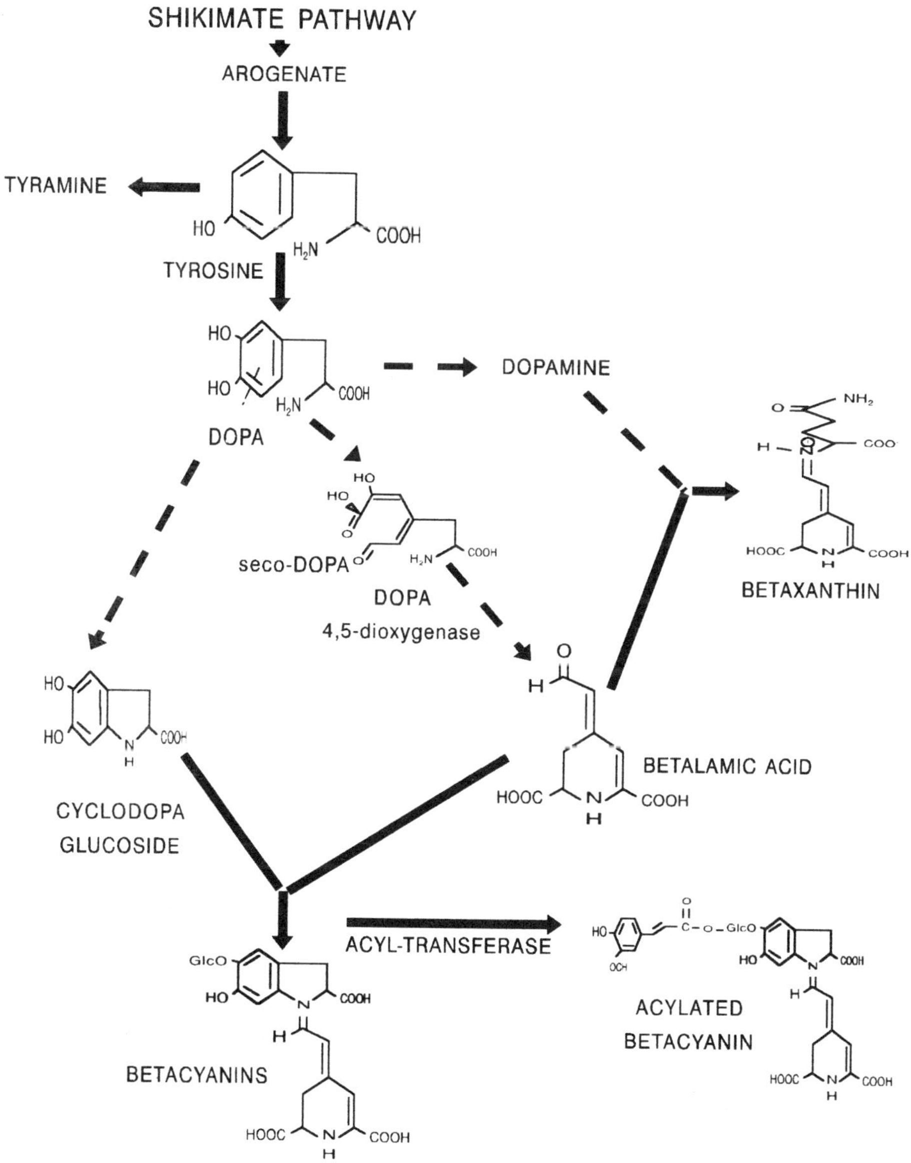

Figure 3. Probable biosynthetic routes leading from shikimic acid pathway to betalains.

Most of the physiologic mechanisms on betalanin formation and their biosynthetic mechanism are not known. However, studies of the influence of light and hormones (cytokinins) on biosynthesis of these pigments demostrate the inductive role of both factors, due to a synergistic effect between them. It is speculated that light and cytokinins act independently along the biosynthetic pathway and might be involved in photoregulation and hormone control (Bianco-Colomas and Hugues, 1990).

PRODUCTION OF BETALAINS BY PLANT CELL AND TISSUE CULTURE

General Aspects

Plant cell and tissue culture has been successfully applied in the production of natural pigments. In the case of betalains, different species of at least six of the nine families of Caryophyllales (which produce pigments *in vivo*), can also produce and accumulate the pigment *in vitro* in considerable amounts (Stafford, 1991; Leathers *et al.*, 1992). The following strategies for production processes are interesting and some of them have been used as models for biochemical and genetic studies (Böhm and Rink, 1988).

The protocol for production of betalanins is similar to that for obtaining secondary metabolites by vegetable cell culture. Betalains can be detected visually and quantified spectrophotometrically. This carry advantages over systems in which gas chromatography, hplc, etc. have to be used (Schwitzguébel *et al.*, 1991).

Culture of callus in semisolid media and of cells in suspended media have been used for betalain production (Weller and Lasure, 1982; Berlin *et al.*, 1986; Böhm *et al.*, 1991; Girod and Zrÿd, 1991; Leathers *et al.*, 1992). Also, culture of transformed roots (Kino-Oka *et al.*, 1992a; 1996) and production in a bioreactor have been reported (Schwitzguébel *et al.*, 1991; Kino-Oka *et al.*, 1992b; Khlebnikov *et al.*, 1995; Jiménez *et al.*, 1996).Cell immobilization in different supports (Knorr and Berlin, 1987) and recovery systems including membrane permeation with DMSO (Brodelius and Pedersen, 1993), electroporation (Dörnenburg and Knorr, 1993) and sonication have been studied (Kilby and Hunter, 1990).

In the above systems, isolation and characterization of tissues which are phenotypically different to those used for the production of betalains have been performed. Callus with a color ranging from yellow (betaxantin) to red-purple (betacyanins) and colorless ones can be found (Girod and Zrÿd, 1992; Leathers *et al.*, 1992; Rodríguez *et al.*, 1996).

Factors Affecting Betalanin Biosynthesis in Plant Cell Culture (PCC)

To carry out PCC at commercially exploitable levels, it is necessary to increase the current level of productivity from 6 to 15 times (Humphrey, 1994). The commercial success of some PCC processes, has been achieved by knowing key stages of the secondary metabolic route involved (Schwitzguébel *et al.*, 1991; Dörnenburg and Knorr, 1997), like in the production of shikonine from *Lithospermum erythrorhizon*. In this system, the productivity increased more than 800 fold as compared with that from the root of the plant (Dörnenburg and Knorr, 1997).

The production and preservation of high yield cellular lines are very important (Girod and Zrÿd, 1991). This is partially achieved by inhibiting permanent damages to the cell which could be passed on to the next generations and also irreversible and potentially irreversible genetic changes. The frequency of appearance which increases with the age of the culture and the intensity of the change is related to the composition of the media and the species of the plant (Leathers *et al.*, 1992). Some of the factors affecting *in vitro* biosynthesis of betalains are provided below.

Nutrients. Development of pigmented cellular lines from plants depends on the method for induction of callus tissue. Individual phenotypes must be maintained on specific culture media (Böhm and Rink, 1988).

Biosynthesis of betalains in species such as *P. grandiflora, B. vulgaris* y *C. argentea* can be achieved by changing the composition of the traditional culture media, like the MS one developed by Murashige and Skoog (1962), the B5 reported by Gamborg et al. (1968) and the SH by Schenk and Hildebrandt (1971). Attention has been drawn to carbon and nitrogen sources, some micronutrients and the addition of precursory substances, such as tyrosine, DOPA or specific amino acids (Elliott, 1983; Rink and Böhm, 1985; Trezzini and Zrÿd, 1991a).

Practical experiences, more than fundamental rules, are followed to manipulate the composition of the culture media. For example, an increase of the ammonia and nitrate levels in M5 and B5 media, increased the biomass but not the amount of pigment produced by *C. rubrum* (Berlin et al.,1986) and in *B. vulgaris* (Trejo *et al.*, 1997).

As far as phosphorus is concerned, it has been reported that in cultures of *C. Rubrum* (Berlin *et al.*,1986), *P. americana* (Sakuta *et al.*, 1986) and *O. microdasys* (Jiménez *et al.*, 1992) the amount of betalanin production increased with an increase of phosphorus above the amounts normally found in ms media.

Cultures of *B. vulgaris* showed a significant decrement of pigment when the amount of this nutrient was increased in the MS media (Leathers *et al.*, 1992). Otherwise this increase had no effect on pigment production when B5 media was used (Trejo *et al.*, 1997).

Carbon source can seriously affect betalain biosynthesis. In cultures of *C. Rubrum,* the maximum production was achieved with 2% sucrose and the maximum cellular growth was reached with 6% sucrose (Berlin et al., 1986). Maximum levels of amaranthin in *A. tricolor*, where achieved when the carbon source concentration was decreased 2 to 3 times as compared with that required for maximum cellular concentration.

In the case of *B. vulgaris,* no differences in betalanin production where found when using glucose, fructose or lactose as the carbon source, and the maximum levels of betalains where achieved when the sucrose concentration was decreased to 30% (Rodríguez *et al.*, 1994). In fact, it is suggested that in red-beet root cell culture, it is not necessary to have different growing and production culture media (Leathers *et al.*, 1992).

Phytohormones. In the case of betalains, selection of high producing cell lines, strongly depends on their hormone treatment for stabilization and maintenance by periodical re-inoculation (Böhm and Rink, 1988). Reports, have shown that if the rates of auxin/cytokinin is not altered, the proportion of betacyanins and betaxanthins does not changes significantly along the cell growing cycle (Girod and Zrÿd, 1991; Rodríguez *et al.*, 1996). So it has been possible to produce cellular lines which can present pigmentation varying from green, yellow and orange to red and purple (Figure 4).

By changing the proportion of phytohormones present in the cell, it is possible to control the direction and frequency of the interconversion of the quimeric phenotypes (Girod and Zrÿd, 1991; Rodríguez *et al.*, 1996). In the case of betalains, and contrary to the case of anthocyanins, the sensitivity to the auxin/cytokinin rate may indicate that the hormone-induced phenotypic transformation may be associated with the replication of cellular DNA and thus be a function of the proliferation process (Hirano and Komamine, 1994).

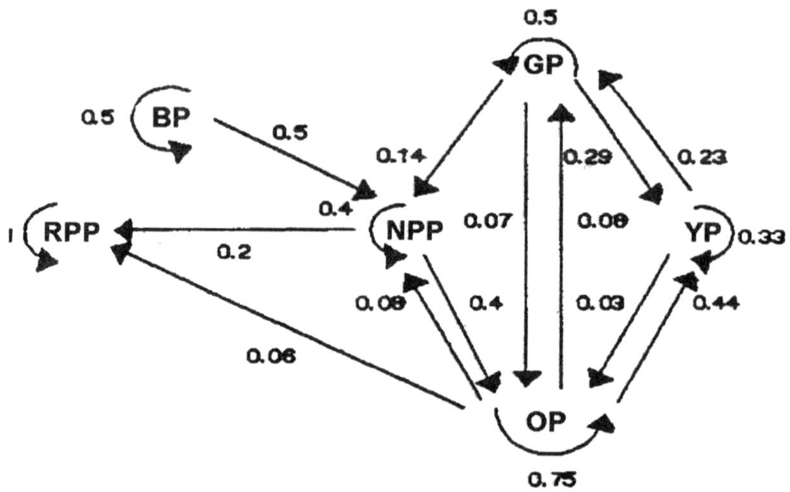

Figure 4. Stability and interconversion of different phenotypes of *Beta vulgaris*. The arrows represent transition occurring, curved-arrows the conversions and the numbers are stability of each phenotype itself. (BP = green; OP = orange; RPP = red-purple; BP = brown; YP = yellow; NPP = non-pigmented phenotype).

Light. Light is an important factor influencing the production of pigments which has not been properly understood (Piatelli, 1981; Böhm and Rink, 1988). Accumulation of pigments is faster under the action of light. For example, it has been reported that *O. microdasys* and *B. vulgaris* lose their capacity to produce pigment in absence of light (Jiménez *et al.*,1992; Leathers *et al.*, 1992).

A synergetic effect between light and presence of cytokinins has been observed in *A. Tricolor* (Bianco-Colomas and Hugues, 1990; Kishima *et al.*, 1991). In some *B. vulgaris* strains, the production of cytokinins increased three fold under condition of illumination and presence of kinetin. When cultures were deprived of light and citokinin action was inhibited, betalanin production decreased at a greater extent than that shown by cultures in the absence of light alone (Leathers *et al.*, 1992). This behavior may be due to the regulatory effect of kinetin on tyrosine availability, probably by activating or regulating phenoloxydase and DOPA-dioxigenase (Böhm and Rink, 1988). In addition, betalains production in some species of *C. Rubrum* can be increased under white or red light. This shows that a system different to phytochrome may play a role in this metabolic route (Berlin et al., 1986).

Long exposure to white light increases synthesis of chlorophyll and betalain (Böhm and Rink, 1988). When levulonic acid (chlorophyll synthesis inhibitor) was added to the media, betalain concentration decreased significantly. This may suggest that the photosynthetic system also plays an important role on betalain production. Actually, stimulus by near infrared light are only perceived by the photosynthetic system of the phosphorylation cycle (Leathers *et al.*, 1992).

Betalain production has been promoted, up to 30%, by short periodic exposure to blue light, as compared with a control system exposed to white light (Berlin *et al.*, 1986). On the other hand, alternating exposure to red and blue lights had no important effect in betalains production.

Suspended Cultures

Production of betalains in suspended cultures has been reported. The proportion of betaxanthins and betacyanins is constant (Böhm and Rink, 1988; Hunter and Kilby, 1990). Also, there are reports on cultures in which betacyanins (Bianco-Colomas and Hugues, 1990; Rodríguez *et al.*, 1994) or betaxanthin concentration does not change (Böhm *et al.*, 1991). Production of betalains is associated with cellular growth in logarithmic phase, contrary to production during stationary phase in most suspended cultures (Komamine *et al.*, 1993).

As mentioned before, suspended cells producing betalains suffer from a drawback in that the pigments are not execrated into the culture media and remain inside the vacuoles (Brodelius and Pedersen, 1993).

Production capacity and productivity are limited by the storing capacity of vacuoles and so downstream recovery is a limiting factor in production (Humphrey, 1994). Diffusion through cell walls controls most recovery schemes. Attempts to achieve an execration of metabolites must be re-directed to increase the diffusion coefficient of the pigmented substances, considering the multiplicity of factors which affect this parameter. The reversible modification of cellular permeability must also be emphasized (Dörnenburg and Knorr, 1995).

PRODUCTION OF BETALAINS IN BIOREACTORS

General Aspects

Most PCC systems have not been scaled-up to commercial levels due to the lack of engineering strategies for designing such processes (Scragg, 1992; Meijer *et al.*, 1993; Meijer *et al.*, 1994). Economical feasibility of this technology is however important for increasing the productivity of the system. This could be achieved by following one or more of the next strategies (Pras, 1992; Humphrey, 1994; Dörnenburg and Knorr, 1995):

a. Decreasing of fermentation time (decrease of cellular duplication time).
b. Increasing the specific cell productivity (addition of precursors, elicitors and control of amount and kind of nutrients and hormones).
c. Achieving high cell concentrations and expression of metabolites during stationary phase.

As far as the two first aspects are concerned, cell metabolism and genetic regulation are involved. The third aspect involves the operation of the reactor. Batch culture represents a good option for increasing productivity (Hooker et al., 1990; Meijer *et al.*, 1993) since a high cell concentration is achieved and the metabolite is mainly produced during stationary phase (Taticek *et al.*, 1991).

Vegetable cells are susceptible to damage due to shear during agitation, which affects rate of growing and viability (Meijer *et al.*, 1994). *B. vulgaris* cells have been used for betalain production in stirred tank reactors (Jiménez *et al.*, 1996) and also in bubble-free, Taylor couette tanks (Janes *et al.*, 1987), fluidized-bed reactors (Khlebnikov *et al.*,1995) and air-lift fermenters (Schwitzguébel, 1997). *C. Rubrum* was grown in a cell-immobilization system for amaranthin production (Knorr *et al.*, 1990).

Hydrodynamic Stress

Since PCC processes are based on culture conditions for bacteria and fungi, they have also been regarded as susceptible to damage by hydrodynamic stress (Hooker *et al.* 1989; Taticek *et al.*, 1991). However, this effect is not always present since vegetable cells can successfully grow in stirred tanks at 200 rpm, and shear rate values of about 1000 min^{-1} for five hours (Scragg *et al.*, 1988). Jiménez *et al.* (1997), when working with a 3 liters stirred tank and at an equivalent rotational speed at the stirrer-end of 3.5 $m \cdot s^{-1}$, evaluated the resistance to shear of a betacyanins producing strain of *B. vulgaris*. The culture reached concentrations of up to 15 of dry weight per liter with no adverse effect on cell viability which remained at above 80%.

The above works have shown that cell fragility depends mainly on the strain, but do not allow the understanding and quantification of the same under specific hydrodynamic conditions. Vegetable cells during culture differ from micoorganisms in size (40 a 100 μm) and in the tendency to grow in clusters. This creates sedimentation problems and generation of several microenvironment fractions in the tank (Takeda *et al.*, 1994).

Problems related to cell fragility are also related with mixing (Thomas, 1990). Hydrodynamics of a bioreactor, and kinetic energy transferred to the fluid contribute to determine the presence of bubbles. Thomas,(1990) has pointed out that shear stress and shear rate should be related to hydrodynamic stress. For very sensitive systems, *i.e.* animal cell culture, shear stress values of 1 - 5 N m^{-2} may significantly decrease the extent of growth (Goosen, 1992; Tramper *et al.*, 1993; Tan *et al.*, 1994).

Rheological Aspects

In a stirred tank, a velocity profile is generated around the stirrer (Curtis and Emery, 1993). Shear stress is the product of shear rate and effective viscosity (Dunlop and Namdev, 1993) and has an influence on mixing, micromixing and gas dispersion (Joshi *et al.*, 1996). There are few reports in the literature on the rheological behavior of PCC and even less on the flow properties of betalains culture media. Pseudoplasticity, Bingham plastic behavior and thixotrophy have all been observed (Tanaka, 1988; Zhong *et al.*, 1992).

B. vulgaris cultures in a stirred tank showed a newtonian behavior at the beginning of the process and changed to a pseudoplastic one when cell concentration and cell shape were modified (Jiménez *et al.*, 1996). Also, the accumulation of extracellular glycoprotein compounds plays an important role in flow behavior (Rodríguez and Galindo, 1997).

The lack of rheological studies has impaired the understanding of the mechanism of damage by shear on vegetable cells cultures. In a bioreactor, shear rate decreases exponentially with distance from the stirrer and for non-Newtonian fluids, Metzner and Otto (1957) have proposed the use of an average deformation gradient (γ_{av}) as given in equation 1.

$$\gamma_{av} = k \cdot N \quad (1)$$

In which N = stirring speed, and k = constant depending on the stirrer geometry.

In this equation, the average shear rate depends only on the type of stirrer and on the rotating velocity and is valid only for streamline flow. For PCC, Meijer *et al.*, (1994) used the maximum deformation rate defined by equation 2.

$$\gamma_{max} = 9.7\ N\ [D(T^{-1})]^{0.3}\ (D(W^{-1}) \quad (2)$$

Table 1. Examples of pigment production by plant cell and tissue cultures (NA means data not available)

Species	Pigment	Yield	Culture type	Reference
Beta vulgaris (red beet)	Betaxanthins and Betacyanins	0.1% dry weight to 0.5 % dry weight	Suspension	Leathers et al., 1992
	Red pigment	NA	Hairy root	Kino-Oka et al., 1996
	Betacyanins	134 µg / dry weight	Bioreactor	Jiménez et al., 1996
Bixa orellana (annatto)	Bixin	3-4% , dry weight	Callus	Callebaut, A. unpublished
Ajuga reptans	Cyanidin-class anthocyanins	1.0-2.5%, dry weight	Suspension	Callebaut, A. unpublished
Vitis vinifera (grape)	Anthocyanins	3-4%, dry weight	Suspension	Pepin, M.F., unpublished
Gardenia jasminoides	Carotenoids including crocin	8 times less than in-vivo levels	Callus	Nawa, T. and Othani, T., unpublished
Portulaca grandiflora	Betaxanthin	NA	Suspension	Böhm, et al., 1991
Phytolacca americana	Betacyanin	Abs 535 nm: 2.33	Suspension	Hirano and Komanine, 1990.
Amaranthus tricolor	Amaranthin	29 µmol/ g dry weight	Suspension	Bianco-Colomas and Hugues, 1990

In which D = stirrer diameter, T = tank diameter and W = thickness of the stirrer pallet.

From equation 2, it has been possible to evaluate a maximum shear rate of 589 s^{-1} for a rushton stirrer with six pallets, in which strong differences in viability and growth of culture were also observed.

The critical value for shear stress for vegetable cells is about 2 $N \cdot m^{-2}$ and for animal cells is around 0.65 $N \cdot m^{-2}$; this value of insect cells is in the range of 1 $N \cdot m^{-2}$. With these data, it is possible to define some operating conditions and design parameters of the bioreactor (Tramper *et al.*,1993).

FINAL CONSIDERATIONS

The hydrodynamic effect of agitation on vegetable cells, has been studied in bioreactors (see Table 2). However, there are insufficient data to understand the susceptibility of the cells under culture conditions since only global parameters are evaluated during production. Different geometric characteristics of the tank and control mechanisms are also reported in the literature.

An alternative way to evaluate the sensibility of vegetable cells to hydrodynamic stress is by using rheometers, which might be used as model systems for establishing controlled shear rate and stress values (Kieran *et al.*, 1995). Dunlop and Namdev (1993) showed that *D. carota* cultures had a decrease of biological activity as a function of the absorbed energy by the culture, after subjected to shear rate values of 450 to 3800 min^{-1}. The same authors observed that in order to decrease 50% of the cell capacity to reinitiate growth, it was necessary to add 10^5 $J \cdot m^{-3}$, while for decreasing mitochondrial activity (electron transport) or membrane integrity, it was necessary to increase input energy values 4 to 5 times (10^9 to 10^{10} $J \cdot m^{-3}$) of the above mentioned figures. Kieran *et al.* (1995)

Table 2 . Rheological characterization of vegetable cell cultures

Culture	Rheological behavior	Rheological parameters	Age of culture	X^a (g/l)	Size of clusters	Shape of cells	Viscometer	Reference
M. citrifolia	Thixotropic	---	16 days	15 DW	n.r.	n.r.	Haake Rotovisco RV3 system MV1-50	Wagner and Vogelman, 1977
N. tabacum	Culture: pseudoplastic Filtrate: Newtonian	0 = 0.69-0.74 := 0.9-1.2 cP	Along Kinetics	>7.7 DW	350-400	n.e.	Brookfield B. type BL Capilar viscometer	Kato et al., 1978
C. tricuspidata	Culture: pseudoplastic	0 = 0.53 K = $(x)e^{6-7}$		10-18 DW	40-2000 44-500 44-2000	n.e.	Brookfield B type BL	Tanaka, 1988
V. rosae		0= 0.70						
A. githago		0 = 0.70						
	Filtrate: Newtonian	: = 1.0-1.4 cP						
P. frutescens	Culture: Bingham Plastic	: = $(x)^4$: = Q cell shape	12 days	5-12.8 DW	37-2000	n.e.	Brookfield B. type BL	Zhong et al., 1992
N. tabacum	Pseudoplastic	: = Q cell morphology	Along Kinetics	250 FW	0.15-260	Elongated	Brookfield R1 Stormer	Curtis and Emery, 1993
D. carata		0 = 0.6						
B. vulgaris	Pseudoplastic	0 = 0.6 : = Q cell morphology	21 days	10 DW	250	Elongated	Haake RV12	Jimenez et al., 1996

a = concentration of cells. DW = dry weight. FW = fresh weight.
n.e. non evaluated
n.r. non reported
Q = function

measured the susceptibility to stress of *M. citrifolia* cultures in a capillary viscometer. They found a first order equation for cellular death, equation 3.

$$X\,(Xo)^{-1} = e^{-kt} \tag{3}$$

Where X_0 = cell viability before subjected to stress, X = cell viability after being subjected to stress and t = time.

From the above model, a linear correlation between the rate of average energy dissipation ($J.kg^{-1}\ s^{-1}$) and the specific rate of cell death is observed within values of shear stress of 25 to 350 N m^{-2}. This is important when cultivating at concentrations higher than 10 g/l, since the rheological behavior is related to mass and heat transfer effects such as oxygen transfer and mixing (Tanaka *et al.*, 1988).

ACKNOWLEDGEMENTS

This work was partially supported by the National Polytechnic Institute of Mexico (DEPI 970069), the National Council for Science and Technology of Mexico (26405N) and the Iberoamerican Network of Physical Properties of Foods (RIPFADI- CYTED).

REFERENCES

Berlin, J.; Sieg, S.; Strack, D.; Bokern, M.; Harms, H. Production of betalains by suspension cultures of *Chenopodium rubrum* L. *Plant Cell, Tiss. Org Cul.* **1986,** *5(2)*,163–174

Bianco-Colomas, J.; Hugues, M.; Establishment and characterization of a betacyanin producing cell line of *Amaranthus tricolor*: inductive effects of light and cytokinin. *J. Plant Physiol.* **1990,** *136*,734–739

Böhm, H.; Böhm, L.; Rink, E.; Establishment and characterization of a betaxanthin producing cell cultures from *Portulaca grandiflora. Plant Cell Tiss. Org. Cult.* **1991,** *26,* 75–82

Böhm, H.; Rink, E.; Betalains. Cap. 26, p.419 463. En "Cell culture and somatic cell genetics of plants" Vol 5 Academic Press, N.Y., 1988

Bokern, M.; Strack, D.; Synthesis of hydroxyccinamic acid esters of betacyanins via 1-*O*-acylglucosydes of hydroxyccinamic acids by protein preparations from cell suspension cultures of *Chenopodium rubrum* and petalas of *Lampranthus sociorum. Planta* **1988**, *174*, 101–105

Brodelius, P.; Pedersen, H.; Increasing secondary metabolites production in plant cell culture by redirecting transport. *TIBTECH.***1993**, *11*, 30–36

Curtis, R.; Emery, A.; Plant cell suspension culture rheology. *Biotechnol. Bioeng.* **1993,** *42*, 520–526

Dörnenburg, H.; and Knorr, D.; Cellular permeabilization of cultured plant tissue cultures by high electric field pulses or ultra high pressure for the recovery of secondary metabolites. *Food Biotechnol.* **1993**, *7(1)*, 35–48

Dörnenburg, H.; and Knorr, D.; Strategies for the improvement of secondary metabolite production in plant cell cultures. *Enzyme Microb. Technol.* **1995**, *17,* 674–684

Dörnenburg, H.; Knorr, D. Challenges and Opportunities of metabolite production from plant cell and tissue culture. *Food Technol.* **1997**, *51(11)*, 47–54

Dunlop, E.; Namdev. P.; Effect of fluid forces on plant cell suspension. *Abstracts of Papers.* 3rd International Conference on Bioreactor and Bioprocess Fluid Dynamics; Nienow, A. (Ed.); Mechanical Eng. Pub. Ltd., London, *1993;* pp. 447–455

Elliott, C. D.; The pathway of betalain biosynthesis: Effect of cytokinin on enzymatic oxidation and hydroxylation of tyrosine in *Amaranthus tricolor* seedlings. *Physiol. Plant.* **1983**, *59*, 428–437

Francis, J.; Food colorants today. *World Ingred.* **1994**, *Oct./ Nov.* 8–9, 11

Gamborg, O. L.; Miller, R. A.; Ojima, K.; Nutrient requirements of suspension culture of soybean cells. *Exp. Cell Res.* **1968**, *50*, 151–156

Girod, P.A.; Zrÿd, J.; Secondary metabolism in cultured red beet (*Beta vulgaris*) cells: differential regulation of betaxanthin and betacyanin biosynthesis. *Plant Cell Tiss Org. Cult.***1991**, *25*, 1–12

Goosen, M.; Large-scale insect cell culture. *Current Opinion Biotechnol.* **1992**, *3*: 99–104

Gullett, E.A.; Color and Food. In *Encyclopedia of food science and Technology.* Vol.1. (Ed) Y.H. Hui; John Wiley and Sons, New York, 1991; pp.452–460

Havkin-Frenkel, D; Dorn, R.; Leustek, T.; Plant tissue culture for production of secondary metabolites. *Food Technol.* **1997,** *51(11)*, 56–59

Hayward, P. M.; Colorantes (Food colorings). *Tecnol. Aliment.* **1992**, *27*(4–5), 34–43

Hempel, J.; Böhm, H.; Betaxanthin pattern of hairy roots from *Beta vulgaris* var. Lutea and its alteration by feeding of amino acids. *Phytochem.* **1977**, *44(5)*, 847–852

Heuer, S.; Strack, D.; Synthesis of betanin from betanidin and UDP-glucose by protein preparation from cell suspension cultures of *Dorotheanthus bellidiformis* (Burm, f). *Planta,* **1992**, *186*: 626–628

Hirano, H.; Komamine, A.; Correlation of betacyanin synthesis with cell division in cell suspension cultures of *Phytolacca americana. Physiol. Plant.* **1994**, *90,* 239–245

Hooker, B.; Lee, J.; An, G.; Response of plant tissue culture to high shear environment. *Enzyme Microb. Technol.* **1989**, *11,* 484–490

Hooker, B.; Lee, J.; An, G.; Cultivation of plant cells in a stirred vessel: effect of impeller design. *Biotechnol. Bioeng.* **1990**, *35*, 296–304

Humphrey, A.; Plant cells as chemical factories: control and recovery of valuable products. Advances in Bioprocess Engineering; Galindo, E.; Ramírez T. (Eds.); Kluwer Academic Pub., *1994,* 103–107

Hunter, C. S.; Kilby, N. J.; Betanin production and release *in vitro* from suspension cultures of *Beta vulgaris.* Cap. 49.; Methods in molecular biology. Pollard, W.; Walker, J.M.; Humana Press. *1990*, 545–554

Ilker, R.; *In vitro* pigment production: an alternative to color synthesis. *Food Technol.* **1987**, *41(4)*, 70–72

Janes, D. A.; Thomas, N.; Callow, J.; Red beet batch culture demonstration of a bubble-free taylor-couette bioreactor. *Biotechnol. Technol.* **1987**, *1(4),* 257–262

Jiménez, A. A.; Dávila, O. G.; Villegas, G. T.; Del Villar, M. A.; Obtención de colorantes de interes alimentario por cultivo de células de *opuntia microdasys* (Lehm) Pfeiff. *Rev. Latinoamer. Quim.* **1992**, *23(1)*, 5–8

Jiménez, A. A.; Gutiérrez, L. G.; Rodríguez, M. M.; Caracterización reológica e hidrodinámica del cultivo de células de betabel (*B. vulgaris* L.) productoras de betalainas en un biorreactor tipo tanque agitado. Propiedades físicas de los alimentos, Vol. 1; Hübinger, M.; Murr, F. X.; Aguilera, J. M. (Eds).; CYTED-UPV, Vlencia, España, 1996; 105–112

Jiménez, A. A.; Rodríguez, M. M.; Gutiérrez, L. G. *Unpublished data.* **1998**

Joshi, J.; Elias, C.; Patole, M.; Role of hydrodynamic shear in the cultivation of animal, plant and microbial cells. *Biochem. Eng. J.* **1996**, *62*, 121–141

Kato, A.; Kawazoe, S.; Soh, Y.; Viscosity of the broth of tobacco cells in suspension culture. *J. Ferment. Technol.* **1978**, *56(3),* 224–228

Khlebnikov, A.; Dubois, B.; Kut, O. M.; Prenosil, J. E.; Growth and productivity of *Beta vulgaris* cell culture in fluidized bed reactors. *Bioprocess Eng.* **1995**, *14*, 51–56

Kieran, P,; O'Donnell, H.; Malone, D.; MacLoughlin, P.; Fluid shear effects on suspension cultures of *Morinda citrifolia. Biotechnol. Bioeng.* **1995**, *45*, 415–425

Kilby, N. J.; Hunter, C.S.; Repeated harvest of vacuole-located secondary product from in vitro grown plants cells using 1.02mhz ultrasound. *Appl. Microbiol. Biotechnol.* **1990**, *33*, 448–451

Kino-Oka, M.; Mine, K.; Taya, M.; Tone, S.; Bioreactor-based cultured of plant hairy roots for production and recovery of pigments. *Biochem. Eng. 2001* **1992a**, 296–298

Kino-Oka, M.; Hongo, Y.; Taya, M.; Tone, S.; Culture of red beet hairy root in bioreactor and recovery of pigment release by oxygen starvation. *J. Chem. Eng. Jpn.* **1992b**, *25(5)*, 490–495

Kino-Oka, M.; Tone, S.; Extracellular production of pigment from red beet hairy roots accompanied by oxygen preservation. *J. Chem. Eng. Jpn.* **1996**, *29 (3),* 488–493

Kishima, Y.; Nozaki, K.; Akasi, R.; Adachi, T.; Light inducible pigmentation in *Portulaca* callus; selection of high betalain producing cell line. *Plant Cell Rep.* **1991**, *10*, 304–307

Knorr, D.; Berlin, J.; Effects of immobilization and permeabilization procedures on growth of *Chenopodium rubrum* cells and amaranthin concentration. *J. Food Sci.* **1987**, *52(5),* 1397–1400

Knorr, D.; Beaumont, M.; Caster, C.; Dörnemburg, H.; Gross, B.; Pandya, Y.; Romagnoli, L. G.; Plant tissue culture for the production of naturally derive food ingredients. *Food Technol.* **1990**, *44(6)*, 71–79

Komamine, A.; Hirano, H.; Kagewata, K.; Suda, J.; Hirose, M.; Roy, R.; Sugiyama, M.; Sakuta, M.; Regulation mechanisms of biosynthesis of betacyanin and anthocyanin in relation to cell division activity in suspension cultures. *Proceedings,* Primary and Secondary Metabolism of Plants and Plant Cell Cultures, Schripsema, J.; Verpoorte, R. (Eds).; Leiden, The Netherlands, 1993, p.105

Leathers, R.; Davin, C.; Zrÿd, J.; Betalain producing cell cultures of *Beta vulgaris* L. (Red beet). *In Vitro Cell Dev. Biol.* **1992**, *28*, 39–45

Mabry, T. J.; Taylor, A.; Turner, B. L.; The betacyanins and their distribution. *Phytochem.* **1963**, *2*, 61–64

Mabry, T. J.; Dreiding, A. S.; The betalaines. Recent advances in phytochemistry, Vol.1. Mabry, R.; Alston, E.; Runeckles, V. C. (Eds).; Appleton Century Crofts, New York, 1968; pp. 145–160

Meijer, J.; ten Hoopen, H.; Luyben, K.; Libbenga, R.; Effects of hydrodynamic stress on cultured plant cells: A literature survey. *Enzyme Microb. Technol.* **1993**, *15,* 234–238

Meijer, J.; ten Hoopen, H.; Jan-Gameren, Y.; Luyben, K.; Libbenga, R.; Effects of hydrodynamic stress on the growth of plant cells in batch and continuous culture. *Enzyme Microb. Technol.* **1994**, *16,* 467–477

Metzner, A.; Otto, R.; Agitation of non-Newtonian fluids. *A.I.Ch.E.J.* **1957**, *3(1),* 3–9

Minale, L.; Piatelli, M.; De-Stefano, S.; Nicolaus, R. A.; Pigments of Centrospermae VI. Acylated betacyanins. *Phytochem.* **1966**, *5,* 1037–1052

Müeller, L. A.; Hinz, U.; Zrÿd, J. P.; The formation of betalamic acid and muscaflavin by recombinant DOPA-dioxygenase from *Amanita. Phytochem.* **1997**, *44(4),*567–569

Müeller, L. A.; Hinz, U.; Uze, M.; Sautter, C.; Zrÿd, J. P. Complementation of betalain biosynthesis in *P. grandiflora* by a fungal DOPA-dioxigenase. *Experientia* **1996**, *52(2):*A20

Murashige, T.; Skoog, F.; A revised medium for rapid growth-promoting activity of various auxins. *Plant Physiol.* **1962**, *57,* 35–41

Piatelli, M.; Minale, L.; Pigments of Centrospermae I. Betacyanins from *Phyllocactus hybridus* Hort. And *Opuntia ficus-indica. Phytochem.* **1964**, *3,* 307–311

Piatelli, M.; Betalains. Chemistry and biochemistry of plant pigments. Vol. 1. Goodwin, T.W. (Ed); Academic Press, New York, 1976; pp.560–596

Piatelli, M.; The betalains: structure, biosynthesis and chemical taxonomy. The biochemistry of plants: A comprehensive treatise. Vol. 17. Secondary plant products. Conn, E.E. (Ed).; Academic Press, New York, 1981; pp.557–575

Pras, N.; Bioconversion of naturally occurring precursors and related synthetic compounds using plant cell cultures. *J. Biotechnol.* **1992**, *26,* 29–62

Rink, E.; Böhm, H. Changed betaxanthin pattern in violet flowers of *Portulaca grandiflora* after the feeding of DOPA. *Phytochem.* **1985**, *24(7),* 1475–1477

Rodríguez, M.; Jiménez, A.; Dávila, G.; Sepúlveda, G.; Effect of carbon source in cell suspension culture of *Beta vulgaris. Biotechnol Lett.* **1994**, *16(8),* 853–858

Rodríguez, M M.; Jiménez, A. A.; Sepúlveda, J. G.; Development of an experimental model for the study of the biosynthesis of betalains (natural colorants) from tissue culture of *Beta vulgaris. Papers* 2nd International Symposium on Natural Colorants. Hereld, P. C. (Ed).; S.I.C. Publishing Company, CT, 1996; pp. 409–416

Rodríguez, M.; Galindo, E.; Efecto del estrés hidrodinámico sobre cultivos de células de *Beta vulgaris. Abstracts of Papers*, Proc. 2nd International Engineering Bioprocesses of Sociedad Mexicana de Biotecnología y Bioingeniería; Mazatlán, México, 1997; p.155

Rosendal-Jensen, S.; Juhl-Nielsen, B.; Dahlgren, R.; Use of chemistry en plant classification. *Rev. Latinoamer. Quim.* **1989**, *Suppl 1,* 66–89.

Sakuta, M.; Takagi, T.; Komamine, A.; Growth related accumulation of betacyanin in suspension cultures of *Phytolacca americana* L. *J. Plant Physiol.* **1986**, *125,* 337–343

Schenk, U. R.; Hildebrandt, C. A.; Medium and techniques for induction and growth of monocotyledoneus and dicotyledoneus plan cell cultures. *Can. J. Bot.* **1972**, *50,* 199–204

Scragg, A.; Allan, E.; Leckie, F.; Effect of shear on the viability of plant cell suspension. *Enzyme Microb. Technol.* **1988**, *10,* 361–367

Scragg, A.; Large-scale plant cell culture: methods, applications and products. *Current Opinion Biotechnol.* **1992**, *3,* 105–109

Schwitzguébel, J. P.; Zrÿd, J. P.; Leathers, R.; From plant cells to biotechnology. *Swiss Biotech.* **1991**, *9(1),* 17–24

Schwitzguébel, J. P; Biosynthesis of betalains by cultivation of red beetroot cells in bioreactors. *Abstract* at WEB site *http://dgrwww.epfl.ch/GB/DP/JPS-1.html,* 1997

Sobkowska, E.; Czapski, L.; Kaczmarek, R.; Red table beet pigment as food colorant. *Int. Food Ing.* **1991**, *3,* 24–28

Spears,K.; Developments in food colorings: the natural alternatives. *TIBTECH* **1988**, *6(8),* 1973–1980

Stafford, A.; The manufacture of food ingredients using plant cell and tissue cultures. *Trends Food Sci. Technol.* **1991**, *5(2),* 116–122

Steghis, W.; Strack, D.; Betalains. The alkaloids. Vol. 39; Brossi, A. (Ed.); Academic Press, Orlando, FL, 1991; pp. 1–62

Strack, D.; Steglich, W.; Wray, V.; Betalains. Methods in plant biochemistry. Vol. 8; Academic Press, Orlando, FL, 1993; pp.421–450

Takeda, T.; Seki, M.; Furusaki, S.; Hydrodynamic damage of cultured cells of *Carthamus tinctorius* in a stirred tank reactor. *J. Chem. Eng.* **1994**, *27(4),* 466–471

Tan, W.; Dai, G.; Chen, Y.; Quantitative investigations of cell-bubble interactions using a foam fractionation technique. *Cytotechnology* **1994**, *15(1–3)*, 321–328

Tanaka, H.; Technological problems in cultivation of plant cells at high density. *Biotechnol. Bioeng.* **1988**, *23*, 1203–1218

Tanaka, H.; Semba, H.; Jitsufuchi, T.; Harada, H.; The effect of physical stress on plant cells in suspension cultures. *Biotechnol Lett.* **1988**, *10(7)*, 485–490

Taticek R.; Moo-Young, M.; Legge, L.; The scale-up of plant cell culture: engineering considerations. *Plant Cell Tiss. Cult.* **1994**, *24*, 139–158

Thomas, C.; Problems of shear in biotechnology. Chemical engineering problems in biotechnology. Winkler, M. (Ed.); Elsevier, Sci. Pub. England, 1990; pp. 25–93

Tramper, J.; Gooijer, D.; Vlak, J.; Scale-up considerations and bioreactor development for animal cell cultivation. Insect cell culture engineering. Goosen, M.; Daugulis, A.; Faulkner, P. (Eds.); Marcel Dekker, Inc., New York, 1993; pp. 139–177

Trejo, T. G.; Rodríguez, M. M.; Jiménez, A. A.; De Jesus, S. A.; Martínez, B. B.; Sepúlveda, J. G.; Salcedo, M. G.; Estudios para la producción de betalainas en cultivos de *Beta vulgaris*: Efecto de la fuente de carbono y nitrógeno. *Abstracts of Papers*, Proc. 2nd International Engineering Bioprocesses of Sociedad Mexicana de Biotecnología y Bioingeniería; Mazatlán, México, 1997; p.206

Trezzini, G. F.; Zrÿd, J. P.; *Portulaca grandiflora:* A model system for the study of the biochemistry and genetics of betalain synthesis. *Acta Hort.* **1990**, *280*, 581–585

Trezzini, G. F.; Zrÿd, J. P.; Two betalains from *Portulaca grandiflora. Phytochem.* **1991a**, *30(6)*, 1897–1899

Trezzini, G. F.; Zrÿd, J. P.; Characterization of some natural and semi-synthetic betaxantins. *Phytochem.* **1991b**, *30(6)*, 1901–1903

Wagner, F.; Vogelman, H.; Cultivation of plant tissue cultures in bioreactors and formation of secondary metabolites. Plant tissue culture and its biotechnological application, Barz, W.; Reinhard, R.; Zenk, H. (Eds.), Springer-Verlag, Berlin, 1977; pp. 245–255

Weller, A. T.; Lasure, L. L.; Betalains in beet root tissue culture. *J. Food Sci.* **1982**, *47*, 162–163

von-Elbe, J. H.; The betalaines. Current aspects of food colorants, Chap. 3, Furia, E.T. (Ed).; CRC Press Inc., Boca Raton, FL, 1977; pp. 29–39

Zhong, J.; Seki, T.; Kinoshita, S.; Yoshida, T.; Rheological characteristics of cell suspension and cell culture of *Periella frutescens. Biotechnol. Bioeng.* **1990**, *40*,1256–1262.

PRODUCTION OF RECOMBINANT BLOOD FACTORS IN TRANSGENIC PLANTS

Manfred Theisen*

Meristem Therapeutics
8 rue des Freres Lumiere
63100 Clermont-Ferrand, France

ABSTRACT

The production of recombinant proteins for therapeutic use in transgenic plants has become an alternative to the isolation of these molecules from natural or other recombinant sources. Plant bioreactors offer the possibility of an inexpensive large scale production of high volumes of recombinant proteins with increased safety concerning contaminations with human pathogens. Several laboratories are involved in the commercial production of the first recombinant products, such as antibodies or recombinant human hemoglobin, from transgenic plants.

INTRODUCTION

Todays increasing demand for purified safe pharmaceutically active human proteins for application in medical treatments can no longer be satisfied by purification from original sources. The supply of human blood, placenta or other tissues and organs as starting material for the purification of proteins is limited. In addition several examples have shown in the past that the extraction of proteins from human tissue is not free of risk of contamination with pathogens such as HIV, Hepatitis B virus or prions and may thus lead to the transmission of infectious diseases to the treated patients. Recombinant DNA technology has allowed to circumvent these problems by adapting the cloned gene encoding a human protein of interest to an expression in heterologous host organisms (Dodet, 1990). Bacterial and yeast expression systems have been well established within the last decade.

* Tél.: 04 73 986810, Fax: 04 73 986819, e-mail: mtheisen@meristem-therapeutics.com

Chemicals via Higher Plant Bioengineering, edited by Shahidi *et al.*
Kluwer Academic / Plenum Publishers, New York, 1999.

Fermentation of these unicellular organisms in large scale fermentation units has allowed the production of several recombinant proteins, some of which have in the meantime reached the marketplace. Although the number of recombinant proteins produced in these expression systems is quite impressive, several limitations of the systems hamper their general application for the production of recombinant proteins of pharmaceutical interest. First, most mammalian proteins go through rather complex post-translational modification processes, which are necessary for their biological function. This includes not only post-translational cleavage of the polypeptide chains, formation of covalent or non-covalent complexes of homo- or heteromeric polypeptide chains, but also glycosylation, carboxylation, lipidation or other chemical modifications of the primary translation product. Bacteria and yeast cells do not necessarily possess the enzymatic machineries that are needed to catalyse these modifications. This leads in some cases to the result, that although the total amount of recombinant protein produced in a fermenter is high, the pharmacologial quality of the product is rather poor and its therapeutic value is limited. Second, large scale fermentation is rather cost-intensive and demands large production units.This cost factor is even more important if the recombinant protein has to be produced in eucaryotic (insect or mammalian) cell culture systems. The high costs for tissue culture media and serum are thereby only partly compensated by a higher conformity of the recombinant product.The application of for example fetal bovine serum or serum derived products in these culture systems unfortunately re-introduces in addition the risk of contamination of the recombinant product with pathogens that could potentially cross the human species barrier . After the recent development of transgenic technologies for large farm animals, the production of recombinant human proteins in these organisms has become possible. Several laboratories and companies have therefore chosen the mammary gland of transgenic cattle, sheep, goat or pig as a "bioreactor" for recombinant protein production (Hennighausen, 1990). Expression of the recombinant protein is thereby targeted to milk producing cells of the mammary gland and the secreted protein product can be purified from the milk of transgenic females. Large scale production of proteins from the milk of one transgenic cow or several transgenic sheep or goats is thereby possible. Although this process is rather straight forward, the development of for example one transgenic cow is very cost intensive. The animals have to be kept under strict sanitary conditions and routine health controls are necessary to monitor and prevent their infection with potentialy harmful pathogens.

Within the last couple of years, an alternative approach for the production of recombinant proteins has been developed, that is based on the use of transgenic plants as bioreactors. Meristem Therapeutics (Limagrain Bio-Santé) and others have developed this concept, which allows to introduce the gene of a human protein or for example a potential vaccinating protein (antigen) into the genome of transgenic crop plants. Plants possess all the catalytical capacities that are necessary for the post translational modification and maturation of an eucaryotic protein (although as discussed below, slight differences between for example mammalian and plant glycosylation exist (Vebert, 1993)). Our "Molecular Pharming©" approach allows a low cost production of recombinant proteins in transgenic crops, due to the fact that large quantities of biomass can easily be grown in the field. Expression of the recombinant protein can be targeted to plant organs, which are easy to store and for whom industrial extraction and manufacturing procedures have been established in the agro-industry. This approach minimizes the risk of contamination of the recombinant protein with potential human pathogens, since plant pathogens that cross the species barrier to humans are unknown, while for example human pathogenic viruses can not develop in the plant cells. Besides these biotechnological aspects the agricultural aspect of the "Molecular Pharming©" program should also not be neglected. This approach is directly linked to the development of a new agriculture, that is not only assuring the production of sufficient nutrition for the human population but will also serve as a tool for the creation of a new generation of health products for the treatment of human diseases. The following text will first explain the general process of recombinant protein production in transgenic plants and then describe some examples for its application in the production of pharmacologically active proteins for human or veterinary medicine.

FROM THE TRANSGENE TO THE DRUG: A GENERAL SCHEME

Figure 1 represents a schematic presentation of the process that allows the production of recombinant human proteins from transgenic crop plants. Several aspects (as for example the choice of the host plant or the plant organ to whom expression of the protein is targeted) may influence the details of the approach, but in general the first step of the process consists in adapting the gene or cDNA that encodes the protein of interest to an expression in the plant. For this the coding region of the gene has to be fused to promoter, leader and terminator sequences of plant origin that will allow the machinery of the plant cell to recognize the newly introduced gene and to synthesize the recombinant protein. The so adapted gene (transgene) will be introduced into plant cells, using well established techniques like transformation by *Agrobacterium tumefaciens* based gene transfer or transfection by biolistics ("gene gun"), depending on the chosen plant species. Transformed plant cells can regenerate new complete plants *in vitro*, provided they are growing under carefully selected and controled culture conditions. A selectable marker gene (kanamycin resistance gene) allows thereby to identify and isolate new transgenic plants from those that have not received and integrated the genetic information to produce the recombinant protein. The newly generated plants are then analyzed in the laboratory to determine the level of expression of the recombinant protein. Only those plants that produce satisfying amounts of the recombinant protein are kept in the process. After a passage and multiplication in the greenhouse, that is also used to start developing the large scale extraction and purification protocol on the level of a semi-industrial pilot plant, the offspring of the selected transgenic plants can be grown in the field to assure large scale production of the necessary biomass. At this point the real adventage of transgenic plants becomes obvious, since scaling up production in the field from several acres to several (tens of) hectars is a straight forward not very expensive agricultural process. The newly produced biomass is harvested using existing agricultural tools and the recombinant protein is isolated and purified in the production plant. The first milligrams to grams of the purified recombinant protein will be used to perform the necessary pharmacological and toxicological tests in animal experiments and the necessary clinical trials before the authorization for commercial utilization of the new medication is sought from the competent authorities.

CHOICE OF THE HOST PLANT FOR PROTEIN PRODUCTION

For practical reasons, tobacco is very often used to test the feasibility of producing a given human recombinant protein in transgenic plants. The transformation of tobacco plants with *Agrobacterium tumefaciens* based binary vectors is well established and easy to perform. Regenaration of tobacco plants and development of enough biomass to do first tests on the conformity of the recombinant protein can be done within three to four months

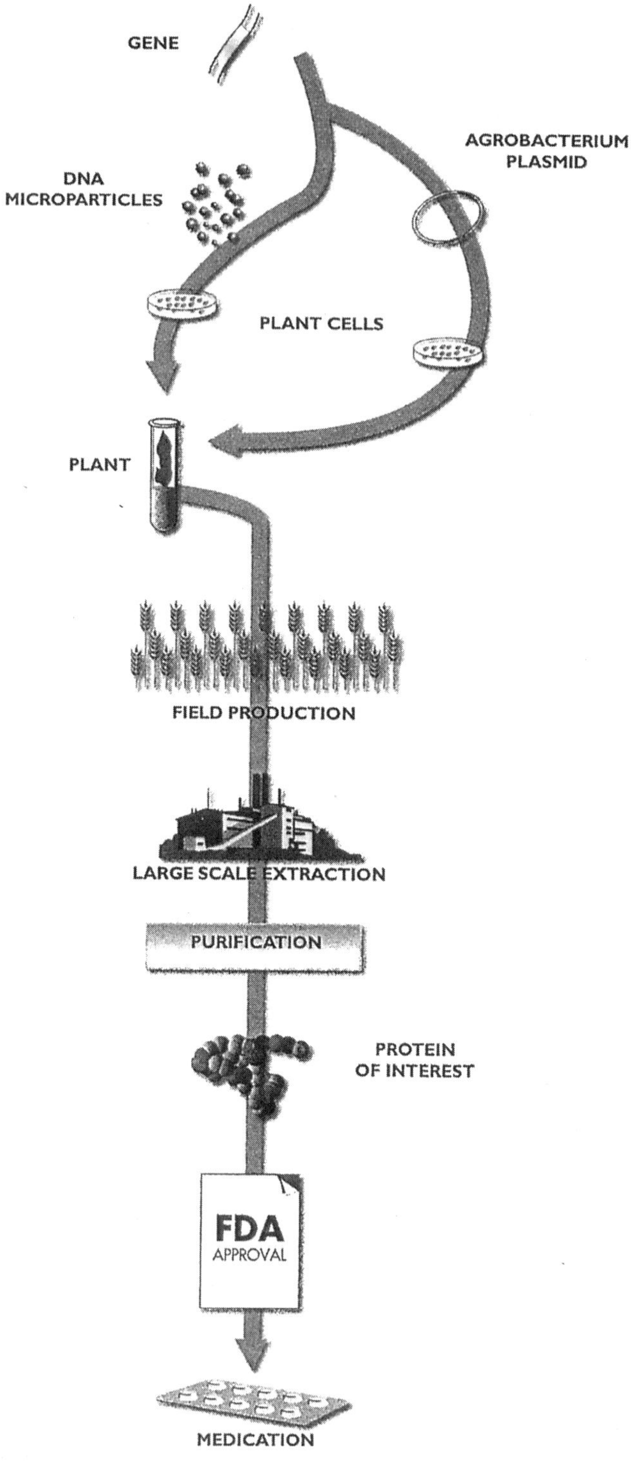

Figure 1. Schematic presentation of the production of a recombinant protein of therapeutic interest in transgenic plants. The gene of interest coding for the recombinant protein is adapted for expression in plant cells. Transgenic plants are generated and selected for the presence of the recombinant protein. Large scale production in the field followed by extraction and purification of the protein on an industrial level allow to produce the quanities of protein needed for controls and development of the new medication.

after the initial transformation. Expression of the transgene in tobacco is generaly targeted to the leaf, since tobacco seeds do not represent enough biomass to obtain large quantities of recombinant protein for an industrial development. To extract large quantities of recombinant proteins from tobacco leaves, the extraction and purification campain either has to be reduced to a very short period during the year, at harvest season, or rather cost intensive storage conditions for the fresh, green biomass have to be developed (freezing, lyophilisation). This represents one of the two major disadvantages of tobacco in a "molecular pharming" approach. The second disadvantage encountered with tobacco is the presence of nicotine and other toxic alkaloids, that have to be eliminated during the purification process. From this point of view edible plants and plant organs, like the seeds of cereals, potato tubers, roots of carrots and beets or the seeds of beans or peas, appear to be more suitable targets for the production of recombinant proteins. Storage of seeds or for example tubers is easy so that a continuous production from stored biomass can be assured throughout the year. In some cases, where the recombinant protein is destined to be used as orally applied medication, it is even thinkable to administer an enriched extract of maize, pea or carrot to the patient, instead of isolating the recombinant protein to its highest possible purity. Rapeseed is another industrial crop plant that can be used to target expression to the seed organs for easy storage of the biomass and industrial extraction of the recombinant product.

CONTROLING EXPRESSION AND TARGETING INTRACELLULAR TRAFFICKING OF THE RECOMBINANT PROTEIN

The principal keypoints for the generation of a transgenic plant that produces a recombinant protein are first, the choice of the appropriate host plant and plant organ for production (see above), second, to obtain the highest possible expression levels in the organ(s) of choice and third to assure the pharmacological activity and conformity of the protein by influencing the intracellular trafficking of the polypeptide chain and thereby its post translational processing. Modern molecular biology and biotechnology allow us to largely influence these parameters. High level expression of the recombinant protein in a relatively ubiquitous way in dicots as tobacco and rapeseed is for example obtained by using the 35S promoter from Cauliflower mosaic virus (CaMV). Several other promoters are available to target expression specificly to seeds especially of monocots like maize. Modern gene technologies allow to generate tailormade expression cassettes for the expression of recombinant proteins by fusing transcriptional control elements of several plant genes, thus creating highly specific and performant tools. The concept of an inducible production of the recombinant protein is hereby of special interest since it allows on the one hand to boost the level of protein produced in the plant shortly before harvest and limits on the other hand the risk that the synthesis of the recombinant protein interferes continuously with the physiology of the transgenic host cell. The optimal inducible system should thereby consist of a well characterized promoter element, coupled to an inducer (hormone, chemical, etc.) that allows a safe, non toxic, inexpensive, and fast induction of recombinant protein production in large quantities of transgenic plants.

As already mentioned, most mammalian or human proteins of pharmacological interest go through an array of post translational modifications that largely influence their biological activity. It is therefore important to not only express the highest possible amount of the protein in the plant cell but also to assure that the cell performs the necessary processing steps. The control of intracellular trafficking of the recombinant protein within the plant cell is therefore of crucial importance. A plant cell is subdivided into different compartments, the nucleus, cytoplasm, endoplasmic reticulum, Golgi system, vacuole or protein body and is surrounded by the extracellular compartment . We know now that the transport of proteins within and through these different compartments is directed by informations contained in the protein sequence in form of leader-, signal-, pro- and pre-peptides either at the N-terminus or the C-terminus of the polypeptide chain (Gomord and Faye, 1990). These peptides are normally cleaved off during the maturation process. In the simplest case the nascent polypeptide chain that is translated by cytoplasmic ribosomes bound to the mRNA molecule does not contain a signal peptide. In this case the growing polypeptide stays in the cytoplasm and the protein is not modified by glycosylation. The presence of a signal peptide will direct synthesis of the protein to the lumen of the endoplasmic reticulum where glycosylation (besides other modifications) is initiated. After its passage through the Golgi system, the protein will finaly end up either in the vacuole or will be anchored into the cell membrane or will be secreted into the extracellular compartment. This passage is coupled to a whole series of modifications as for example the addition of complex glycans, a process that differs in plants and mammalian cells (Gomord et al., 1996). Complex N-glycans of plant origin possess α-1-3 fucose and β-1-2 xylose residues that are missing in mammalian proteins, while sialic acid (N-acetyl-Neuraminic acid), a major component of mammalian N-linked glycans is not found in plant glycoproteins. In this respect plant glycoproteins resemble invertebrate glycoproteins. To avoid a potential influence of this plant specific heterologous glycosylation on the pharmacological activity or immunogenicity of the recombinant protein a simple strategy has been developed. The short peptide sequence H/KDEL fused to the C-terminus of a polypeptide chain leads to its retention in the endoplasmic reticulum, where the nascent glycoprotein is retained in its high mannose form, identical between plants and mammals (Wandelt et al., 1992). This technology, or genetic modifications of the host plant leading to the same result, allow to prevent plant specific glycosylation of recombinant proteins but they do not guarantee the production of glycoproteins with a perfect mammalian glycan pattern in the plant. Future work to genetically adapt glycan biosynthesis pathways in the plant for a mammalian specific glycosylation of recombinant proteins will be crucial for the development of plant production models for human pharmaceuticaly active proteins.

In this context it is also worth mentioning that the expression of a recombinant protein can be targeted to plant organelles. This is either done by using again specific signal-peptides that control intracellular trafficking or by directly transforming the genome of the organelle with a transgene, whose expression is directed by organelle specific promoters. We have demonstrated that this approach is important if the recombinant protein binds specific co-factors that are preferentialy synthesized in the organelle (see below) (Dieryck et al., 1997). Targeting transgene expession to organelles could also be of interest for the expression of procaryotic proteins in plant cultures.

THE TARGET PROTEINS

An increasing number of scientific publications demonstrate the possibility to produce recombinant proteins of human, mammalian or viral origin in transgenic plants. The target proteins are thereby largely the same as for example for production in transgenic animals although a certain preference can be found for the production of oraly applicable proteins and more specificaly oral vaccines (see below).

Sijmons et al. (1990) have shown that human serum albumine (HSA) can be produced in leaves of transgenic tobacco and potato. Expression of the transgenes was driven by the CaMV 35S promoter. The transgenic constructs contained either the authentic HSA signal-peptide or a plant specific pre-pro peptide (PR-S) fused to the coding region of processed HSA. The secreted protein was either isolated from leaf tissue or from plant cell suspension cultures. Chromatographicaly purified HSA was analyzed by N-terminal amino acid sequencing. The recombinant protein containing the human signal peptide was not correctly processed, while the construct with the PR-S peptide allowed the production of correctly cleaved HSA in the plant cells. Expression levels of 0.025 % were so far obtained. The authors conclude that an increase of expression levels to 1 to 2% in potato tubers should allow to produce 12 kg of recombinant HSA per hectare.

Charles Arntzen of the Boyce Thompson Institute in the US is among others developing transgenic plants for the production of "edible" vaccines for human or veterinary medicine (Moffat, 1995). His team could show, that subunit antigens against hepatitis B virus (Thanavala et al., 1995), Norwalk virus (Mason et al., 1996) or *E. coli* heat-labile enterotoxin (LT-B) (Haq et al., 1995) that were produced in tobacco or potato plants induced specific antibodies in mice after either i.p. or oral administration. Targeting expression of the Norwalk virus capsid protein to tobacco or potato leads for example to the assembly of pseudoviral particals (VLP) that could eventually serve as an oral vaccine against this gastro-enteritis causing virus. The choice of the host plant for oral vaccines production is certainly of prime importance. While tobacco and raw potato extracts are sufficient to prove the feasibility of the approach in animal tests, plant derived edible vaccines for human vaccination campains should be based on other host plants. Attempts to transform for example bananas for vaccine production are currently undertaken. Other potential targets for vaccine production in transgenic plants are the rabies virus glycoprotein, whose production in transgenic tomatoes was described (McGravey et al., 1995), or for example cholera toxin subunits that are produced in tobacco. The expression of subunit vaccines for the immunization of farm animals against veterinary pathogens is also of important economical interest. Vaccination of cattle with for example corn, that contains a specific recombinant antigen should be envisioned. The production of vaccines in transgenic plants is not limited to oral immunization approaches. Purification of the antigen to a sufficient level that allows vaccination by i.p., s.c. or i.m. injection appears feasible.

The production of vaccines that was described in the previous examples is based on the expression of full length recombinant proteins in stably transformed transgenic plants. It is worth mentioning that an alternative method for the production of recombinant vaccines has been developed. It consists in the production of recombinant plant viruses that express potentially immunogenic epitopes of human or animal virus surface proteins as fusion proteins within their capsid (Porta et al., 1994). Two different plant viral vectors have been developed, one is based on Tobacco mosaic virus (TMV), the other on cowpea mosaic virus (CPMV). Recombinant virus is generated in the laboratry and used to infect the appropriate host plant (tobacco or cowpea), the recombinant virus is purified and used as such in a vaccination protocol. This method has been used for example to generate a recombinant CPMV virus that contains a 22 amino acid peptide of gp 41 from the human immunodeficiency virus type 1 (HIV 1) (McLain et al., 1998). Dalsgaard et al. (1997) could show that vaccination of mink with a recombinant CPMV expressing a short epitope from the VP2 capsid protein of mink enteritis virus (MEV) protected the animals against viral diarrhea.

If recombinant plant viruses for vaccine production could be complementary to the creation of recombinant vaccines in transgenic plants, a major limitation, the size of the recombinant peptide (and not protein) that can be expressed, hinders so far their use for the production of large complex proteins.

The following examples demonstrate that stable trangenic plants are capable to sythesize complex multimeric proteins of mammalian origin that possess their full physiological activities. One example is the production of antibodies in transgenic plants (Whitelam and Cockburn, 1996), the other the synthesis of functional human haemoglobin in tobacco seeds (Dieryck et al., 1997). The demand for recombinant antibodies for therapeutic or diagnostic use is increasing. Functional antibodies are multimeric proteins that consist of either two heavy and two light chains (for example IgG) or even four heavy and four light chains, that are joined by a J chain and protected against proteolysis by a secretory component (SC) in the case of secretory IgA (SIgA) molecules. SIgA molecules are involved in mucosal immunity. Their application in passive immunization by oral administration is of great therapeutic interest. To achieve production of functional antibodies in plants the genes for the different subunits have to be integrated and co-expressed in the same plant. Ma and colleagues (Ma, et al., 1995) showed that functional SIgA can be produced in transgenic tobacco. Independent transgenic lines were created that expressed either the Ig κ light chain, Ig α heavy chain, the J chain or the SC polypeptide. Cross fertilization of single transgenic plants expressing the light and heavy chain genes led to descendants that produced a monomeric Ig. Breeding of these double transgenic plants with a plant that expressed the J chain allowed the production of dimeric Ig in the triple transgenic offspring. Crosspolynization of the triple transgenic plants with plants expressing the SC protein gave plants with four independent transgenes that produced functional SIgA.

Together with our colleagues at INSERM U 299, Kremlin Bicêtre, Meristem Therapeutic's has generated transgenic tobacco plants that produce functional human hemoglobin (Dieryck et al., 1997). Purified hemoglobin as blood substitute for emergency use is currently prepared from outdated human blood or from animal blood sources which includes the already discussed risks of contamination of the product with potential pathogenic agents. Active hemoglobin is a tetrameric molecule with two alpha and two beta globin chains. The functional molecule contains in addition one heme molecule per globin subunit. To produce functional globin from a recombinant source, it has not only to be assured that the two different globin chains are both synthesized in sufficient quantitiy, but also that the heme co-factor is present and that its integration into the molecule leads to the synthesis of a stable active tetrameric hemoglobin. Dyricks et al. (1997) obtained transgenic tobacco plants that were transfected with a co-expression vector containing both alpha and beta globin genes of human origin under the control of the CaMV 35S promoter. The nascent polypeptide chains were directed to the chloroplast to assure their assembly in the presence of heme molecules. The presence of human hemoglobin in extracts of the transgenic tobacco plants was shown on the protein level. Physiological tests show that the recombinant human globin is functionaly active and capable to fulfill its function as oxygen carrier. Transgenic plants can therefore be used as bioreactors for the production of complex multimeric human proteins.

CONCLUSIONS

The scientific work of several laboratories and companies during the last couple of years has shown that transgenic plants are suitable host organisms for the production of recombinant human or mammalian proteins of therapeutical interest. In comparison to other organisms like bacteria, yeast, mammalian tissue culture cells or even transgenic animals, plants possess certain advantages. These include an increased safety of production, due to the lack of known pathogens that are common to plants and humans, the possibility to produce proteins with a greater conformity than for example procaryotic systems and the fact that large scale production costs can be minimized by agricultural exploitation of the transgenic plants. Several examples have shown (see above) that complex recombinant proteins from plant sources are physiologically active and may soon be used for medical applications. The steps that will largely determine the future of this production technology will be the development of feasible large scale industrialization processes, the proof of efficiency and safety of plant derived proteins in medical trials, the acceptance of plant derived medical proteins by the regulatory authorities and finally the acceptance of these recombinant proteins by the medical community and the patients. Plant based medication has accompanied man since its origins. We are now able to use modern biotechnology to specifically produce targeted medical products in plants.

REFERENCES

Dalsgaard, K., Uttenthal, A., Jones, T.D., Zu, F., Merryweather, A., Hamilton, W.D.O., Langeveld, J.P.M., Boshuizen, R.S., Kamstrup, S., Lomonossoff, G.P., Porta, C., Vela, C., Casal, J.I., Meloen, R.H. and Rodgers, P.B. Plant-derived vaccine protects target animals against a viral disease. *Nature Biotechnology* 1997, *15*, 248–252.

Dieryck, W., Pagnier, J., Poyart, C., Marden, M.C., Gruber, V., Bournat, P., Baudino, S. and Merot, B. Human haemoglobin from transgenic tobacco. *Nature (London)* 1997, *386*, 29–30.

Dodet , B. Production de protéines recombinantes: quel systèm choisir? *Biofutur* 1990, *91*, 20–29.

Gomord, V. , Sailer J-P et Faye L. Plante et médicament, un nouveau départ. *Biofutur* 1996, *154*, 26 30.

Gomord, V. and Faye, L. Signals and mechanisms involved in intracellular transport of secreting proteins in plants. *Plant Physiol. Biochem.* 1996, *34,* 165–181.

Haq, T.A., Mason, H.S., Clements, J.C. and Arntzen, C.J. Oral immunization with a recombinant bacterial antigen produced in transgenic plants. *Science* 1995 *268*, 714–716.

Hennighausen, L., Ruiz, L. and Wall, R. Transgenic animals - production of foreign proteins in milk. *Curr. Opin. Biotechnol.* 1990, *1*, 74–78.

Ma, J.K.C., Hiatt, A., Hein, M., Vine, N.D., Wang, F., Stabila, P., Van Dolleweerd, C., Mostov, K. and Lehner, T. Generation and assembly of secretory antibodies in plants. *Science* 1995, *268*, 716–719.

Mason, H.S., Ball, J.M., Shi, J.J., Jiang, X., Estes, M.K. and Arntzen, C.J. Expression of Norwalk virus capsid protein in transgenic tobacco and potato and its oral immunogenicity in mice. *Proc. Natl. Acad. Sci. USA* 1996, *93*, 5335–5340.

McGravey, P.B., Hammond, J., Dienelt, M.M., Hooper, D.C., Fu, Z.F., Dietzschold, B., Kprowski, H. and Michaels, F.H. Expression of the rabies virus glycoprotein in transgenic tomatoes. *Bio/Technology* 1995, *13,* 1484–1487.

McLain L., Durrani, Z., Wisniewski, L.A., Porta, C., Lomonossof, G.P. and Dimmock, N.J. Stimulation of neutralizing antibodies to human immunodeficiency virus type 1 in three strains of mice immunized with a 22 amino acid peptide of gp 41 expressed on the surface of a plant virus. *Vaccine*1996 *14*, 799–809.

Moffat, A.S. Exploring transgenic plants as a new vaccine source. *Science* 1995, *268*, 658–660.

Porta, G., Spall, V., Loveland, J., Johnson, J., Barker, P. and Lomonossoff, G. Development of cowpea mosaic virus as a high-yield system for the presentation of foreign peptides. *Virology* 1994, *202*, 949–955.

Sijmons, P.C., Dekker, B.M.M., Schrammeijer, B., Verwoerd, T.C., Van den Elzen, P.J.M. and Hoekema, A. Production of correctly processed human serum albumin in transgenic plants. *Bio/Technology* 1990, *8*, 217–221.

Thanavala, Y., Yang, Y.F., Lyons, P., Mason, H.S. and Arntzen C. Immunogenicity of transgenic plant-derived hepatitis B surface antigen. *Proc. Natl. Acad. Sci. USA* 1995, *92*, 3358–3361.

Verbert, A. La glycosylation des protéines recombinantes. *Biofutur* 1993, *125*, 40–44.

Wandelt, C.I., Rafiqul M., Khan, I., Craig, S., Schroeder, H.E., Spencer D. and Higgins, T.J.V. Vicilin with carboxy-terminal KDEL is retained in the endoplasmic reticulum to high levels in leaves of transgenic plants. *Plant J.*1992, 181–192.
Whitelam, G.C. and Cockburn, W. Antibody expression in transgenic plants. *Trends Plant Sci.* 1996, *8*, 268–272.

17

RECENT PROGRESS IN BIOTECHNOLOGY OF MEXICAN MEDICINAL PLANTS

Ma. Luisa Villarreal,[1,4] Pilar Nicasio,[1] Gabriela Rojas,[1] Laura Alvarez,[2] and Rodolfo Quintero[3,4]

[1]Centro de Investigación Biomédica del Sur
Instituto Mexicano del Seguro Social Argentina No1
Xochitepec, Morelos Mexico
[2]Facultad de Ciencias Químicas e Industriales
Universidad Autónoma del Estado de Morelos
Cuernavaca, Morelos, Mexico
[3]Instituto de Biotecnología
UNAM, Av. Universidad 2001
Cuernavaca 62210, Morelos, Mexico
[4]Centro de Investigación en Biotecnología
Universidad Autónoma del Estado de Morelos
Cuernavaca, Morelos, Mexico

ABSTRACT

Based on traditional medicinal knowledge, it was possible to identify the plant species *Solanum chrysotrichum* as the source of a new antimycotic agent designated SC-1. Cell suspension batch cultures from this plant were established in shake flasks, in which the production of SC-1 was optimized, reaching values fifty times higher than those registered in field grown plants. Large-scale cultivation of the active biomass from *S, chrysotrichum* was established in 10 l airlift bioreactors, and productivity levels of SC-1 were increased by 60% when using a draw-fill mode in the bioreactors.

INTRODUCTION

The continuous use of plant-derived products for health care problems, has encouraged man to consider as important priorities the knowledge on the pharmacological properties of its surrounding flora, and the design of appropriate technologies to exploit the medicinal plant resources. In science, as well as in industry, there is a great interest for in-

Chemicals via Higher Plant Bioengineering, edited by Shahidi *et al.*
Kluwer Academic / Plenum Publishers, New York, 1999.

troducing modern technologies which involve biotechnological procedures to properly establish the optimized production of active plant selected compounds.

It is evident that the biosynthetic repertory of higher plants for the production of plant drugs is quite large, and that many of the already identified molecules are novel chemical structures. Most of these products are complex molecules, and their synthesis represent a permanent challenge for the chemical industry. This last problem has never succeeded due to very low yields, difficulties in obtaining chemical transformations; or to the need of obtaining high purity isomers from very complex mixtures. In this way, plants continue to be the most effective source for the biosynthesis of chemical structures with important biological activities.

One viable approach for obtaining these compounds is represented by plant cell and tissue culture systems. These procedures are gaining good reputation in the last few years due to two main reasons: one is that these techniques could actually be applied to most plant species, and the other is that the potential of some culture systems for production of useful medicines had already been demonstrated. It can be surely said that *in vitro* cultures represent a future commercial alternative for the production of high value plant-derived medicines.

Some important drugs used in the clinic are actually manufactured by plant cells and tissue culture procedures. There are many benefits that the use of these techniques could offer. Some outstanding utilities could be the controlled and homogeneous production of plant compounds, securing high yields of bioactive products, and the conservation of high value genetic pools.

The commercial exploitation of cell culture techniques requires scaling-up facilities with the use of bioreactors, in order to provide better conditions for growth and biosynthesis. The development of plant cultures in bioreactors could be a very sophisticated technology. Nevertheless, it is actually possible to grow plant suspension cultures up to volumes of 75,000 l when using appropriate systems. Some important scaled-up cultures have been developed in Japan. In this country, natural plant derived products play a main role in health care problems, and the Japanese industries have been supporting research and development programs for these type of products. Until now, three large-scale commercial processes for the production of medicinal compounds have been established in Japan by Mitsui Petrochemical Industries. In North America, there are two processes that have been developed for commercialization. Taxol, an antitumorigenic compound, has been produced in 75,000 l bioreactors in Germany.

The Antifungal Medicine from *Solanum chrysotrichum*

In México, the knowledge on the curative properties of plants is permanently used among indigenous populations, and medicinal plants constitute the most valuable and large resource of Mexican Traditional Medicine (Lozoya, 1976; Zolla, 1979). The benefits that this resource could offer are enormous. The fact that medicinal plants could be used to solve health priority problems in a large percentage of the population, has generated an increasing interest in various academic and health areas for investigating the local flora. The permanent use of certain plants for medicinal purposes, is an important criteria that in some way validates their therapeutical use, and offers incentive for development and production of new medicines (Meckes, 1993).

Among the higher plants most frequently used for medicinal purposes by the highland Mayas of Chiapas, México, *Solanum chrysotrichum* (popularly known as *sosa*) is of outstanding importance to treat dermatological infections such as *athlete foot* and skin mycosis.

According to popular nosologies about skin infections, and from the ethnobotanical information that emerged from the program IMSS-COPLAMAR, which correlate Traditional Medicine with Institutional Medicine, in 200 rural communities of the state of Chiapas; it was reported that aqueous extracts from the leaves of *S. chrysotrichum* provided the treatment of choice for mycotic skin diseases (Zurita and Zolla,1986). The subsequent studies on the genus *Solanum*, allowed the confirmation of the widely use of these plants as antifungal remedies; being *Solanum chrysotrichum* the species selected for starting specific pharmacological evaluations (Lozoya and Aguilar,1987).

Clinical and Pharmacological Studies

In preliminary pharmacological evaluations, *S. chrysotrichum* organic extracts from the leaves of *sosa* were prepared. These extracts were evaluated in *in vitro* bioassays using different pathogens which include Gram+ and Gram- bacteria, yeast such as *Candida albicans,* and dermatophytes such as *Microsporum gypseum.* The Minimal Inhibitory Concentration (MIC) values of these extracts against the last two microorganisms were considered important, once that they show some degree of resistance to the commercial medicines presently prescribed. In another series of studies we evaluated the antimycotic activity of the extracts against the most common dermatophytes that cause athlete foot and skin mycosis; these included *Trycophyton mentagrophytes*, *Trycophyton rubrum* and *Microsporum gypseum.* The obtained results showed that the mentioned extracts did not provoke any toxicity against the selected bacteria at a concentration of 30 mg ml^{-1}, which is considered the limit concentration needed to register activity using this model. The inhibition of growth of *Candida albicans* showed MIC value of 20 mg ml^{-1} and the antifungal action against *T. mentagrophytes*, *T.rubrum* and *M.gypseum* showed antimycotic activity with MIC values equal or less than 15 mg ml^{-1} (Lozoya *et al*,1991)

As a result of these studies a preliminary study was started using organic extracts from *sosa* in patients from Hospital Regional No 1 of the Instituto Mexicano del Seguro Social at Cuernavaca, Morelos. A group of 28 patients with a previous diagnosis of *tiñae pedis* were selected, using the following exclusion criteria: *diabetis mellitus*, onicomycosis, peripheric neuropathy, heart and renal insufficiency, and hemophilia. A group of 14 patients were treated with the local application of a cream containing 5% of a methanolic extract from the leaves of the plant. An equal number of patients were considered as a control group, to which the usual miconazol nitrate medicine was applied. After four weeks of treatment, 42% of the patients which received the *sosa* extracts ameliorated, showed disappearance of most of the signs and symptoms of the disease as edema, ache, itch and vesicles (Lozoya *et al*,1991).

MATERIALS AND METHOD

Clonal Propagation

The therapeutic value of this natural resource, and the fact that this species grows in a very limited area of the state of Chiapas, were two main reasons to start *in vitro* tissue cultures for its clonal propagation. The microprogation was carried out using axillary buds as explants, which were obtained from germinated plantlets. The explants were grown in the nutrient media of Murashige and Skoog (1962), supplemented with growth regulators. Induction of differentiated plants was obtained when indol-3 acetic acid (IAA) was present as an auxin, in combination with one of two citoquinins: kinetin (KN) or benciladenine (BA). Nevertheless, the addition of 0.1 mg $l^{-1}$of IAA + 0.2 mg l^{-1} BA was optimum to induce morphogenesis (Villarreal and Muñóz, 1991). The specimens obtained using this procedure were rooted *in vitro*, and totally adapted to live in the soil at the state of Morelos (CIBIS-IMSS). The material obtained through tissue culture showed biological activities which were comparable to those obtained from wild plants. (Villarreal *et al*, 1997)

Chemical Isolation and Identification of Bioactive Sc-1 from *Solanum chrysotrichum*

The procedures carried out to isolate, purify and elucidate the molecular structure of the compound responsible for fungicidal activity of this plant species were established using the leaves as a raw material. The bioactive molecule was denominated SC-1, and it corresponds to the same structure obtained from the *in vitro* cultures. The knowledge on the molecular structure of SC-1, allowed the development of proper analytical procedures for its identification and quantification in tissue cultures.

Chemical Extraction

The leaves were obtained from adult specimens of *S. chrysotrichum* collected in the state of Chiapas, and dried at room temperature. The material was powdered and macerated at room temperature with methanol.

Bioguided Fractionation for Preparation of SC-1

In order to direct isolation, a bioguided fractionation of methanolic extracts from the leaves of this plant species was carried out using *Trycophyton mentagrophytes* as the biological monitor, and obtaining minimal inhibitory concentration (MIC) values throughout the whole procedure. Once the methanolic extract was prepared, it was further evaporated to dryness under reduced pressure, followed with the addition of a mixture of hexane-water (1:1). Complete partition was carried out until two fractions were obtained: an organic fraction without activity, and an aqueous one which kept the bioactivity of the extract (MIC *T.m.* 15 mg ml^{-1}). A second partition of the aqueous fraction was done using a mixture of chloroform:water (1:1), obtaining an active chloroformic fraction (MIC *T.m.* 2.5 mg ml^{-1}), and and aqueous one devoided of antifungal activity (MIC *T.m.*> 15 mg ml^{-1}). The chloroform fraction was separated using silica gel chromatographic columns (70–230 mesh Merck), with an elution system of chloroform-methanol (94:6), obtaining several fractions from which two exhibited important biological activity. The active fractions were combined, and rechromatographed using chloroform-methanol (94:6). From this column a mixture of products was obtained which were separated using silica gel thin layer chromatography (TLC) (60 F 254, 5x10 cm, 0.25 mm with, Merck), and with a solvent system of chloroform-methanol-acetone (5:1:3); by carrying out the Lieberman reaction, a big brown spot was visualized along with several contaminants.

Purification of SC-1

The isolated fraction was purified through silica gel chromatography using as eluent a clroroform-methanol solvent system(9:1), and obtaining one active product (MIC *T.m.*

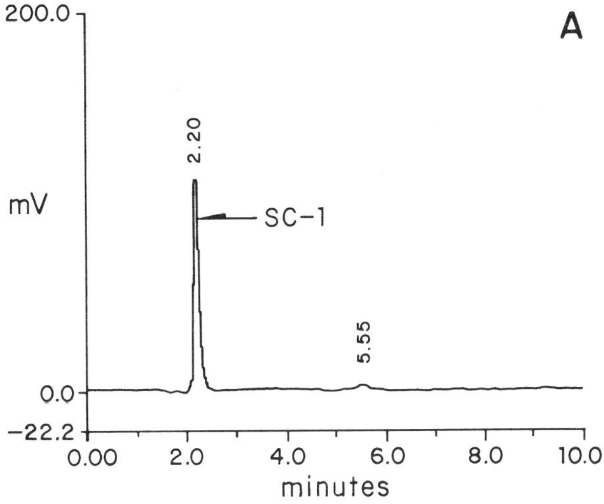

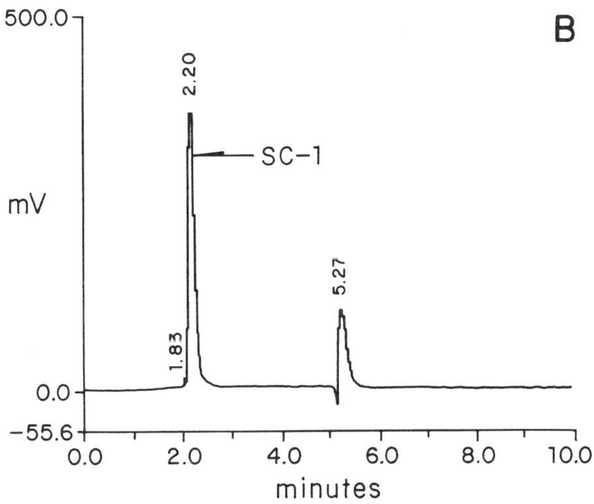

Figure 1. High Performance Liquid Chromatography patterns from leaves (A) and 20 days old cell suspension (B) extracts of *Solanum chhrysotrichum*.

100μg ml^{-1}) . The isolated product was washed with chloroform-methanol (9:1), obtaining an apparently pure solid, which was again analyzed by TLC using the already described system, and showing important antifungal activity (MIC *T.m.* 50 μg ml $^{-1}$). Through a third chromatographic separation, the products were again purified (MIC *T.m.* 20 μg ml^{-1}) and subjected to TLC using a system of chloroform:methanol:acetone (5:1:3), and identifying two spots with Rf_s values of 1) 0.07 and 2) 0.10; and with strong activities: MIC values 1) *T.m.* 10 :g ml^{-1}, 2) *T.m.* 10 :g ml $^{-1}$.

High Performance Liquid Chromatography (HPLC)

The chromatographic separation of SC-1 was carried out from product 2) obtained through TLC separation. Purification was done using a normal silica gel chromatographic column Lichrosfer Si (5μ) 4 × 250 mm, with a Merck-Hitashi program and a Refractive Index RI-71 at a temperature of 40°C. The samples were eluted using a flux of 1.2 ml min^{-1} and a solvent system of methanol-chloroform-water (29:70:1). Figure 1 shows the corresponding peaks of SC-1 obtained from leaves and from 20 days old cell suspensions, respectively.

Structural Elucidation of SC-1

The mixture of the semipurified products was analyzed; the infrared spectrum showed an intense absorption band at 3435 cm^{-1}, typical of hydroxyl groups, and an enlarged carbon-hydrogen band at 2926 cm^{-1}. The NMR spectra (^{1}HNMR) of the active mixture and the purified products were performed using a 300 MHZ machine with an ASPECT 2000 computer (Instituto de Química UNAM). The samples were dissolved in chloroform. The ^{1}HNMR of the compound showed in the carbonyl region (3.0 and 4.0 ppm), a series of signals which indicates a glycosylated compound. The acid hydrolysis, using 2N HCl, allowed to deduce the presence of glucose as a component of the mixture, which was further identified by direct comparison (CO-TLC) with an authentic sample. Another sugar was also identified by TLC. Three characteristic doublets of anomeric carbons,were also observed, which indicated a triglycosylated compound, glucose being one of the sugars. The ^{13}C NMR spectra at 500 MHZ showed 45 carbons, 18 of which corresponded to three sugars that, according to the chemical displacement theory (Levy and Nelson, 1972) and with comparison with some reported data (Agrawal et al,1985), were assigned to two molecules of glucose and one of fucose. The rest of the 27 signals corresponded to a steroidal molecule with a spirostanol backbone, which was hidroxylated in position C6. Therefore, it was possible to establish the steroidal skeleton as well as the sugars. The compared SC-1 corresponded to a new spirostanic saponin-type molecule, triglycosilated with two glucoses and one fucose (Pérez *et al.*,1996).

Cell Suspension Cultures

The next step was the establishment of a biotechnological process that allowed the production of the antimycotic medicine SC-1 using *in vitro* cultures of *Solanum chrysotrichum*, which could be used for large scale growth.

Establishment and Characterization of Cell Suspension Cultures

Cell suspensions were initiated from previously cultivated undifferentiated callus tissue. The calluses were derived from hypocotils and cotyledons of plantlets obtained from the germination of aseptic seeds of *Solanum chrysotrichum*. Since the first fifteen days of hormonal treatments, these tissues showed undifferentiated growth, reaching values of 92% calluses formation at 30 days in culture. These calluses did not show any organogenesis, and proliferated mostly into a mass of undifferentiated tissue of friable consistency. The cell suspension cultures were established in Murashige and Skoog (1992); (MS) complete media, supplemented with different combinations of growth regulators, and using kinetin as the citoquinin, but adding one of four different auxins: MS1 (2 mg l^{-1} 2,4-dichlorophenoxyacetic acid), MS2 (0.5 mg l^{-1} naphtalene acetic acid), MS3 (1.3 mg l^{-1} indoleacetic acid), MS4 (1 mg l^{-1} 2,4,5 triclorophenoxyacetic acid). The cell suspensions were grown in batch in 250 ml shake flasks for 25 days. The cultures were incubated at 28°C with a photoperiod of 16 hr light in an orbital rotary shaker (New Brunswick), at

Table 1. Antimycotic activity of tissues and purified compounds from *Solanum chrysotrichum* (minimal inhibitory concentrations, MIC [mg ml^{-1}])

Extracts and products	*Microsporum gypseum*	*Trycophyton mentagrophytes*	*Trycophyton rubrum*
Leaves	15.00	5.00	5.00
Stems	4.00	4.00	5.0
Callus	10.00	4.00	NT
Cell suspension	NT	6.00	6.0
Active fraction	0.02	0.01	NT
SC-1	0.04	0.04	NT
Nystatin	0.16	0.01	NT

130 rpm. The kinetics of growth and accumulation of the antifungal SC-1 was established using the following variables measured every five days: dry weight, fresh weight, pH, cell viability, sugar concentration in the culture media, as well as accumulation of SC-1. The four analyzed media that were used to initiate the suspension cultures observed good growth, but MS1 was selected since it produced a finer and homogeneous suspension compared to the other media.

Production of SC-1 in Tissue Cultures

The antifungal activities shown by the products obtained through *in vitro* cultures, in comparison with tissues and fractions from the collected wild material, were evaluated through the assay usually reported for the study of natural products with antimycotic activities (Paxton,1991). The dermatophytes used for this purpose were *Trycophyton rubrum* NRRL A-2846, *Trycophyton mentagrophytes* NRRL-1942 and *Microsporum gypseum* NRRL A-2605 (Table 1).

SC-1 Quantification

Identification and quantification of the antimycotic products in cell suspension cultures of this plant were carried out by HPLC analysis throughout the growth suspension cycle. Retention times of SC-1 peaks from cell cultures and from wild plant material, were compared using co-chromatography. The calibration curves were constructed using the leaves' active product as a standard. Authentication of SC-1 in the cultures was corroborated by infrared comparison of the *in vitro* spectra, with the ones obtained from wild plants.

Optimization of Growth and SC-1 Accumulation

In order to improve biomass growth as well as accumulation of the antifungal SC-1 in batch suspension cultures of *S. chrysotrichum*, different concentrations of several selected nutrients were assayed, as well as variations in inoculum size.

1. Inoculum concentration: In spite that plant cell cultures required an optimum inoculum in order to initiate fast growth, the effect of inoculum size in biomass and product accumulation was evaluated using 2, 3 and 5 g l^{-1} dry weight.
2. Effect of sucrose concentration: The utilization of sucrose in suspension cultures of plant cells shows individual patterns, and the initial concentrations of the carbon source exerts a fundamental influence in the course of the culture

(Fowler, 1982; Merillon, 1994). The effect of various sucrose concentrations (30(control), 45,60 and 90 g l^{-1}) was evaluated.

3. Effect of phosphate concentrations: Different phosphate concentrations of 0, 0.25(control) and 3.75 mM, were evaluated in their ability to influence biomass growth and SC-1 accumulation.
4. Effect of nitrate and ammonium concentrations: The concentration of nitrogen from nitrate and ammonium has a central role in the optimum growth of *in vitro* plant cultures. In this investigation, concentrations three fold and three times lower than the usual concentrations of the nutrients in the MS medium, were assayed.

Large-Scale Cultivation of *S. chrysotrichum* Cells

Once optimized conditions for biomass growth and accumulation of SC-1 were reached in the shake-flask experiments, the scale-up of the cultures was established. Large-scale cultivation of cell suspensions was performed in air-lift 10 l bioreactors with an 8.5 l working volume. The bioreactors used in this study were in-house designed draught tube internal loop air-lifts (Villarreal *et al.*, 1997b). Cell suspensions of *S. chrysotrichum* grown in 2 l Erlenmeyer flasks and of 15 days old, were used to obtain inocula of 20 and 30 g l^{-1} fresh weight. Three to four-weeks old batch-suspension cultures in bioreactors were implemented, using the same MS nutrient medium, but substituting sucrose with sucrose. Culture temperature was maintained at 25°C, and air was supplied at various rates from 0.11 l min^{-1} (0.013 vvm) to 0.8 l min^{-1} (0.09 vvm) according to cell growth, and in order to maintain culture homogeneity. A draw-fill batch culture mode was also investigated by harvesting 50% of the cell culture and replacing it with fresh medium.

RESULTS AND DISCUSSION

The results of the biomass growth and SC-1 accumulation of batch suspension cultures of *S.chrysotrichum* in shake flasks are shown in Figure 2. The maximum biomass obtained for this cultures was 12.9 g l^{-1} dry weight, which was 5.6 times higher than the initial inoculum. The culture reached a stationary phase after 10 days, even though a high concentration of total sugars still remained in the culture medium (*c.a* 22 g l^{-1}). Duplication times based on dry and fresh weights were 4.5 and 5.0 days, respectively. The maximum concentration of SC-1 was 14.6 mg g^{-1} dry weight; this value was 50 times higher than that obtained from wild plants, in which the accumulation of SC-1 was of 0.282 mg g^{-1} dry weight. The production of SC-1 was non-growth associated, as it was observed from a relatively low accumulation of the product before reaching a biomass concentration of 10 g l^{-1} dry weight and a relatively high accumulation of SC-1 afterwards. Yields of the product from biomass (Yp/x) were 6.14 and 17 mg g^{-1} dry weight on days 15 and 25, respectively. Productivity values obtained for the experiments in shake-flasks were 0.01 g l^{-1} day^{-1} (Villarreal *et al*, 1997 a).

Sugar utilization by cell cultures was directly associated with biomass increment. Sucrose was completely hydrolyzed into glucose and fructose by day 10, and consequently glucose and fructose was increased early in the cultures. The consumption of glucose started after day 5, and the concentration of fructose began to decrease after day 10. Glucose was completely consumed after 25 days in culture, while fructose concentration was above 7 g l^{-1} at such a time.

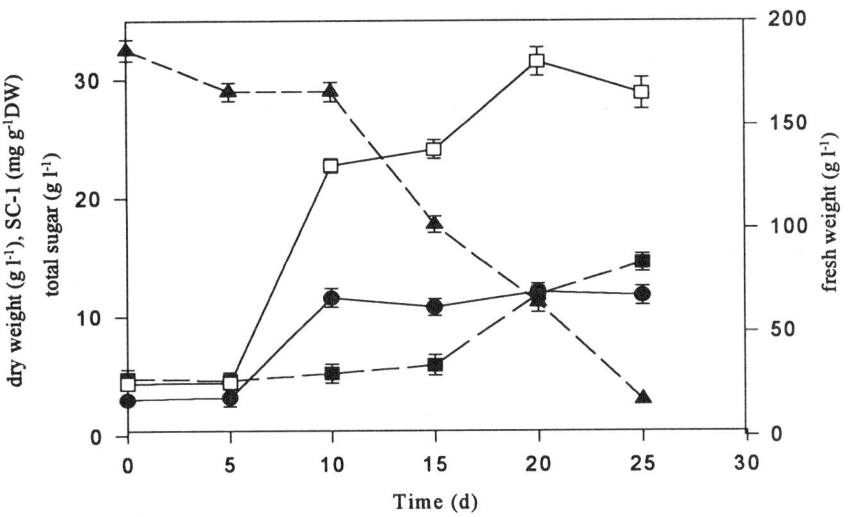

Figure 2. Typical growth, SC-1 accumulation and carbohydrate uptake by *S. chrysotrichum* cell suspensions cultures. (□) fresh weight, (●) dry weight, (▲) total sugars, and (■) SC-1. Reproduced from Plant Cell Tissue and Organ Culture Vol, 50, pp. 39–44, 1997.

Results using different inocula sizes indicated that the highest concentration of biomass was reached with the 5% inoculum, while the highest SC-1 accumulation was obtained when using a 2% inoculum (Villarreal *et al.*, 1997 a). Results with various sucrose concentrations (30,45,60 and 90 g l^{-1}) after 20 days in culture over biomass growth and production of SC-1, indicated that maximum accumulation of the bioactive product was obtained with sucrose concentrations between 30 and 45 g l^{-1}, while maximum cell growth was registered using a concentration of 60 g l^{-1}. The results with different concentrations of nitrates, ammonium, phosphates and 2,4-dichlorophenoxyacetic acid (2,4-D) that were analyzed by an ortogonal design using four factors, indicated that neither variations in nitrate, phosphate and 2,4D produced any positive results over cell growth and SC-1 accumulation, and that only the highest ammonium concentration (3x the usual concentration), showed important increments in order to reach 14 mg l^{-1} biomass, and 15 mg g^{-1} dry weight of SC-1 after 20 days in culture.

Results with experiments in the shake flasks demonstrated that cell suspensions of *Solanum chrysotrichum* maintain their capacity to synthesize and store significant levels of SC-1, obtaining concentrations that were 50 times higher than the maximum concentrations observed with wild plants (Villarreal *et al,* 1997 a). The registered duplication times based on dry and wet weights, were near the typical values that had been reported for cell suspension cultures of some other species of *Solanum* (Alvarez *et al.*, 1993), as well as other botanical species (Kutsal and Caglar, 1992). The inverse relationship between accumulation of SC-1 and biomass growth (non-growth associated), has been reported with some plant species that accumulate terpenoids (Charlwood and Charlwood,1991) The use of sugars by the growing cells was directly related to dry weight increment, nevertheless, cell growth stopped even though the sugar concentration remained high in the medium, suggesting that cell growth was probably limited by the lack of some other nutrients, or by the accumulation of toxic material. The continuos decrease in the specific growth rate in the cultures, is in accordance with this hypothesis. The results obtained with inoculum size are in accordance with some earlier reports (Alvarez *et al*,1993; Fujita *et al*, 1981), which demonstrated that the maximum yields for the secondary metabolites were obtained using the smallest inoculum size. A high level of sucrose (60 g l^{-1}) provoked an increase in biomass, even though when the highest concentration of this sugar was used (90 g l^{-1}), a decrease in biomass was observed, presumably due to changes in the osmolarity of the cultivated cells. Nevertheless the accumulation of the product did not show big differences in relation to the sucrose levels utilized. These results suggest that the synthesis of SC-1 was not limited by the carbon source concentration in the medium.

The cell suspension grew in bioreactors as an homogeneous whitish suspension of growing cells. Microscopic inspection of the suspension cultures used in this study showed a variety of cell types with phenotype differences with respect to cell size, shape and vacuolar development. Batch-cellular growth and accumulation of SC-1 during 21 days in culture of both cell lines (*ccvx* and *ccvz*) using 2g l^{-1} dry weight as inoculum showed a maximum biomass concentrations of 14.1 and 5.9 g l^{-1} dry weight, for *ccvx* and *ccvz* respectively. Cell suspensions displayed different growth profiles with doubling times of 3.15 and 6.9 days for *ccvx* and *ccvz* respectively, which for *ccvx,* represented a lower value than the one obtained in the shake-flasks experiments for the same cell line (T_d = 4.15 days). Likewise, accumulation of SC-1 was non growth associated, and reached maximum values of 21 and 19 mg.g^{-1} for *ccvx* and *ccvz*, respectively, versus 14 mg g^{-1} obtained for *ccvx* in shake-flask experiments (Villarreal *et al.*, 1996). Figure 3 shows results of growth and product accumulation for both cell lines when the inoculum was equal to 3 g l^{-1} dry weight, and using the same culture conditions described above. Maximum biomass concentrations were 14.6 and 7.7 g l^{-1} dry weight for *ccvx* and *ccvz* showing doubling times of 5.3 and 5.8 days, respectively. Maximum SC-1 concentrations for *ccvx* and *ccvz* were 23 and 20 mg g^{-1} dry weight at 17 and 24 days in culture, respectively.

From the above results it can be seen that even though SC-1 accumulation reached approximately the same value for the two clones, *ccvx* exhibited a higher biomass concentration. On the other hand, inoculum size at the two tested concentrations did not apparently influence biomass accumulation or SC-1 concentration in the bioreactor cultures. This result is in contrast with that found in shake-flask suspensions of this (Villarreal *et al.*, 1997a), as well as other plant species (Alvarez *et al*, 1993; Fujita *et al*, 1981), where the maximum product yield was obtained when the lowest inoculum size was used. Sucrose uptake in the bioreactor cultures was relatively fast; at 12 days practically all the sucrose was hydrolyzed. The concentration of glucose and fructose initially increased but later decreased. While glucose consumption initiated after day 6, fructose consumption occurred only after 10 days of culture. Having established the feasibility of cell suspension growth and bioactive SC-1 production of *S. chrysotrichum* on a large-scale basis, a change in the culture mode was investigated in order to increase productivity. Figure 4 represents the growth profile and SC-1 production of *S. chrysotrichum* cells when a draw-fill batch culture mode was introduced. Interestingly, a steady state in specific SC-1 concentration of about 25 mg g^{-1} dry weight was obtained during the second stage of the culture.

As productivity of any process, including plant cell cultures, depends on growth rate yield and biomass concentration, improvement in all three values should be considered. Plant product accumulation has received increasing attention; however, only few studies have reported yields above 10% dry weight which are considered to be very high. Values around 2% dry weight, as those obtained for SC-1, or other compounds like serpentine (Zenk *et al.*, 1985), are considered promising.

Although biomass concentrations between 10–40 g l^{-1} dry weight can be obtained in cell cultures, the usually obtained biomass concentration ranges between 5 and15 g l^{-1} dry

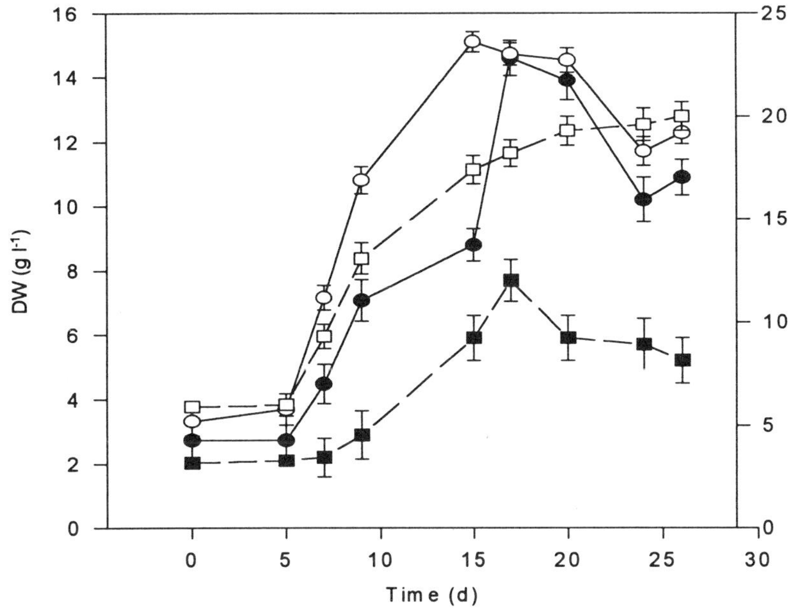

Figure 3. Typical biomass (● ccvx, ■ ccvz) and SC-1 accumulation (o *ccvx*, *ccvz*) in cell suspension cultures of *Solanum chrysotrichum* grown in 10 l airlift bioreactors. Reproduced from Plant Cell Reports Vol 16 pp 653–656,1997.

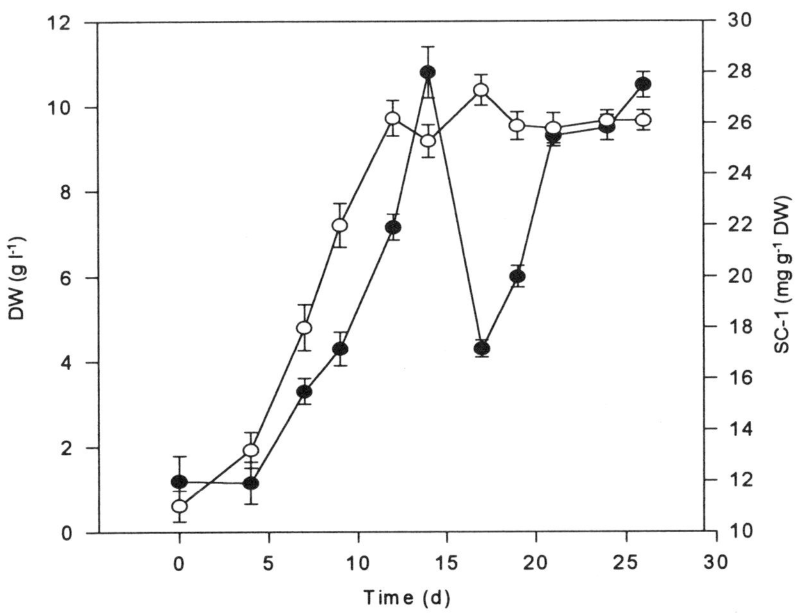

Figure 4. Cell concentration (●) and SC-1 accumulation in (O) *S. chrysotrichum* (ccvx) in draw-fill cell culture mode. Reproduced from Plant Cell Reports. Vol.16, pp653–656,1197.

Table 2. Productivity of *Solanum chrysotrichum* cell suspensions (*ccvx* cell line)

	Maximum productivity	
Volume	(g l^{-1} day^{-1})	T*(days)
250 ml	0.017	16t
10 l batch	0.025	9
10 l draw-fill		
1st stage	0.018	14
2nd stage	0.041	4

T* time in culture
t Villarreal *et al.*, 1997a

weight, when initial carbohydrate concentrations of 1–3% are used (Scragg, 1995). The biomass concentrations reached in this investigation are within such a range, being the concentration for the *ccvx* cell line near the highest reported values. SC-1 productivity from *ccvx* cell suspension cultures grown in batch mode was around 0.025 g l^{-1} day^{-1} (Table 2), a value similar for some other plant species such as *Thalictrum minus* for the production of berberine (Kobayashi *et al.,*1988), and anthocyanin from *Vitis* spp. (Yamakawa *et al.,* 1993). The productivity reached in the bioreactors was between 2.33 and 2.0 times higher than in the previous report using shake flask cultures (Villarreal *et.al.,* 1997a). This situation was probably due to a better oxygen supply. It was also remarkable that the use of a draw-fill mode in the bioreactor culture, allowed an increased productivity value. Namely, a 2.25 higher productivity was obtained in the second stage of the culture, compared to the first stage. Such a high productivity was reached as early as four days after the addition of fresh medium. These results suggest that the use of the draw-fill batch culture mode can increase productivity, eliminating dead periods, such as the time required to sterilize the bioreactor and the initial lag phase of the culture.

CONCLUSIONS

According to ethnobotanical information, the plant species *Solanum chrysotrichum* was selected as the most commonly used plant by the highlands Mayas of Chiapas, México, to treat skin mycosis. Through *in vitro* pharmacological bioassays and with a preliminary clinical study, the efficiency of the methanolic extracts from the leaves of this species was tested. The isolation and structural elucidation of a new antimycotic molecule designated SC-1 from this species was carried out; and the micropropagation of the plant was conducted. Cell suspension batch cultures in shake flasks were established, in which the production of SC-1 was optimized, reaching values fifty times higher than those registered in field grown plants. Large-scale cultivation of the active biomass from *S, chrysotrichum* was conducted in 10 l airlift bioreactors, and productivity levels of SC-1 were increased by 60% when using a draw-fill mode in the bioreactor.

Through the use of plant cell culture technology we were able to produce a new antimycotic from a plant popularly used in Mexican Traditional Medicine, and the potential for clinical application and commercial production of SC-1 is under evaluation.

ACKNOWLEDGMENTS

We wish to thank Dr. Humberto Granados for his helpful discussion and revision of the manuscript. Financial support was provided by CONACYT 3413P-B9608.

REFERENCES

Agrawal, PK.; Jain, DC.; Gupta, RK.; Thakur, RS. Carbon-[13] NMR spectroscopy of steroidal sapogenins and steroidal saponins. *Phytochem.* **1985**, *24*, 11, 2479–2496.

Charlwood, BV. & Charlwood , KA .Terpenoid production in plant cell cultures. In: Ecological Chemistry and Biochemistry of Plant Terpenoids; Harborne JB., Tomas , FA, Eds. Clarendon , Oxford, **1991.**

Fowler, MW. Substrate utilization by plant cell cultures. *J. Chem. Tech. Biotec.*,**1982**, *132*, 338–346.

Kutsal,T.; Caglar, A. Biotechnological advances in developing countries. In: Recent Advances in Biotechnology. Vardar, F.-Sukan, S. Eds. Kluwer Academic. Pub. Dordrecht, **1992**

Levy, G.; Nelson, GL Resonancia magnética Nuclear de Carbono 13. Ediciones Bellaterra S.A. John Wiley and Sons Eds. New York,**1972**

Lozoya, X. Estado Actual del Conocimiento en Plantas Medicinales Mexicanas. IMEPLAN De. México,**1976.**

Lozoya, X.; Aguilar, A. Encuesta sobre el uso actual de plantas en la Medicina Tradicional Mexicana. *Rev. Med. IMSS* (México) **1987**, *25*, 283–291.

Lozoya, X.; Navarro, V.; García, M.; Zurita, M. *Solanum chrysotrichum* (Schdl) a plant used in Mexico for treatment skin mycosis. *J of Ethnopharm.* **1991,** *36,* 127–132.

Meckes, M. Investigación de las Plantas Medicinales en México. Avances y Perspectivas. In: La Investigación Científica de la Herbolaria Medicinal Mexicana. Secretaría de Salud Eds. México,**1993.**

Merillon, JM.; Rideau, M.; Chenieux, JC. Influence of sucrose on levels of ajmalicine serpentine and tryptamine in *Catharantus roseus* cells in vitro. *Plant. Med.*, **1984**, *48,* 497–501.

Murashige,T.; Skoog, E. A revised medium for rapid growth and bioassays with tobacco tissue cultures. *Physiol. Plant.* **1992**, *15*, 473–497.

Paxton, JD. Assays for antifungal activity. In Methods in Plant Biochemistry Dog PM.,Harborne, JD. Eds. Academic Press, Toronto,**1991.**

Pérez, MA.; Villarreal, ML.; Navarro, V.: González,JL.; Alvarez,L.; Delgado, G. A triglycosilated spirostanol with antifungal action from *Solanum chrysotrichum* (sosa). Abstract 9a Jornada de Química. Universidad Autónoma del Estado de Morelos. **1996**,21.

Scragg, A The problems associated with high biomass levels in plant cell suspensions. Plant *Cell Tissue & Org. Cult.* **1995**, *43*,163–170.

Villarreal, ML.; Muñoz, J. Studies on the medicinal properties of *Solanum chrysotrichum* tissue culture. Callus formation and plant induction from axillary buds. *Arch. of Med. Res.* **1991**, *22*: 128–133.

Villarreal, ML.; Arias, C.; Feria-Velasco , A.; Ramírez ,O.; Quintero, R. Cell suspension culture of *Solanum chrysotrichum* (Schldl.). A plant producing an antifungal spirostanol saponin. *Plant Cell Tissue & Org. Cult.* **1997(a)**, *1*,39–44

Villarreal, ML; Arias, C.; Vega, J.; Feria-Velasco, A., Ramírez OT., Nicasio, P., Rojas, G., Quintero, R. Large-scale cultivation of *Solanum chrysotrichum* cells: Production of the antifungal saponin SC-1 in 10 l airlift bioreactors. *Plan. Cell. Rep.***1997(b)**, *16*,653–656.

Yamakawa,T.; Kato, S.; Ishida, K..;Kodama, T.; Minoda, Y. Production of anthocyanins by *Vitis* cells in suspension culture. *Agric.Biol. Ghem.*, **1993**, *47*, 2185- 2191.

Zenk, MH.; El-Shagi, H.; Shulte, U. Anthraquinone production by cell suspension cultures of *Morinda citrifolia. Planta Med.* **1985** Suppl. 79–101.

Zolla C. La Etnobotánica en el Estudio de la Medicina Tradicional. Alternativa para la Salud. CEESTEM. Vol 2, México, **1979.**

Zurita,M.; Zolla,C. Enfermedades Dermatológicas en la Medicina Tradicional de México. *Boletin de la Oficina Sanitaria Panamericana* **1986**, *101,* 339–344.

18

CHEMICALS FROM ROOTS, HAIRY ROOTS, AND THEIR APPLICATION

B. Canto-Canché and V. M. Loyola-Vargas*

Centro de Investigación Científica de Yucatán
Unidad de Biología Experimental
Apdo. Postal 87, CP 97310, Cordemex, Yucatán, México

ABSTRACT

Plants produce thousands of different compounds through the secondary metabolism pathways. Since many of these products are obtained by direct extraction from plants that are cultivated in the field or some times even collected in their original habitat several factors can alter their yield. The use of plant cell cultures has overcome several inconveniences for the production of secondary metabolites. Organized cultures, and especially root cultures, can make a significant contribution to our understanding of secondary metabolism. Furthermore, a new alternative has arisen: transformed root cultures. Until now, hairy roots have been obtained from more than 100 different species. The products that they are able to produce range from alkaloids to aromatic compounds and dyes. These kinds of cultures have turned out to be an invaluable tool to study the biochemistry and the gene expression of the metabolic pathways in order to elucidate the intermediaries and enzymes involved in the biosynthesis of secondary metabolites.

PLANT ROOTS

Plant kingdom includes a very large group of living organisms, ranging from single cells to gigantic trees. Although there is high diversity, the vegetative body of seed plants is basically composed of three organs: leaf, stem and root. From a very simplistic point of view, the leaf's function is photosynthesis, while the stem provides support and transport of the nutrients. The function for the root is to anchor the plant and function for the uptake

* Email: vmloyolacicy.cicy.mx

water and nutrients from the soil. Terrestrial plant roots are vegetative organs which attach the plant to the ground for support.

On the bases of their origin, roots are classified as primary, adventitious and laterals. Primary roots emerge directly from the plant stem, while the roots emerging from internods are called adventitious. Secondary and tertiary roots growing from other roots are called lateral. This classification has been called developmental classification. Typically, a plant produces one or more primary roots and these can then produce secondary lateral roots and then, they later can ramify considerably, forming the functional root system.

The size of the roots is variable; an average root diameter is less than 0.5 mm, roots with diameters from 20 mm or more are considered large. In grasses, the root system has very fine terminal branches (less than 100 mμ diameter). Common root system length could vary from few centimeters to 4–5 meters, as is the case of *Eucalyptus calophylla*. Their color is also variable. Young roots can be unpigmented (white or pale yellow), pink or orange. While older roots are brown. Roots of *Datura* and others plant genera are able to form chlorophyll when exposed to light. (Fitter, 1991; Glinski and Lipiec, 1990).

PLANT ROOT ANATOMY

Anatomically, a typical root is composed of root cap, epidermis, cortex, vascular stele and endodermis (Figure 1A) (Glinski and Lipiec, 1990). The root cap is attached to the head of the root. It consists of a conical mass which produces and secretes mucigel, allowing the soft root to migrate into the hard soil protecting the apical meristems. The root cap cells regulate the geotropic curvature of the root serving as a gravi-sensing tissue (Taiz and Zeiger, 1991). The cap cells are filled with starch which acts both as nutrient source for production of gel that lubricates the root tip, and allows the root to detect the direction of earth's gravitational field and then direct the root to grow downward. These gravity-sensitive bodies are called statoliths. The cap tissue cells are continuously renewed. Statocytes, gravity perceiving cells-, arise from meristematic division and are transformed into secretion cells. Although, the cycle time depends on a number of factors, typically, in a few days, the cap has been renewed completely. When the cells reach the outer layer of the cap, their middle lamellae is degraded. Cells separate each other and from the root. This event defines a kind of cells called border cells; their characteristic is that they are transcriptionally active after separation (Bringham, 1995). Cells are loosely adhered to each other and to the root cap by a water soluble mucilage which can be removed by gentle agitation in water. These cells are able to survive in a wide range of osmotic conditions, including distilled water, without lysis or apparent cellular damage.

Epidermis is formed from the outer layer of thin-walled elongated cells. Aerial epidermis has a waxy cuticle while epidermis root does not have it. The root epidermis is adapted for absorption. Its permeability depends on the suberization extent. One of its particular characteristics is its extension into root hairs, which increases the surface area and contributes to absorption. They arise from the secondary wall and penetrate the primary wall emerging into the rizosphere. Lectins, which are responsible for legume recognition by rhizobium, are secreted at the root hair surface. Vesicular-arbuscular mycorrhization is also through the root hairs.

The cortex lies on the outside of the endodermis. This layer of cells has intercellular spaces which allows longitudinal movement of gases.

Vascular stele is the system specialized for transport. It is composed of xylem, phloem and pericycle. Xylem conducts water and minerals from the soil to the shoots;

A

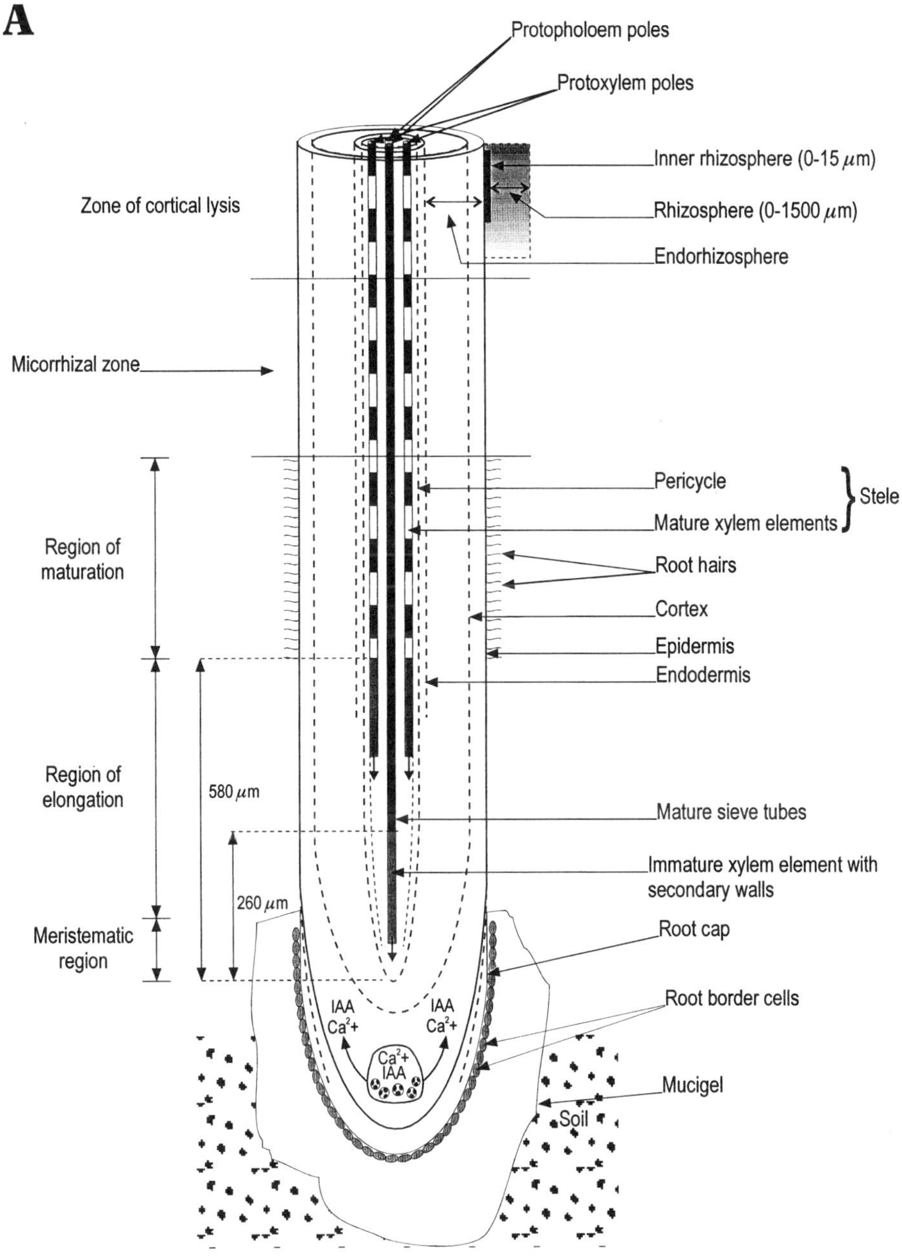

Figure 1. Diagram of a tobacco root tip showing relative order of maduration and major regions of differentiation in the root (A) and rhizosphere (B). Redrawn from Foster et al. (1983) and Kramer (1983) with permission.

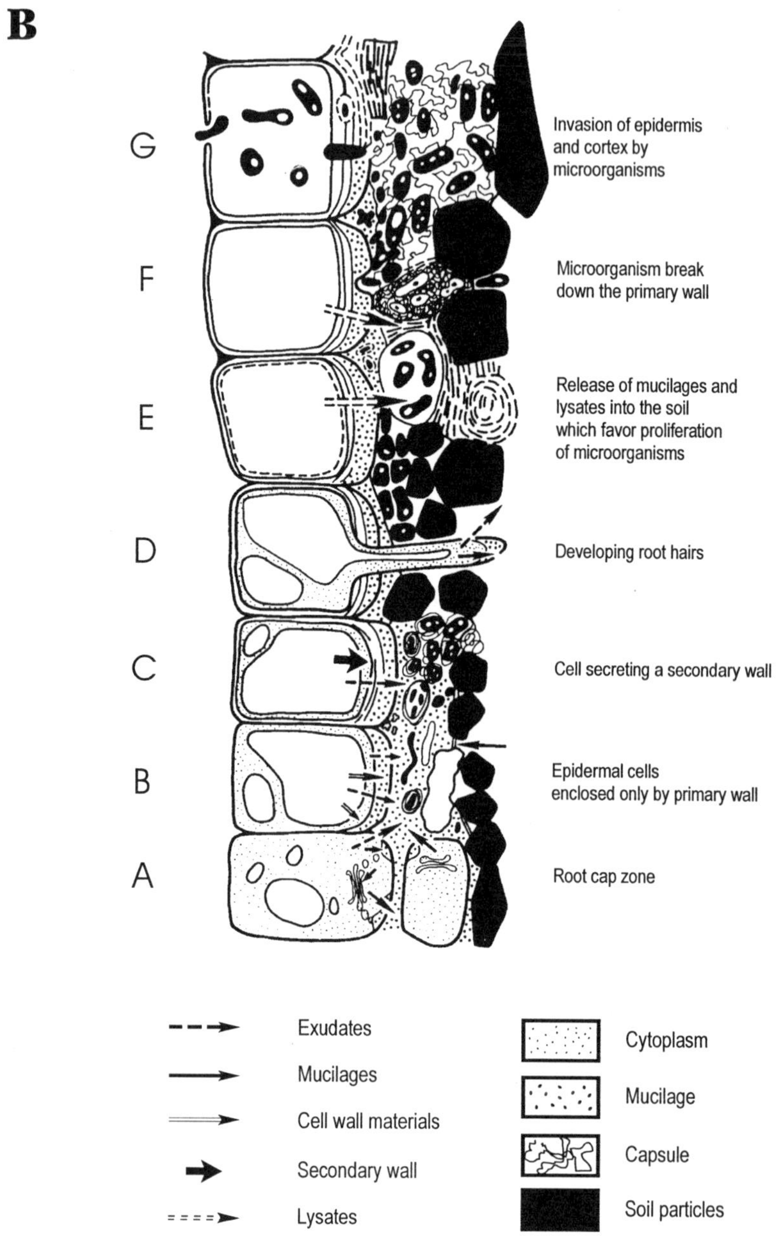

Figure 1. (*Continued.*)

phloem transports assimilate in the opposite direction. These transport systems are surrounded by the pericycle.

Endodermis constitutes a layer of thick cells surrounding the stele. Endodermal cells produce suberin; its deposition forms a specialized stratum which isolates and impermeabilizes the inner area.

Biological FunctionsIn general, main functions related to roots are: anchor the plant and to provide it with minerals like nitrogen, potassium, phosphorus, magnesium, calcium, etc., and water. Inorganic nutrients move by diffusion and mass flow. The rate of input depends on the size of the root system, demain impose by plant tissues and mobility of nutrient in the soil.

Roots synthesize the growth regulators cytokinin, auxin, gibberellin, abscisic acid and ethylene. Some of them are transported via xylem to the shoots, establishing a communication between roots and aerial parts (Itai and Birnbaum, 1991). Their level is a balance between synthesis, degradation, conjugation and transport from the root. Phyto-regulators have a deep influence on the functions and metabolic processes of the roots, including its adaptation to the rounding soil (nutrient stress, salt stress, drought and flood).

Roots contribute to the composition of the soil, both when they are living and when they are decomposing. The organic materials released into the soil ecosystem by roots are classified as exudes, secretions, mucilage, mucigel and lysates (Rovira et al., 1979). These are released from the epidermal and cortical tissues, senescence woody, fine roots, damaged roots due to microbial disease, autolysis, etc. (Vogt et al., 1991). The root-soil interface, where roots have influence, is called rhizosphere. Plant roots can interact, not only with abiotic environment of rhizosphere, but also with the biotic elements. Some microorganisms are attracted by volatile compounds liberate into the soil by roots. Because rhizosphere is very rich in organic matter, microorganisms are stimulated to grow and reproduce (Foster et al., 1983).

Under natural conditions, many plant roots are able to establish a mutualistic symbiotic relationship with some fungi. The composite organs formed as a result of this by the roots and fungi are called mycorrhizae. Mycorrhizae seems to occur in around 83% of dicotyledonous and 79% of monocotyledonous plants (Wilcox, 1991). Root supplies carbohydrates to the fungus, and in return fungus helps in the uptake of immobile inorganic nutrients, specially phosphorous (Mukerji et al., 1991). Members of Brassicaceae and Cucurbitaceae do not show mycorrhizal association, while all gymnosperms appear to establish this association. In addition, legumnous plant establish symbiotic relationship with *Rhizobium,* thus forming root nodules specialized in nitrogen fixation. Some others functions of roots are: vegetative propagation and storage.

From the agronomical point of view, roots are important because they protect soil from erosion and the efficient use of external inputs depends on them. Efficient use of available water and nutrients reduces the need for fertilizer, herbicide, irrigation and drainage with repercussions on economy and ecology (Van Noordwijk and De Willigen, 1991).

Human Application of Plant Roots

Anthropologically, plant roots are very important, not only because they are naturally involved in plant survival but also because of their very complex metabolism and are called chemical factories by Flores (1992). Their biosynthetic value is enormous. For many centuries herbal medicines have been used in the treatment of human diseases. Phytochemical analysis of plants used in folklore medicine, has made it possible to identify and isolate the active principles from a variety of sources. To date, in particular from roots, biological pesticides, herbicides, dyes, flavors, fragrance essences, pharmaceutical active principles, have been recovered, which support giant industries with enormous economical returns.

In addition, some roots have been used as food since immemorial times. In the opinion of Kutschera (1991), a main function of the roots is to take up the excess of assimilates produced by the shoot and store it. In some plant species, roots represent important storage organs. Storage rhizomes and tubers are important food crops. From Food and Agricultural Organization (FAO) records, in 1996 millions of tons of potato (*Solanum tuberosum*), sugar beet (*Beta vulgaris*), Cassava (*Manihot esculenta*), sweet potato (Ipomoea batatas), yam (*Dioscorea* spp.) and carrots (*Daucus carota*) were raised and harvested (Loyola-Vargas and Hernández-Sotomayor, 1997). Potato is the most important non-cereal world food in terms of total food production. It has the capacity to produce more energy and protein per unit land than any other food crop, and contains a better balance of essential amino acids, particularly lysine (Bajaj, 1987). At least, nineteen other nutritious root and tuber crops are used as food as well (for more information, refer to Loyola-Vargas and Hernández-Sotomayor, 1997).

While root for eating come from farming, many metabolites extracted from roots are obtained from native plants cultivated in the field. In worst cases, plants are harvested from their original habitats, as very high amounts of plant source are needed for extraction of these compounds (Loyola-Vargas and Miranda-Ham, 1995), and this represents a serious ecological problem. In some cases, the careless exploitation has pushed some metabolite plant sources toward extinction, as are the cases for *Dioscorea deltoidea* from Mexico and *Rawolfia serpentina* from India.

Harvesting plant from the field has another shortcoming. Metabolite yield is very variable, as the production is under the influence of various environmental factors such as temperature, rain, light, pest and disease. In addition, it could be a time-consuming task as some plants need many years of cultivation before use.

As was mentioned earlier roots are an important site for the synthesis of some important metabolites (Table 1), *in vitro* culture of roots serves as an alternate to overcome the inconveniences mentioned above.

IN VITRO ROOT CULTURES AND HAIRY ROOTS

Plant tissue and organ cultures are techniques which allow growth and maintenance of the material in chemical media, under axenic conditions. *In vitro* cell and tissue cultures began from the ancient idea (in 1800's) that individual plant cells have the ability of totipotency, i.e. they could regenerate a whole plant. Initially, it was easy to probe using vegetative cells, but was more difficult in the case of well differentiated cells organized in tissues and organs as they were unable to grow and divide. However, in 1934 Philip White established a successful and indefinite culture of excised roots of tomato.

Discovery of phytohormones like auxins and cytokinins had great influence on tissue cultures. The knowledge of the role of these hormones, and the critical effect of the cytokinin/auxin ratio had on differentiation, led Steward et al. (1958) to regenerate whole plants from cells obtained from carrot roots (reviewed in Maheshwari, 1990). Later, cell tissue and organ culture techniques have developed significantly (Maheshari, 1990; Thorpe, 1990; Allan, 1991). To date, it is possible to cultivate *in vitro* single cells, anthers, protoplasts, embryos, etc. with multiple biotechnological applications.

Table 1. Some secondary metabolites produced in plant roots (Charlwood and Rhodes, 1990)

Group of compounds	Plant source
Alkaloids	
Indole alkaloids	*Catharanthus roseus*
Isoquinoline alkaloids	*Cinchona ledgeriana, Papaver bracteatum*
Pirrolizidine alkaloids	*Senecio erucifolios, S. squalidus, S. vernalis, S. vulgaris*
Quinoline alkaloids	*Cinchona ledgeriana, C. succirubra*
Steroidal alkaloids	*Solanum chacoense, S. khasianum*
Tropane alkaloids	*Atropa belladonna, Calystegia sepium, Datura innoxia, D. fastuosa, D. leichhardtii, D. stramonium, Duboisia hopwoodii, D. leichhardtii, D. myoporoides, Hyoscyamus albus, H. bohemicus, H. canariensis, H. gyorffi, H. muticus, H. myoporoides, H. niger, Scoplia parviflora*
Phenolics	*Anethum graveolens, Derris elliptica, Linum flavum, Pimpinela anisum,*
Steroids	*Digitalis lanata, D. purpurea*
Terpenoids	*Centranthus macrosiphon, C. ruber, Valeriana officinalis, Zingiber officinale*

In the case of the production of plant secondary metabolites the use of single cells, callus and organ *in vitro* cultures has been explored. In general, organ cultures give better production than less organized cultures, despite the fact that non-organized cultures do not lack all the biosynthetic ability required to produce secondary metabolites.

In vitro root culture consists of initiating and maintaining successfully excised plant roots from seedlings or from roots formed on the surface of callus cells. A complex medium containing a carbon source, vitamins and inorganic salts is used to support the culture survival. Auxins are essential for root growth. Normal root cultures possess lower levels of endogenous auxin than those required to sustain active growth and morphogenesis. All the surface material of the explant from which the tissue culture would start has to be sterilized. Culture manipulations are made in laminar flow sterile cabinets and flask containing the cultures are maintained under specific growth conditions and subcultured at specific time intervals. Roots have been cultured not only on solid agar-containg media but also in liquid media, as in the latter, they usually grow faster.

Since the initial culture of tomato roots was achieved, many other plant root cultures have been established, both from angiosperm and gymnosperm families (Table 2). Although many advantages emerge from root cultures over non-organized cultures, some troubles are associated with them as well (Table 3). To avoid some of the disadvantages of the normal root cultures the *Agrobacteriun rhizogenes*-transformed hairy roots system has been established (Flores and Medina-Bolívar, 1995, Loyola-Vargas and Miranda-Ham, 1995).

Agrobacterium is a Gram negative bacterium from the soil and it is able to infect cells from higher plants (Figure 2). The bacterium contains a large plasmid (200 Kb), part of which is transferred (T-DNA) and incorporated into the plant host nuclear genome (Chilton et al., 1982). The metabolism of the host is modified by the integrated DNA. Plant begins to synthesize constitutively phytohormones, which alter the developmental

Table 2. Plants that has been used to induce non-transformed root cultures

Family	Genus and species	Reference
Apocynaceae	*Catharanthus trichophyllus*	(Cordell and Farnsworth, 1976)
Araliiaceae	*Panax ginseng*	(Furuya et al., 1989; Ushiyama et al., 1989; Ushiyama and Furuya, 1989)
Asteraceae	*Tagetes erecta*	(Mukundan and Hjortso, 1990)
Buxaceae	*Buxus papillosa*	(Atta-ur-Rahman et al., 1992)
Compositae	*Eupatorium cannabinum*	(Siebertz et al., 1989)
Curcurbitaceae	*Trichosanthes kirilowii*	(Wang et al., 1994)
Labiatae	*Coleus forskohlii*	(Sen et al., 1993)
Leguminosae	*Glycine max*	(Ebel et al., 1989)
	Swainsona galegifolia	(Ermayanti et al., 1994)
Linaceae	*Linum flavum*	(Berlin et al., 1988; Van Uden et al., 1991)
Menispermaceae	*Menispermum davricum*	(Sugimoto et al., 1996; Sugimoto et al., 1994)
	Stephania cepharantha	(Sugimoto et al., 1988; Sugimoto and Yamada, 1993; Sugimoto et al., 1989)
Phytolaccaceae	*Phytolacca acinosa*	(Strauss et al., 1995)
Rubiaceae	*Cephaelis ipecacuanha*	(Jha et al., 1991; Yoshimatsu and Shimomura, 1991)
	Cinchona ledgeriana	(Heimer 1975)
Solanacea	*Atropa belladonna*	(Hartmann et al., 1986)
	Capsicum annuum	(Chávez-Moctezuma and Lozoya-Gloria, 1996)
	Datura stramonium	(Olby 1974)
	Hyosciamus niger	(Hashimoto et al., 1986; Matsuda et al., 1991)
	Hyosciamus albus	(Hashimoto et al., 1989)
	Nicotiana alata	(Friesen et al., 1992)
Scrophulariaceae	*Digitalis lanata*	(Lui and Staba, 1979)
Scrophulariaceae? Gesneriaceae?	*Rehmannia glutinosa*	(Matsumoto et al., 1989)
Taxaceae	*Taxus baccata*	(Zocher et al., 1996)

program of the transformed tissue, producing the characteristic phenotype of the disease. Transformed cells produce unusual amino acids derived from arginine; they are called opines and they are used by the bacteria. *A. rhizogenes* is closely related to *A. tumefasciens* (Chilton et al., 1977). The plasmid from *A. rhizogenes* is transferred to the plant genome and is called Ri because of its root-inducing abilities.

The ability of *Agrobacterium* to infect plant cells has been largely used to establish *in vitro* hairy roots. Several protocols have been developed. Initially infections were made by cocultivating plant protoplast with the bacterium. Today, infection is done on surface-sterilized stem fractions, detached leaves, leaf discs, tuber discs, etc. from a plant grown in

Table 3. Non-transformed root cultures characteristics

Advantages	
	Biosynthetic cability to produce secondary metabolites.
	Higher production of secondary metabolites than callus and cell suspension.
Disadvantages	
	Exogenous auxin-dependency.
	Large inoculum is required for adequate growth.
	Low branching degree.
	Long lag phases.
	In many cases yield of product is less than roots of the whole plant.

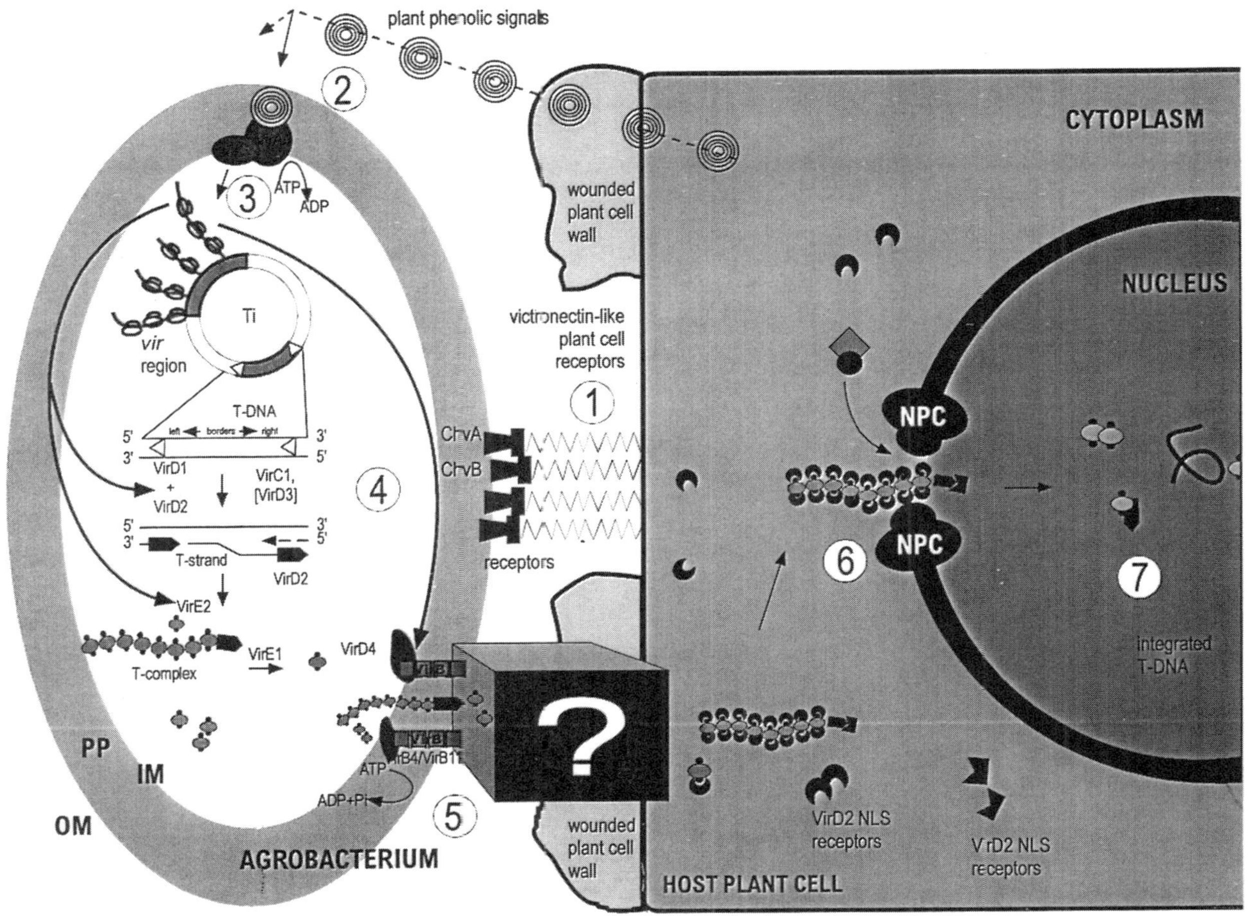

Figure 2. Diagram summarizing all major cellular reactions involved in T-DNA transport. Redrawn from Sheng and Citovsky (1996) with permission.

Table 4. Characteristics of transformed roots derivated from plant cells *Agrobacteriun rhizogenes-infection*

High incidence of lateral branching
Exogenous phytohormones-independent growth
Negative geotropism
Synthesis of opines (agropine, mannopine or cucumopine)
Easier to regenerate into plants
Stable and fast growth
Genetic and biochemical stability

greenhouse or with axenic plants grown under *in vitro* conditions. A wound is made using a sterile syringe needle which contains a concentrated bacterial suspension (few drops). After 1–4 weeks, fast-growing adventitious roots emerge, either from the site of the wounding or from a tumor formed first on the site of inoculation (Chilton et al., 1977; David et al., 1984).

Therefore, the bacterium can be eliminated from the root culture. Emerged roots are excised and put on a solid selective media without phytohormones but containing ampicillin, carbenicillin or cefotaxione in order to eliminate the bacterium. Hairy roots for some genera as *Datura* and *Nicotiana* are relatively easy to establish as independent cultures (Flores and Filner, 1985a; Parr and Hamill, 1987); Although for other genera such as *Catharanthus* and *Cinchona* is more difficult (Parr et al., 1988; Hamill et al., 1989). After transfer to solid media, roots produce callus. Decreasing the ionic strength of the media has shown good results (Toivonen et al., 1989; Ciau-Uitz et al., 1994). Many factors affect hairy roots growth such as pH, media composition and ionic strength, type and concentration of sugars, etc. (reviewed by Loyola-Vargas and Miranda-Ham, 1995). Data from several groups (Flores and Filner 1985b; Yoshikawa and Furuya, 1987, Jaziri et al., 1988 , Kyo et al., 1990; Shimomura et al., 1991; Benjamin et al., 1994) and also from our laboratory has shown that when optimum conditions of growth are established, hairy roots usually grow faster than normal roots (non-transformed) cultivated *in vitro* (Vazquez-Flota et al., 1994; Maldonado-Mendoza et al., 1993; Ciau-Uitz et al., 1994). In general, culture media are based on Gamborg´s B_5 (Gamborg et al., 1968) or Murashige and Skoog (1962) salt without hormones and antibiotics. Hairy root characteristics are summarized in Table 4.

Today the list of plant species from which hairy roots have been obtained comprise more than 30 plant families (Table 5). Asteraceae, Cruciferae and particularly Solanaceae are the families with more species transformed *in vitro* with *A. rhizogenes*.

APPLICATIONS OF HAIRY ROOTS CULTURES

Production of Chemical Products

Because of their high metabolic potential, transformed roots have been largely used to obtain their respective secondary products. The first work focusing on this point was from Flores and Filner in 1985. They transformed a member of the *Solanaceae*, family, *Hyosciamus muticus* and observed that transformed root lines produced similar tropane alkaloid concentrations, as was seen from the roots of the whole plant. Later, Hamil et al. (1986) found similar results in *Beta vulgaris* and *Nicotiana rustica*. Hairy root cultures from both species synthesize their characteristic secondary products at levels comparable with those of *in vivo* roots. The list of chemical products obtained from hairy roots has grown enormously (this subject has been recently review by Loyola-Vargas and Miranda-Ham, 1995). Products recovered from hairy roots comprise alkaloids (indole alkaloids, isoquinoline alkaloids, pyrrolizidine alkaloids, quinoline alkaloids, quinolizidine alkaloids, tropane alkaloids, etc.), terpenoids (monoterpens, sesquiterpens, diterpens, triterpens, steroids, cardenolides), flavonoids, phenol compounds (coumarines, tannins), anthraquinones, quinones, glucosinolates, lignans, proteins (peroxidases, antivirals), etc.

It is generally known that undifferentiated *in vitro* cultures produce lower levels of secondary metabolites than those from more differentiated cultures. In the history of *in vitro* plant culture a correlation between developmental stage of culture and secondary metabolites production has been established (Nessler, 1990), thus suggesting that the fact that secondary metabolism pathways are under developmental regulation (Miranda-Ham, 1990). As a source of chemical products, hairy roots offer a number of advantages. As opposed to the non-transformed root cultures, which need auxins to grow and to maintain their phenotype, hairy roots are largely auxin-independent; they show a fast growth and a high lateral branching in an auxin-free medium. Therefore, antagonic effect of auxins on secondary biosynthetic pathways (Hashimoto et al., 1986; Deno et al., 1987; Rhodes et al., 1994) is eliminated. Furthermore, when exogenous auxin is added to solanaceous transformed root cultures, for example, it produces an intense undifferentiation mass from roots into cell suspension with a concomitant loss of their biosynthetic capacity (Flores and Filner, 1985a; Deno et al., 1987; Ruiter and Kollöffel, 1985; Parr et al., 1990)

The ability of hairy roots to produce similar or higher levels of the characteristic metabolites from the plant is stable during subculture periods (Flores and Filner, 1985b; Maldonado-Mendoza et al., 1993; Ciau-Uitz et al., 1994; Joao and Brown, 1994; O'Keefe et al., 1997). This biochemical stability is related to the chromosomal stability in these cultures. The non-organized cultures show changes in ploidy and also chromosomal instability (Larkin and Scowcroft, 1981), whereas hairy roots retain their euploid chromosomal number from the parent plant (Baíza, 1990; Banerjee-Chattopadhyay et al., 1985; Ambros et al., 1986; Hänish et al., 1987; Aird et al., 1988). However, recently changes in chromosomal number in alfalfa hairy roots during subcultures has been reported (Xu and Jia, 1996; Zu et al., 1997).

In order to increase the secondary metabolites production in hairy root cultures, several strategies have been established. Firstly parental plants should be chosen very carefully, since *in vitro* culture derived from high yielding parent plants maintains higher levels of secondary products (Zenk et al., 1977, Kinnersley and Dougall, 1980). However, this is not always true as the condition of the culture may affect deeply the metabolic behavior of cultures (Galneder and Zenk, 1990). Screening of different species and also of different varieties is a possibility which needs to be explored. In case of *Catharanthus roseus* there are at least four different cultivars commercially used, each of them producing their particular alkaloid spectrum (Kurz et al., 1985). Second, variation arises in cultures derived from explants of the same explant, therefore a screening of secondary metabolites production in the established hairy root lines is suggested (Bhadra et al., 1993). There are a number of different techniques available in order to screen the metabolites content from the starting plant material: thin layer chromatography, high performance liquid chromatography, spectrophotometry, fluorescence, radioimmunoassay, etc. Selection of the most suitable method depends on the characteristics of the compounds to be monitored.

The inorganic salts or vitamin content of the media composition can also be modified. In the case of alkaloids, the content increases either by lowering nitrogen or phosphorous concentration of the medium (Hilton and Rhodes, 1993; Hilton and Rhodes,

Table 5. Hairy root cultures generated for production of secondary metabolites and basic studies

Family	Genus and species	Major metabolite	Reference
Apocynaceae	*Amsonia elliptica*	Indole alkaloids	(Sauerwein et al., 1991a)
	Artemisia annua	Artemisinin	(Jaziri et al., 1995; Weathers et al., 1994; Teoh et al., 1996; Mukherjee et al., 1995; Smith et al., 1997)
	Catharanthus roseus	Ajmalicine, serpentine, catharanthine, vindoline, vindolinine	(Ciau-Uitz et al., 1994; Bhadra et al., 1993; Ho and Shanks, 1992; Pereda-Miranda et al., 1993; Bhadra et al., 1991; Jung et al., 1992; Parr et al., 1988; Toivonen et al., 1991; Toivonen et al., 1990; Toivonen et al., 1989; Vázquez-Flota et al., 1992; Jung et al., 1994; Bhadra and Shanks, 1995; Jung et al., 1995; Jung et al., 1994; Bhadra and Shanks, 1997)
	Catharanthus trichophyllus	Indole alkaloids	(Davioud et al., 1989a; Davioud et al., 1989b)
	Lactuca virosa	Sesquiterpene lactones	Stojakowska et al., 1993)
	Leontopodium alpinum	Hydroxycinnamic acid esters (essential oils)	(Hook 1994)
	Rauwolfia serpentina	Ajmalicine, serpentine	(Benjamin et al., 1994; Falkenhagen et al., 1993)
	Vinca minor	Indole alkaloids	(Tanaka et al., 1994; Tanaka et al., 1995)
Araliaceae	*Panax ginseng*	Biotransformation, polyacetylenes, saponins	(Asada et al., 1993; Hirakura et al., 1991; Yoshikawa and Furuya, 1987; Yoshikawa et al., 1993; Yoshimatsu et al., 1996)
	Panax quinquefolius	Polyacetylens	(Smith et al., 1996)
Asteraceae	*Acmella oppositifolia*	Polyacetylenes	(Flores et al., 1993)
	Artemisia absinthium	Volatile oils	(Kennedy et al., 1993)
	Artemisia annua	Artemisinin	(Paniego and Giulietti, 1996)
	Ambrosia artemisiifolia	Thiarubrine A & B	(Gomez-Barrios et al., 1992)
	Ambrosia trifida	Thiophenes, thiarubrines	(Lu et al., 1993)
	Bidens alba	Polyacetylenes	(Norton and Towers, 1986)
	Chaenactis douglasii	Thiophenes, thiarubrines	(Constabel and Towers, 1989; Constabel and Towers, 1988)
	Coreopsis tinctoria	Phenylpropanoids	(Horz and Reichling, 1993; Reichling and Thron, 1990)
	Echinacea purpurea	Alkalmides	(Trypsteen et al., 1991)
	Echinops pappii	Thiophenes	(Abegaz 1991)
	Lactuca virosa	Sesquiterpene lactones	(Kisiel et al., 1995)
	Stevia rebaudiana	Steviol glucosides	(Yamazaki et al., 1991)
	Tagetes campanulata	Thiophenes	(Talou et al., 1994
	Tagetes erecta	Thiophenes	(Mukundan and Hjortso, 1990b; Kim et al., 1995; Jacobs et al., 1995; Croes et al., 1994; Hjortso and Mukundan, ; Flint-Wandel and Hjortso, 1993)
	Tagetes laxa	Thiophenes	(Talou and Giulietti, 1995; Talou et al., 1994)
	Tagetes minuta	Thiophenes	(Talou et al., 1994)

Table 5. (*Continued*)

Family	Genus and species	Major metabolite	Reference
	Tagetes patula	Thiophenes, biotransformation	(Buitelaar et al., 1993; Croes et al., 1989; Buitelaar et al., 1992; Uesato et al., 1986; Kyo et al., 1990; Mukundan and Hjortso, 1990a; Mukundan and Hjortso, 1990c; Mukundan and Hjortso, 1991a; Mukundan and Hjortso, 1991b; Parodi et al., 1988; Wilson et al., ; Westcott ; Arroo et al., 1995b; Croes et al., 1994; Jacobs et al., 1995; Arroo et al., 1995c; Arroo et al., 1995a)
	Tagetes terniflora	Thiophenes	(Talou et al., 1994)
Boraginaceae	*Lithospermum erythrorhizon*	Shikonin	(Sim and Chang, 1993; Shimomura et al., 1991b; Wilson et al.,)
Campanulaceae	*Campanula medium*	Polyacetylenes	(Tada et al., 1996b; Lloyd 1995)
	Lobelia cardinalis	Polyacetylenes	(Tanford 1983)
	Lobelia chinensis	Polyacetylenes	(Tada et al., 1995b)
	Lobelia inflata	Lobetyolinin, lobetyol, lobeline, polyacetylenes	(Ishimaru et al., 1992a; Ishimaru et al., 1991; Yonemitsu et al., 1990)
	Lobelia sessilifolia	Lobetyolin, biotransformation	(Ishimaru et al., 1994; Yamanaka et al., 1995)
	Platycodon grandiflorum	Polyacetylenes (@lobetiol, lobetyolin and lobetyolinin)	(Tada et al., 1995a; Ahn et al., 1996)
Casuarinaceae	*Casuarina glauca*	Nitrogen fixation	(Diouf et al., 1995)
Chenopodiaceae	*Beta vulgaris*	Betalain	(Benson and Hamill, 1991; Hamill et al., 1986; Dilorio et al., 1993)
Cephalotaxaceae	*Cephalotaxus harringtonia*	Homoharringtonine	(Wickremesinhe and Arteca, 1991)
Compositae	*Carthamus tinctorius*	Pigments	(Cheetham et al., 1989)
	Lactuca floridana	Sesquiterpen lactones	(Song et al., 1995)
	Lactuca virosa	Sesquiterpen lactones	(Kisiel et al., 1995)
	Leontopodium alpinum	Anthocyanin	(Comey et al., 1992)
	Perezia cuernavacana	Sesquiterpene quinone	(Arellano et al., 1996)
	Serratula tinctoria	Dyes	(Corio-Costet et al.,)
Convolvulaceae	*Calystegia sepium*	Alkaloids	(Jung and Tepfer, 1987)
	Ipomoea aquatica	Peroxidases, superoxide dismutase	(Taya et al., 1994)
Cruciferae (Brassicaceae)	*Armoracia lapathifolia*	Peroxidases	(Saitou et al., 1991)
	Armoracia rusticana	Peroxidases	(Parkinson et al., 1990; Uozumi et al., 1992; Taya et al., 1989; Parkinson et al., 1990; Repunte et al., 1993)
	Brassica napus	Nitrogen metabolism, microgravity	(Downs et al., 1994b; Downs et al., 1994a; Slater and Guo, 1996)

(*continued*)

Table 5. (*Continued*)

Family	Genus and species	Major metabolite	Reference
Curcurbitaceae	*Cucurbita pepo*	Peroxidases	(Katavic et al., 1991; Repunte et al., 1993)
	Momordica charantia	Proteins	(Wagner and Flores, 1992)
	Luffa cylindrica	Bioactive proteins	(Di Toppi et al., 1996)
	Trichosanthes cucumerina	Bioactive proteins	(Savary and Flores, 1991)
	Trichosanthes cucumeroides	Bioactive proteins	(Savary and Flores, 1991)
	Trichosanthes kirilowii	Bioactive proteins	(Savary and Flores, 1993; Savary and Flores, 1994; Wang et al., 1994; Thorup et al., 1994; Savary et al., 1997)
Euphorbiaceae	*Euphorbia lathyrins*	Alkaloids	(Follansbee et al., 1995)
	Phyllanthus niruri	Phenolic compounds	(Ishimaru et al., 1992b)
Gentianaceae	*Swertia japonica*	Phenyl glucosides, amarogentin, amaroswerin	(Ishimaru et al., 1990c; Ishimaru et al., 1990b)
Geraniaceae	*Geranium thunbergii*	Tannins	(Ishimaru and Shimomura, 1991)
Gramineae	*Swainsona galegifolia*	Swainsonine	(Ermayanti et al., 1994a)
Juglandaceae	*Juglans regia*	Rooting	(Caboni et al., 1996)
Labiatae (Lamiaceae)	*Ajuga reptans*	Phytoecdysteroids	(Matsumoto and Tanaka, 1991; Uozumi et al., 1993; Yagi et al., 1996; Fujimoto et al., 1996; Tanaka and Matsumoto, 1993)
	Ocimum basilicum	Rosmarinic acid	(Tada et al., 1996a)
	Salvia miltiorrhiza	Diterpenes	(Hu and Alfermann, 1993)
	Scutellaria baicalensis	Flavonoids, phenylethanoids	(Zhou et al., 1997)
Lactuceae	*Lactuca floridana*	Lactucin	(Song et al., 1995)
Leguminosae	*Astragalus boeticus*	Polysaccharides, Triterpenoids	(Ionkova et al., 1992; Ionkova and Alfermann, 1990)
	Astragalus gummifer	Polysaccharides	(Ionkova et al., 1992)
	Astragalus hamosus	Polysaccharides, Triterpenoids	(Ionkova et al., 1992; Ionkova and Alfermann, 1990)
	Astragalus membranaceus	Astragalosides	(Hirotani et al., 1994a; Hirotani et al., 1994b; Zhou et al., 1995; Ionkova et al., 1992)
	Astragalus mongholicus	Saponins, polysaccharides	(Ionkova et al., 1992)
	Cassia obtusifolia	Pigments	(Ko et al., 1995)
	Cassia occidentalis	Pigments	(Ko et al., 1995)
	Cassia torosa	Pigments	(Ko et al., 1995; Taya et al., 1989)
	Glycyrrhiza glabra	Liquiritigenin, isoliquiritigenin	(Toivonen and Rosenqvist, 1995; Toivonen and Rosenqvist, 1995)
	Lupinus hartwegii	Isoflavones	(Berlin et al., 1989)
	Lupinus polyphyllus	Isoflavones	(Berlin et al., 1989)
	Lotus corniculatus	Tannins, phenylpropanoids	(Doolittle 1994; Robbins et al., 1991; Carron et al., 1994; Robbins et al., 1995)
	Psoralea spp.	No furanocoumarins	(Nguyen et al., 1992)
	Robinia pseudoacacia	Regeneration	(Han et al., 1993)
	Spartium junceum	Quinolizidine alkaloids	(Wink and Witte, 1987)
	Swainsona galegifolia	Swainsonine	(Ermayanti et al., 1994b)

Table 5. (*Continued*)

Family	Genus and species	Major metabolite	Reference
	Trifolium repens	Nodulation	(Díaz et al., 1995)
	Trigonella foenum-graecum	Diosgenin	(Rodríguez-Mendiola et al., 1991; Merkli et al., 1997)
Linaceae	*Linum flavum*	5-mehoxypodophyl - lotoxin	(Oostdam et al., 1993)
Malvaceae	*Althaea officinalis*	Polysaccharides	(Ionkova et al., 1991)
Marantaceae	*Thaumatococcus daniellii*	Thaumatin	(Witty 1990a; Witty 1990b; Witty 1989)
Menispermaceae	*Menispermum dauricum*	Bisbenzylisoquinol ine alkaloids	(Sugimoto et al., 1996)
	Stephania cepharantha	Bisbenzylisoquinol ine alkaloids	(Sugimoto et al., 1990)
Papaveraceae	*Papaver somniferum*	Sanguinarine	(Williams and Ellis, 1988; Williams and Ellis, 1993)
Pedaliaceae	*Sesamum indicum*	Anthraquinone, naphthoquinone	(Ogasawara et al., 1993)
Pinaceae	*Larix spp.*	Rooting	(McAfee et al., 1993)
	Pinus spp.	Rooting	(McAfee et al., 1993)
Podophyllaceae	*Podophyllum peltatum*	Podophyllotoxin	(Gussman and Porter, 1989)
Polygonaceae	*Fagopyrum esculentum*	Flavanol	(Trotin et al., 1993)
Rosaceae	*Coluria geoides*	Eugenol	(Olszowska et al., 1996)
	Fragaria x ananassa	Mycorrhizas	(Nuutila et al., 1995; Motomori et al., 1995)
	Sanguisorba officinalis	Tannins	(Ishimaru et al., 1990a)
Rubiaceae	*Cephaelis ipecacuanha*	Emetic alkaloids	(Jha et al., 1991)
	Cinchona leddgeriana	Quinoline alkaloids	(Hamill et al., 1989)
Ruta graveolens	Furanocoumarins	(Knobloch and Berlin, 1981; Eilert and Bohlmann, 1989)	
Scrophulariacea	*Antirrhinum majus*	Regeneration	(Senior et al., 1995)
	Digitalis purpurea	Digitoxin	(Saito et al., 1990b; Saito et al., 1990a)
	Scopolaria dulcis	6-metoxybenzoxazolinone	(Hayashi et al., 1994; Yamazaki et al., 1996)
Solanaceae	*Anisodus luridus (Scopolia Jacq.)*	Alkaloids	(Jovanovic et al., 1991)
	Atropa belladonna	Hyoscyamine, scopolamine	(Hashimoto et al., 1993; Jung and Tepfer, 1987; Kamada et al., 1986; Knopp et al., 1988; Sharp and Doran, 1990; Walton et al., 1990; Nyman 1994; Santana et al., 1995; Kwok and Doran, 1995; Chumsri et al., 1990; Falk and Doran, 1996; Hilton and Wilson, 1995; Subroto et al., 1996; Kanokwaree and Doran, 1997; Aoki et al., 1997)
	Datura candida (B. candida) = Brugmansia aurea X B. versicolor	Hyoscyamine, scopolamine	(Christen et al., 1990; Christen et al., 1989a; Giulietti et al., 1993; Pita-Alvarez and Giulietti, 1995; Pitta-Alvarez and Giulietti, 1997)
	D. candida x D. aurea	Hyoscyamine, scopolamine	(Robins et al., 1990; Hilton and Wilson, 1995)
	D. candida x D. candida	Hyoscyamine, scopolamine	(Homeyer et al., 1991; Kephalas et al., 1989; Christen et al., 1989b; Christen et al., 1989a)

(*continued*)

Table 5. (*Continued*)

Family	Genus and species	Major metabolite	Reference
	Datura ferox	Hyoscyamine, scopolamine	(Knopp et al., 1988; Parr et al., 1990)
	Datura innoxia	Hyoscyamine, scopolamine	(Knopp et al., 1988; Ohkawa et al., 1989; Parr et al., 1990; Shimomura et al., 1991a; Flores et al., ; Ionkova et al., 1994; Doerk et al., 1989; Ionkova et al., 1989; Boitel Conti et al., 1995)
	Datura mentel	Hyoscyamine, scopolamine	(Knopp et al., 1988)
	Datura quercifolia	Hyoscyamine, scopolamine	(Dupraz et al., 1994; Dupraz et al., 1993)
	Datura stramonium	Hyoscyamine, scopolamine	(Hilton and Rhodes, 1993; Dräger et al., 1992; Flores et al., 1993; Furze et al., 1991; Maldonado-Mendoza et al., 1993; Jaziri et al., 1988; Hilton and Rhodes, 1993; Knopp et al., 1988; Payne et al., 1987; Parr et al., 1990; Robins et al., 1991c; Robins et al., 1991b; Robins et al., 1991a; Wilson et al., ; Walton et al., 1990; Walton et al., 1994; Rhodes et al., ; Hilton and Rhodes, 1990; Burbidge et al., 1990; Hilton and Wilson, 1995; Rhodes et al., 1988; Sáenz-Carbonell et al., 1991; José-Yacamán et al., 1996; Ford et al., 1996)
	Datura wrightii	Hyoscyamine, scopolamine	(Parr et al., 1990; Hilton and Wilson, 1995)
	Duboisia leichhardtii	Hyoscyamine, scopolamine	(Kitamura et al., 1992; Kitamura et al., 1993; Mano et al., 1989; Leete et al., 1990; Muranaka et al., 1992)
	D. leichhardtii x D. myoporoides	Hyoscyamine, scopolamine	(Subroto et al., 1996)
	Duboisia myoporoides	Hyoscyamine, scopolamine	(Deno et al., 1987; Yukimune et al., 1994b; Yukimune et al., 1994a)
	Hyoscyamus albus	Hyoscyamine, scopolamine, hyalbidone	(Doerk-Schmitz et al., 1994; Sauerwein et al., 1993; Sauerwein and Wink, 1993; Christen et al., 1992; Hashimoto et al., 1986; Hashimoto et al., 1989; Knopp et al., 1988; Parr et al., 1990; Sauerwein et al., 1992; Sauerwein and Shimomura, 1991; Sauerwein et al., 1991b; Shimomura et al., 1991a; Doerk et al., 1991; Ishimaru and Shimomura, 1989; Doerk et al., 1989)
	Hyoscyamus canariensis	Hyoscyamine, scopolamine	(Hashimoto et al., 1986)
	Hyoscyamus desertorum	Hyoscyamine, scopolamine	(Parr et al., 1990)
	Hyoscyamus gyorffi	Hyoscyamine, scopolamine	(Hashimoto et al., 1986)

Table 5. *(Continued)*

Family	Genus and species	Major metabolite	Reference
	Hyoscyamus muticus	Hyoscyamine, scopolamine	(Flores and Filner, 1985b; Hashimoto et al., 1986; Parr et al., 1990; Flores et al., ; Jaziri et al., 1994; Oksman-Caldentey et al., 1994; Vanhala et al., 1995; Medina-Bolívar and Flores, 1995; Ramakrishnan et al., 1994; Curtis et al., 1995; Hilton and Wilson, 1995; Flores and Filner, 1985c; Sevón et al., 1997)
	Hyoscyamus niger	Scopolamine, solavetivone	(Hashimoto et al., 1992; Corry et al., 1993; Flores and Filner, 1985b; Dunlop and Curtis,1991; Hashimoto et al., 1986; Jaziri et al., 1988; Knopp et al., 1988; Shimomura et al., 1991a; Flores and Filner, 1985c)
	Hyoscyamus pusillus	Hyoscyamine, scopolamine	(Hashimoto et al., 1986)
	Hyoscyamus niger x H. albus = *Hyoscyamus györffyi*	Hyoscyamine, scopolamine	(Ionkova et al., 1994)
	Hyoscyamus turcomanicus	Hyoscyamine, scopolamine	(Doerk et al., 1989)
	Lycopersicon esculentum	Genetic stability	(Joao and Brown, 1994)
	Nicandra physaloides	Hygrine	(Parr, 1992)
	Nicotiana africana	Nicotine	(Parr and Hamill, 1987)
	Nicotiana cavicola	Nicotine	(Parr and Hamill, 1987)
	Nicotiana glauca	Anabasine, nicotine	(Fecker et al., 1992; Green et al., 1992)
	Nicotiana hesperis	Nicotine	(Parr and Hamill, 1987; Walton and Belshaw, 1988)
	Nicotiana rustica	Nicotine, N'-ethyl-S-nornicotine	(Benson and Hamill, 1991; Boswell et al., 1993; Furze et al., 1987; Hamill et al., 1986; Parr and Hamill, 1987; Rhodes et al., 1986; Robins et al., 1987; Walton et al., 1988; Wilson et al., ; Aird et al., ; Rhodes et al., 1988)
	Nicotiana tabacum	Nicotine, cadaverine, anabasine, phytoalexins (capsidiol, debrieyol)	(Fecker et al., 1993; Flores and Filner, 1985b; Flores and Filner, ; Parr and Hamill, 1987; Wink and Witte, 1987; Walton and McLauchlan, 1990; Wibberley et al., 1994; Ramakrishnan et al., ; Palazón et al., 1997)
	Nicotiana umbratica	Nicotine	(Parr and Hamill, 1987)
	Nicotiana velutina	Nicotine	(Parr and Hamill, 1987)
	Scopolia carniolica	Hyoscyamine, scopolamine	(Knopp et al., 1988)
	Scopolia japonica	Hyoscyamine, scopolamine	(Mano et al., 1986; Nabeshima et al., 1986)
	Scopolia stramonifolia	Hyoscyamine, scopolamine	(Parr et al., 1990)
	Scopolia tangutica	Hyoscyamine, scopolamine	(Shimomura et al., 1991a)
	Solanum aculeatissimum	Aculeatiside	(Ikenaga et al., 1995)
	Solanum mauritianum	Solasodine	(Drewes and Van Staden, 1995)
	Solanum aviculare	Solasodine	(Subroto and Doran, 1994; Yu et al., 1996)

(continued)

Table 5. (*Continued*)

Family	Genus and species	Major metabolite	Reference
	Solanum dulcamara	Steroidal alkaloids	(Kumar et al., 1992; Kumar et al., 1991a; Kumar et al., 1990; Kumar et al., 1991b)
	Withania somnifera	Withanolides	(Banerjee et al., 1994)
Umbelliferae (Apiaceae)	*Daucus carota*	Flavonoids, anthocyanins, peroxidases	(Bel Rhlid et al., 1993; Taya et al., 1989; Kim et al., 1994; Uozumi et al., 1991; Kim and Yoo, 1996; Kim et al., 1992)
Valerianaceae	*Centrathus ruber*	Valepotriates	(Gränicher et al., 1995c)
	Valeriana officinalis	Valepotriates	(Gränicher et al., 1992; Gränicher et al., 1995a; Gränicher et al., 1995b)
Verbenaceae	*Lippia dulcis*	Hernandulcin	(Sauerwein et al., 1991c)
Zygophyllaceae	*Peganum harmala*	Serotonine, ß-carbolide alkaloids	(Berlin et al., 1993; Berlin et al., 1994; Berlin et al., 1992)
Ephedraceae (Gymnosperm)	*Ephedra gerardiana*	Ephedrine	(O'Dowd and Richardson, 1994)
	Ephedra minima	Ephedrine	(O'Dowd and Richardson, 1994)
	Ephedra minima hybrid	Ephedrine	(O'Dowd and Richardson, 1994)
	Ephodra saxatilis	Ephedrine	(O'Dowd and Richardson, 1994)

1994; Hilton and Wilson, 1995; Sáenz-Carbonell and Loyola-Vargas, 1996; Sikuli and Demeyer, 1997). Not only the amount of nitrogen modifies the secondary metabolite synthesis but also the relationship between ammonium and nitrate. For example, changes in the ammoniun/nitrate ratio from 1/64 to 1/16 increased the anthocyanin synthesis in *Euphorbia millis* (Yamamoto, 1989). Similarly, Hagimori et al. (1982) reported that tissue cultures of *Digitalis purpurea* showed an increase in the digitoxin production from 2.5 to 45 mg g^{-1} dry weight when the ammonium/nitrate ratio was changed from 1/10 to 2/3.

The type and concentration of sugar is also important (Uozumi et al., 1993); generally, sucrose is used as the only carbon source. Although, high concentrations of sucrose leads to a decrease in growth, thus has a beneficial effect on the production of metabolites (Payne et al., 1987; Jaziri et al., 1988; Parr et al., 1988; Sato et al., 1991).

As mentioned above, phytohormones, type and concentration, have one of the most dramatic effects on secondary metabolites production. Addition of indolbutyric acid to *Panax ginseng* hairy root resulted in an increase of 84% in the saponins content, but with the addition of kinetin, saponins concentration increased by 179% (Yoshikawa and Furuya, 1987). We found that in *Catharanthus roseus* transformed root cultures the addition of methyl jasmonate induces a higher accumulation of indole alkaloids, particularly in catharanthine and ajmaline (Vázquez-Flota et al., 1994), while the addition of acetylsalicylic acid causes the accumulation of indole alkaloids in tumor suspension cultures (Godoy-Hernández and Loyola-Vargas, 1997).

Light (Mukundan and Hjortso, 1991a), pH (Mukundan and Hjortso, 1991b) and temperature (Toivonen et al., 1992) play critical roles in the production of secondary metabolites. There is an antagonism between light and temperature necessities to grow or to produce secondary metabolites. Low temperatures provoke lower growth rates of plant cultures, but induce accumulation of secondary metabolites. Similarly, some cultures need to be maintained under dark regime to grow, although some metabolites are only synthesized when cultures are grown in the light. For example, quinoline alkaloids from *Cin-*

Table 6. Factors which affect secondary metabolites production in hairy root cultures

Media composition	Inorganic salts, Nitrogen source: type and concentration *Phosphorous level *Carbohydrate source: type and concentration **Phytohormones
pH of the media	Effect both on growth and secondary metabolites production
Osmotic stress	Sugar, salt, polymers. Generally decrease growth rate with benefical effect on secondary metabolites production
Light regime	Sometimes light is required to accumulate a secondary product while some cultures only grow in the dark
Temperature	Low temperature generally decrease growth rate with benefical effect on secondary metabolites production

In general, decrease content of nitrogen or phosphorous (*) in the media, has a positive effect on secondary metabolites production. In opposite, in order to increase metabolites accumulation, sucrose (**) should be high, although it reduces the growth rate of the culture.

chona ledgeriana are induced by the change of 2,4-dichlorophenoxyacetic acid (2, 4-D) by naphthalene acetic acid (NAA) in the presence of light as in this case the development of green structures is necessary (Hoekstra et al., 1990). Also, chlorophyll accumulation is essential to produce gingenosides in different lines of *Panax ginseng* (Odnevall and Björk, 1989). In the particular case of the indole alkaloid serpentine we found a direct correlation between chlorophyll and the alkaloid content, both in normal and transformed cultures of *Catharanthus roseus* root growth in the presence of light (Loyola-Vargas et al., 1990).

One of the most frequently used strategies to increase the production of secondary metabolites is the use of elicitation. Cultures respond to plant pathogens dramatically (Mukundan and Hjortso, 1990; Mukundan and Hortso, 1990b; Neill et al., 1994; Singh et al., 1994). The response can be provoked either by the whole fungus or by sterilized mycelium. We also found fungus-derived hydrolase enzymes, such as cellulases and macerozymes resulted in an increase in the content of some indole alkaloids (Vázquez-Flota et al., 1994). Not only biotic elicitation, but abiotic elicitation can also modify the secondary metabolite production. In *Lotus corniculatus* the addition of reducing agents increased the production of phenylpropanoids (Goodwin and Britton, 1988). Also, cadmium, a heavy metal, at 1 mM concentration, induced the production of lubimin and 3-hydroxylubimin in *Datura stramoniun* (Furze et al., 1991). Similarly, tropane alkaloids were accumulated when copper was added to *Hyosciamus albus* hairy roots (Gränicher et al., 1992).

Many efforts have been focused to improve the production of secondary products from *in vitro* hairy root cultures, and that many strategies to improve the yield have been established. However, to date, none of those secondary metabolites obtained from hairy roots have been used for commercial purposes.

BASIC STUDIES

Hairy roots offer a proper experimental model to carry out metabolic studies: they grow fast, are axenic cultures growing under specific conditions, produce the secondary products in a similar pattern, and their metabolism is altered in similar manner when the tissue is invaded by plant pathogens. For these reasons hairy roots have been used as a biological system to perform a number of basic studies.

A. Elucidation of Biosynthetic Pathways: Basically, metabolic pathways which take place in plant roots could be studied in either normal or transformed root cultures. Identification and purification of enzymes, regulatory mechanisms, enzyme localization, etc. are part of the enzymological approaches followed to elucidate a biosynthetic route. Also, the use of tracers allow the identification of precursors and intermediates in the biosynthesis of a final product. Despite these advantages of hairy roots as biological model to elucidate metabolic pathways, this approach has not been extensively exploited.

The best example of a metabolic pathway elucidated in hairy root cultures is the tropane alkaloid route produced in members of family solanaceae. From all the *in vitro* cultures available from solanaceous members, only root cultures are able to produce these alkaloids, cell suspension cultures have failed to synthesize them. Tropane alkaloids, as hyoscyamine and scopolamine, are derived from pyridine. Also, pyridine is a precursor for nicotine, which is present in *Nicotiana* species. Several genera of solanaceae have been studied in order to elucidate the biosynthetic pathway of tropane alkaloids and several enzymes have been identified and characterized from hairy root cultures: i.e. diamino oxidase from *Hyosciamus niger* (Hashimoto et al., 1990); putrescine N-methyltransferase from *Hyosciamus albus* (Hashimoto et al., 1989; Hashimoto et al., 1989b; Hibi et al., 1992), *Atropa belladona* (Walton et al., 1990) and *Datura stramoniun* (Walton et al., 1994); acetyl-CoA-dependent acyltransferase from *D. stramoniun* (Robins et al., 1991); tropinone reductase from *D. stramoniun* (Dräger et al., 1992; Portsteffen et al., 1992) and *H. niger* (Hashimoto et al., 1992); hyoscyamine 6ß-hydroxylase (Yamada and Hashimoto, 1988). Also, the use of tracers has allowed identification of some tropane alkaloids precursors (Sauerwein et al., 1993). Using this strategy, Robins et al. (1994) have shown that phenylactic acid is a true intermediate in tropane biosynthesis while tropic acid was excluded.

The most important among the tropane alkaloids is scopolamine. Scopolamine is an anticholinergic agent that acts on the parasympathetic nerve system. The enzymology of scopolamine biosynthesis in *Hyoscyamus* has been studied (Yamada and Hashimoto, 1988; Matsuda et al., 1991; Hibi et al., 1994; Nakajima et al., 1994). A key enzyme in scopolamine biosynthesis is the hyoscyamine 6ß-hydroxylase, a 2-oxoglutarate-dependent dioxygenase. It catalyzes the final two steps in the scopolamine biosynthesis. This dioxygenase first hydroxylates hyoscyamine at position 6ß and subsequently proceeds with the epoxide formation (Matsuda et al., 1991). This enzyme is located in the root pericycle (Hashimoto et al., 1991) but is not found in stems, leaves, or cultured cells from *Duboisea*, *Atropa* or *Hyoscyamus* cultures. Location of hyoscyamine 6ß-hydroxylase explains why cell cultures failed to produce tropane alkaloids.

Fujimoto's laboratory is currently studying the biosynthesis of 24ß-alkyl-δ25- sterols in *Ajuga reptans* hairy roots. They have identified biosynthetic precursors and intermediates for some steroids (Fujimoto et al., 1996; Fujimoto et al., 1997; Lee et al., 1994; Yagi et al., 1996).

In our group we are studying the indole alkaloid biosynthetic pathway in *Catharanthus roseus* hairy roots. The glycoalkaloid strictosidine, the universal precursor of all indole alkaloids in *Catharanthus* and other apocynaceas, is synthesized by a condensation reaction of tryptamine -the decarboxylated product from tryptophane- and secologanine, a monoterpene iridoid derived from mevalonate. We purified tryptophane decarboxylase (TDC) to homogeneity (Islas et al., 1994) and partially purified strictosidine synthase (SS) (Galaz-Avalos et al., unpublished results). We also found that TDC is largely induced in *C. roseus* hairy roots by macerozyme, cellulase and *Aspergillus* homogenates (Vázquez-Flota et al., 1994). We also have a model in which *C. roseus* hairy roots were undifferentiated into cell suspension and then redifferentiated back into roots. Cell suspensions showed only 2 to 5% of the original TDC activity found in roots and alkaloids content was similar to those of suspension cultures (Moreno-Valenzuela et al., in preparation). When hairy roots were redifferentiated from the suspension cultures, they recovered their full biosynthetic capacity, although TDC activity was only 10% from the activity found in the initial hairy root cultures. These results support data from the literature which suggest that secologanine biosynthesis is the limiting step in the indole alkaloid synthetic pathway.

Actually, we are focusing our efforts on the biosynthesis of secologanine. We are purifying the enzymes 3-hydroxy-3-methylglutary CoA reductase (Gutiérrez-Pachecho et al., published) and iridodial cyclase (Sánchez-Iturbe and Loyola-Vargas, unpublished). Both of these enzymes could play important roles in the biosynthesis of secologanine. In addition, we are also studying the biochemical regulation of geraniol 10-hydroxylase (G10Hase) (Canto-Canche and Loyola-Vargas, unpublished), a cytochrome P-450-dependent monooxygenase. Its activity has been correlated with indole alkaloid content in all indole alkaloid-producing cultures screened until now, and it has been proposed as being a key limiting enzyme in the secologanine biosynthesis.

B. Genetic Manipulation of Metabolic Pathways: Genetic engineering includes isolation, characterization and reorganization of genetic material and its transfer to foreign organisms (Dörnenburg and Knorr, 1995). These techniques have recently been used to manipulate metabolic secondary pathways. Manipulation of metabolism comprises the following genetic strategies:

a) Increase copy number of selected genes. This could increase the enzyme activity (Holden, 1991). b) Addition of developmental regulation genes encoding key enzymes, but under a plant constitutive transcriptional promoter. c) Addition of novel genes. d) Down regulation of specific genes by the use of antisense RNA techniques (Yeoman and Yeoman, 1996).

Some of the goals in increase gene expression include increasing the rate of synthesis, extend the synthetic period, obtain synthesis in other organs, and to obtain synthesis of secondary metabolites in plants of others species (Holden, 1991).

In the case of Ri plasmid, border sequences and vir region have been identified as the essential elements of t-DNA for transfer process. DNA sequences between the t-DNA borders can be transferred and incorporated into plant cell genome. Foreign plant or other alien genes fused to plant or plant virus transcription promoters have been successfully introduced into plant cell by *Agrobacterium rhizogenes*-based vectors.

In *Nicotiana glauca* (Fecker et al., 1992) and *N. tabacum* (Fecker et al., 1993) hairy root cultures, bacterial lysine decarboxylase gene under the regulation of constitutive 35S CaMV promoter was introduced, resulting in an increase of two to three fold of anabasine. The introduced lysine decarboxylase showed little or no negative regulation by lysine, allowing cultures that were fed with lysine to enhance the anabasine and cadaverine production (Fecker et al., 1993). Likewise, Hashimoto et al. (1993) introduced the hyoscyamine 6ß-hydroxylase gene from *H. niger* to *A. belladona* by a binary vector system using *A. rhizogenes*. *A. belladona* hairy roots, which originally accumulate hyoscyamine but not scopolamine, changed their spectrum to scopolamine as the major alkaloid produced (Hashimoto et al., 1993). Similar results have been obtained in the past when *A. belladona* was transformed by *A. tumefasciens*. The resultant transgenic plant contained elevated levels of scopolamine (Yun et al., 1992). On the other hand, repression of a specific gene expression has been reported in *Lotus corniculatus*. In this hairy root culture the accumulation of undesirable tannins was decreased by introduction of the RNA antisense of the dehydroflavonol reductase (*dfr*) gene from *Antirrhinum majus* (Carron et al., 1994).

C. Genetic Engineering to Acquire New Capabilities: Incorporation of foreign genes into hairy roots opens a wide range of new possibilities. The natural biological barrier has been taken care of and, in theory, whatever gene from any biologic source can be introduced into a new biological host, conferring new biological capabilities.

Resistance to Herbicides. Phosphinotricin is a non-selective herbicide which acts on the enzyme glutamine synthase (GS). This enzyme plays a key role in nitrogen assimilation. In the presence of phosphinotricin, toxic ammonium ions are accumulated in the plant cells. It was thought that cell death is provoked by ammonium accumulation, but it seems that glutamine depletion is one of the mechanisms that the herbicide uses. Addition of 10 mM of glutamine to *Brassica napus* hairy roots culture in the presence of phosphinotracin overcomes the herbicide effect (Downs et al., 1994). The cytosolic glutamine synthase gene (*gs*) from *Glycine max* under the control of the CaMV promoter, was introduced into *Brassica napus* leaves by a binary vector system with a strain of *A. rhizogenes*. The emerged hairy roots showed higher GS activity, lower ammonium levels and faster growth than phosphinotricin-treated controls (Wibberley et al., 1994).

Resistance to Pathogen Organisms. Introduction of the coat protein gene from beet necrotic yellow vein virus into *Beta vulgaris*, conferred protection against the virus attack (Ehlers et al., 1991).

Novel Products. Recently, monoclonal antibodies were produced in *N. tabacum* hairy roots against a 185 kDa cell surface protein from *Streptecoccus mutants* (Wongsamuth and Doran, 1997). The transformed *N. tabacum* roots expressed both heavy and light immonoglobulin chains. By Western Blots analysis of root extracts fully assembled antibodies were confirmed. The antibody was able to aggregate *S. mutants* bacteria, showing its bivalent antigen-binding capacity and confirming its functionality. Maximum accumulation of the antibody was related with the end of exponential growth phase. Energy requirements to express, assemble and secrete the antibody appear to have negligible effect on hairy root growth. Manipulation of the media with polyvinypyrrolidone and gelatin, results in the improvement of the antibody secretion into the media from 14% up to 43%. These results establish hairy roots as an alternative source to produce animal proteins which have a direct and large commercial market.

There is no doubt that genetic engineering will be extensively exploited in future to manipulate secondary metabolites. Some criteria have been suggested in order to obtain successful biotransformation results (Stepan-Sarkissian, 1991): a) The accumulated product must not be toxic to the culture. b) The rate of production must exceed the rate of its further metabolism. c) The product must reach its appropriate cellular compartment or organelle without effective degradation.

Study of Processes that Take Place in Roots

Study of Cadmium Uptake and Assimilation. Cadmium is an important heavy metal contaminant in wastewater, aquatic environment and in the soil. It is liberated as waste product of metal refining, as a major contaminant of household waste (button batteries) and also, it is a frequent contaminant of fertilizers. Cadmium is very toxic and is neither degraded, nor converted in other less toxic forms. Plants acidify the soil surrounding in order to solubilize required minerals, particularly iron. In these conditions, cadmium is taken up from the soil as well. Cadmium uptake by plants and its subsequent accumulation, not only represents danger to the plant, but to the subsequent organisms in the food chain, including man. The mechanism of the cadmium uptake has been difficult to study using whole plants. Initially, Tepfer et al. (1989) proposed the use of hairy roots as an alternative model to study the cadmium assimilation process. Later, Macek et al. (1994) studied physicochemical factors which influence the metal accumulation by roots. Cadmium uptake rate was dependent on cadmium concentration, temperature, pH and type of buffer. Citrate buffer resulted in a lower assimilation by *Solanum nigrum* hairy roots. Similar results have been reported with aluminum (Kochian, 1995).

Understanding Microorganism-Root Symbiosis. A system as complex as the rhizosphere is difficult to study experimentally without simplification. Hairy roots simplify the study of plant roots and its interaction with the outside surrounding. Large quantities of roots under axenic conditions are obtained but the principal problem, the difficulty to access the rhizosphere habitat, is not addressed. Also, hairy roots allow the study of obligate parasites which can not be cultured away from the root *in vitro* (Tepfer et al., 1989). Hairy roots are currently used to study nodulation and mycorrhization processes.

The promoter of nodule-specific *gs* gene of *Phaseolus vulgaris* was characterized in *Lotus corniculatus* hairy root cultures. Through a series of deletion and fusion with the ß-glucuronidase reporter gene, the cis-acting elements responsible for developmental and tissue specific regulation were identified (Shen et al., 1992).

Lectins are nonenzymatic sugar-binding glycoproteins. Root lectins are plant factors involved in nodulation; they are characterized by their sugar-binding specificity. For example, *Pisum sativum* lectin is glucose/mannose specific (Díaz et al., 1990) while white clover (*Trifolium repens*) root lectin is 2-deoxyglucose specific (Sherwood et al., 1984). *P. sativum* lectin gene (*psl*) was expressed and targeted into white clover hairy root cells (Díaz et al., 1995) and this confers white clover susceptibility to *Rhizobium leguminosarum* which normally does not nodule *T. repens*. Wild type *psl* yielded true nodules, while a *psl* mutant, which encodes a non-sugar-binding lectin yielded pseudonodules (Van Eijsden et al., 1995).

The first mycorrhization of a hairy root culture was initially established by Mugnier and Mosse (1987) who infected *Convolvulus sepium* hairy roots with *Glomus mosseae*. The potential of mycorrhizae hairy roots as a model to understand the fungus-root interaction was recognized by Bécard laboratory (Bécard and Piché, 1988; 1989a; 1989b; 1992). They compared behavior of transformed roots of *Daucus carota* and *Beta vulgaris*, a host and a non-host plant respectively. Exudes from both root systems differed in their ability to stimulate the hyphal growth of germinating spores of *G. margarita*. It has been suggested that nonmycorrhizal plants do not produce factors required to promote the infection rather they produce inhibitory factors. Analysis of exudes revealed the presence of flavonoids. Bel Rhlid et al. (1993) isolated six flavonoids of mycorrhizal carrot hairy roots, of which three were flavones and three flavonols. One of them, quercetin, exists in either the free form or glycosylated, and is widely distributed in plants in which mycorrhizal symbiosis occurs.

It has been observed that endomycorrhizal infection enhances the resistance to pathogens infection, but root protection by arbuscular mycorrhizal fungi is little understood. Possible induction of defense genes is proposed (Bonfante-Fasolo et al., 1991). Benhamou et al. (1994) are the pioneers in the study of the effect of mycorrhizal infection to induce resistance to pathogen *in vitro* cultures. They claimed that *in vitro* mycorrhizal hairy roots offers the possibility to monitor the events related with pathogen attack and thus give the advantage of excluding undesirable natural contaminants which produce confusing effects. Their observations with *Fusarium*-infected mycorrhizal roots show that the signal produced by the pathogen is essential to trigger synthesis and accumulation of defense gene products. This result suggests that mycorrhization sensitizes the plant to respond more rapidly to microbial attack rather than accumulate defense gene products which would require extensive loss of energy.

CONCLUSIONS AND FUTURE RESEARCH NEEDS

The roots of plants are important biological factories (Flores, 1992). Thousands of natural products are recovered from plant roots. The new antiviral proteins, antineoplassic and other important pharmaceutical products which are discovered from plants, and also the real possibility of manipulation of metabolic pathways through genetic engineering (up or down regulation of key enzymes or introducing novel genes) are helping in the recent increase of interest and perspectives in the secondary metabolite applications. Accumulation of biochemical and molecular biology information and identification of key enzymes is required.

Emphasis has been placed on biosynthetic-type biotransformation. Degradative-type transformation is a field only marginally explored with plants. Regenerated plants expressing enzymes involved in xenobiotics degradation could be used as bioremediation tools to eliminate pollulants from the environment. Contamination of water and soil with heavy metals is a problem which requires immediate attention. Unlike organic pollulants, they are not degraded into less toxic substances. Some plants are tolerant to heavy metals and this characteristic seems to be related to phytochelatin induction (Maitani et al., 1996). Plant tolerance to heavy metals mechanism needs to be better understood. The heavy metal cadmium tends to be retained in plant roots after absorption. Some hairy roots could be used as biological arms to remove cadmium and thus prevent its entry into the food chain (reviewed by Macek et al., 1994).

Calystegia sepium hairy roots, for example, are able to survive for long periods after their removal from organ cultures. When they are placed in the soil they stop growing but remain metabolically active (Tepfer, 1989). This culture and others with similar characteristics could be tested as tools to remove heavy metals from the soil. Introduction of pathogen-resistant crops will eliminate the necessity of application of chemical biocides which are not environmentally friendly.

The use of hairy roots as biological models is largely recognized. They help to elucidate some processes which take place in roots, such as absorption, assimilation and release of substances, and also root interaction with pathogens and symbiotic organisms. Hairy roots will help us to better understand these processes.

Understanding symbiosis interaction would provide both the basic knowledge and biotechnological applications. Mycorrhizal hairy roots produced in large scale can be used as inocule in agriculture in order to optimize the use of water and mineral supplies and diminish the necessity of fertilizers.

It is generally accepted that plants respond to pathogen attack activating defense genes. However, how plants "sense" the presence of pathogen and how they respond is still not fully understood. Some elements involved in the mechanisms of response to external stimuli are being identified using hairy roots (De los Santos et al., 1997; Rodríguez-Zapata and Hernández-Sotomayor, 1997). As Benhamou et al. (1994) claimed, axenic hairy roots offer the valuable advantage of absence of undesirable contaminants and their confounding effects. This constitutes a very good system to study root-rhizosphere (bacteria, fungi, etc.) interaction. Further investigation is necessary to shed light on these relationships.

The outer layer of root cap, the border cells, are themselves useful in several investigations. Some of their application comprises: a) studies of cell structure and biology, b) study of chemotaxis, agglutination and attachment of bacteria and fungi to host cells, and c) plant resistance or susceptibility assays to fungal toxins (Brigham et al., 1995). Since hairy roots grow fast, they would be considered as good starting material to obtain border cells.

In summary, hairy roots have been used for several approaches, particularly to obtain chemical products. Actually, they are being used as biological systems to study several physiological, biochemical and molecular processes in order to manipulate their metabolic pathways. In the near future they will be used to produce new products as is evidenced by the increasing number of available literature on hairy roots.

ACKNOWLEDGMENTS

The authors thanks Drs. Teresa Hernández and Suveena Sharma for the revision of the English version of the manuscript. The work of the authors has been supported by CONACYT, México, grants 0429N9108, P122CCOT894672 and 4023-N9403.

REFERENCES

Abegaz, B. M. Polyacetylenic thiophenes and terpenoids from the roots of *Echinops pappii*. *Phytochem.* **1991**, *30*, 879–881.

Ahn, J. C.; Hwang, B.; Tada, H.; Ishimaru, K.; Sasaki, K.; Shimomura, K. Polyacetylenes in hairy roots of *Platycodon grandiflorum*. *Phytochem.* **1996**, *42*, 69–72.

Aird, E. L. H.; Hamill, J. D.; Rhodes, M. J. C. Cytogenetic analysis of hairy root cultures from a number of plant species transformed by *Agrobacterium rhizogenes*. *Plant Cell Tiss. Org. Cult.* **1988a**, *15*, 47–57.

Aird, E. L. H.; Hamill, J. D.; Robins, R. J.; Rhodes, M. J. C. Chromosome stability in transformed hairy root cultures and the properties of variant lines of *Nicotiana rustica* hairy roots. In Manipulating Secondary Metabolism in Culture; Robins, R. J., Rhodes, M. J. C. Eds.; Cambridge University Press: Cambridge, **1988b**; pp. 137–144.

Allan, E. Plant cell culture. In Plant cell and tissue culture; Stafford, A., Warren, G. Eds.; Open University Press: London, **1991**; pp. 1–24.

Ambros, P. F.; Matzke, M. A.; Matzke, A. J. M. Detection of a 17 kb unique sequence (T-DNA) in plant chromosomes by in situ hybridization. *Chromosoma* **1986**, *94*, 11–18.

Aoki, T.; Matsumoto, H.; Asako, Y.; Matsunaga, Y.; Shimomura, K. Variation of alkaloid productivity among several clones of hairy roots and regenerated plants of *Atropa belladonna* transformed with *Agrobacterium rhizogenes* 15834. *Plant Cell Rep.* **1997**, *16*, 282–286.

Arellano, J.; Vázquez, F.; Villegas, T.; Hernández, G. Establishment of transformed root cultures of *Perezia cuernavacana* producing the sesquiterpene quinone perezone. *Plant Cell Rep.* **1996**, *15*, 455–458.

Arroo, R. R. J.; De Brouwer, A. P. M.; Croes, A. F.; Wullems, G. J. Thiophene interconversion in elicitor-treated roots of *Tagetes patula* L. *Plant Cell Rep.* **1995a**, *15*, 133–137.

Arroo, R. R. J.; Develi, A.; Meijers, H.; Van de Westerlo, E.; Kemp, A. K.; Croes, A. F.; Wullems, G. J. Effect of exogenous auxin on root morphology and secondary metabolism in *Tagetes patula* hairy root cultures. *Physiol. Plant.* **1995b**, *93*, 233–240.

Arroo, R. R. J.; Jacobs, J. J. M. R.; De Koning, E. A. H.; De Waard, M.; Van de Westerlo, E.; Van Galen, P. M.; Swolfs, A. E. M.; Klunder, A. J. H.; Croes, A. F.; Wullems, G. J. Thiophene interconversions in *Tagetes patula* hairy-root cultures. *Phytochem.* **1995c**, *38*, 1193–1197.

Asada, Y.; Saito, H.; Yoshikawa, T.; Sakamoto, K.; Furuya, T. Biotransformation of 18ß-glycyrrhetinic acid by ginseng hairy root culture. *Phytochem.* **1993**, *34*, 1049–1052.

Atta-ur-Rahman; Asif, E.; Ali, S. S.; Nasir, H.; Arshad Jamal, S.; Ata, A.; Farooq, A.; Choudhary, M. I.; Sener, B.; Turkoz, S. New steroidal alkaloids from the roots of *Buxus papillosa*. *J. Nat. Prod.* **1992**, *55*, 1063–1066.

Baíza, A. M. Cytogenetics and transformed hairy root cultures. In Production of Secondary Metabolites from Plant Tissue Cultures and its Biotechnological Perspectives; Loyola-Vargas, V. M. Ed.; CICY: Merida, Yucatan, **1990**; pp. 244–261.

Bajaj, Y. P. S. *Potato. Biotechnology and 21st century of potato. Biotechnology in agriculture and forestry.* Springer Verlag: **1987**; pp. 1 - 700

Banerjee, S.; Naqvi, A. A.; Mandal, S.; Ahuja, P. S. Transformation of *Withania somnifera* (L.) dunal by *Agrobacterium rhizogenes*: infectivity and phytochemical studies. *Phytoth. Res.* **1994**, *8*, 452–455.

Banerjee-Chattopadhyay, S.; Schwemmin, A. M.; Schwemmin, D. J. A study of karyotypes and their alterations in cultured and *Agrobacterium* transformed roots of *Lycopersicon peruvianum* mill. *Theor. Appl. Genet.* **1985**, *71*, 258–262.

Bel Rhlid, R.; Chabot, S.; Piché, Y.; Chênevert, R. Isolation and identification of flavonoids from Ri T-DNA-transformed roots (*Daucus carota*) and their significance in vesicular-arbuscular mycorrhiza. *Phytochem.* **1993**, *33*, 1369–1371.

Benhamou, N.; Fortin, J. A.; Hamel, C.; St-Arnaud, M.; Shatilla, A. Resistance responses of mycorrhizal Ri T-DNA-transformed carrot roots to infection by *Fusarium oxysporum* f. sp. *chrysanthemi*. *Phytophatol.* **1994**, *84*, 958–968.

Benjamin, B. D.; Roja, G.; Heble, M. R. Alkaloid synthesis by root cultures of *Rauwolfia serpentina* transformed by *Agrobacterium rhizogenes*. *Phytochem.* **1994**, *35*, 381–383.

Benson, E. E.; Hamill, J. D. Cryopreservation and post freeze molecular and biosynthetic stability in transformed roots of *Beta vulgaris* and *Nicotiana rustica*. *Plant Cell Tiss. Org. Cult.* **1991**, *24*, 163–172.

Berlin, J.; Bedorf, N.; Mollenschott, C.; Wray, V.; Sasse, F.; Höfle, G. On the podophyllotoxins of root cultures of *Linum flavum*. *Planta Med.* **1988**, *54*, 204–206.

Berlin, J.; Rippert, M.; Mollenschott, C.; Maywald, F.; Strack, D.; Wray, V. Formation of isoflavonoid glucosides in normal and transformed cell cultures of Lupinus. *Planta Med.* **1989**, *55*, 685–686. (Abstract)

Berlin, J.; Kuzovkina, I. N.; Ruegenhagen, C.; Fecker, L.; Commandeur, U.; Wray, V. Hairy root cultures of Peganum harmala. II. Characterization of cell lines and effect of culture conditions on the accumulation of beta-carboline alkaloids and serotonin. *Z. Naturforsch. C.* **1992**, *47*, 222–230.

Berlin, J.; Rügenhagen, C.; Greidziak, N.; Kuzovkina, I. N.; Witte, L.; Wray, V. Biosynthesis of serotonin and ß-carboline alkaloids in hairy root cultures of *Peganum harmala*. *Phytochem.* **1993**, *33*, 593–597.

Berlin, J.; Rügenhagen, C.; Kuzovkina, I. N.; Fecker, L. F.; Sasse, F. Are tissue cultures of *Peganum harmala* a useful model system for studying how to manipulate the formation of secondary metabolites. *Plant Cell Tiss. Org. Cult.* **1994**, *38*, 289–297.

Bécard, G.; Fortin, J. Early events of vesicular-arbuscular mycorrhiza formation on Ri T-ADN transformed roots. *New Phytol.* **1988**, *108*, 211–218.

Bécard, G.; Piché, Y. New aspects on the acquisition of biotrophic status by a vesicular-arbuscular mycorrhizal fungus, *Gigaspora margarita*. *New Phytol.* **1989a**, *112*, 77–83.

Bécard, G.; Piché, Y. Fungal growth stimulation by CO_2 and root exudates in vesicular-arbuscular mycorrhizal symbiosis. *Appl. Environ. Microbiol.* **1989b**, *55*, 2320–2325.

Bécard, G.; Piché, Y. Establishment of vesicular-arbuscular mycorrhiza in root organ culture: review and proposed methodology. *Methods Microbiol.* **1992**, *24*, 89–107.

Bhadra, R.; Ho, C.-H.; Shanks, J. V. Growth characteristics of hairy root lines of *Catharanthus roseus*. *In Vitro Cell. Dev. Biol.* **1991**, *27*, 109A

Bhadra, R.; Vani, S.; Shanks, J. V. Production of indole alkaloids by selected hairy root lines of *Catharanthus roseus*. *Biotechnol. Bioeng.* **1993**, *41*, 581–592.

Bhadra, R.; Shanks, J. V. Statistical design of the effect of inoculum conditions on growth of hairy root cultures of *Catharanthus roseus*. *Biotechnol. Technol.* **1995**, *9*, 681–686.

Bhadra, R.; Shanks, J. V. Transient studies of nutrient uptake, growth, and indole alkaloid accumulation in heterotrophic cultures of hairy roots of *Catharanthus roseus*. *Biotechnol. Bioeng.* **1997**, *55*, 527–534.

Boitel Conti, M.; Gontier, E.; Laberche, J. C.; Ducrocq, C.; Sangwan Norreel, B. S. Permeabilization of *Datura innoxia* hairy roots for release of stored tropane alkaloids. *Planta Med.* **1995**, *61*, 287–290.

Bonfante-Fasolo, P.; Tamagnone, L.; Peretto, R.; Esquerré-Tugayé, M.-T.; Mazau, D.; Mosiniak, M.; Vian, B. Immunocytochemical location of hydroxyproline-rich glycoproteins at the interface between a mycorrhizal fungus and its host plant. *Protoplasma* **1991**, *165*, 127–138.

Boswell, H. D.; Watson, A. B.; Walton, N. J.; Robins, D. J. Formation of *N'*-ethyl-*S*-nornicotine by transformed root cultures of *Nicotiana rustica*. *Phytochem.* **1993**, *34*, 153–155.

Brigham, L. A.; Woo, H.; Hawes, M. C. Root border cells as tools in plant cell studies. In Methods in cell biology Vol. 49; Anonymous Academic Press: 1995; pp. 377–387.

Bringham, L. A.; Woo, H. H.; Hawes, M. C. *Root border cells as tools in plant cell studies*. **1995**; pp. 377–387.

Buitelaar, R. M.; Cesário, M. T.; Tramper, J. Elicitation of thiophene production by hairy roots of *Tagetes patula*. *Enzyme Microb. Technol.* **1992**, *14*, 2–7.

Buitelaar, R. M.; Leenen, E. J. T. M.; Geurtsen, G.; De Groot, Æ.; Tramper, J. Effects of the addition of XAD-7 and of elicitor treatment on growth, thiophene production, and excretion by hairy roots of *Tagetes patula*. *Enzyme Microb. Technol.* **1993**, *15*, 670–676.

Burbidge, A.; Gartland, K. M. A.; Jenkins, R. O.; Woolley, J. G.; Elliot, M. C. Accumulation of tropane alkaloids by hairy root cultures of *Datura stramonium*. *J. Exp. Bot.* **1990**, *41*, P5–13. (Abstract)

Caboni, E.; Lauri, P.; Tonelli, M.; Falasca, G.; Damiano, C. Root induction by *Agrobacterium rhizogenes* in walnut. *Plant Sci.* **1996**, *118*, 203–208.

Carron, T. R.; Robbins, M. P.; Morris, P. Genetic modification of condensed tannin biosynthesis in *Lotus corniculatus*. 1. Heterologous antisense dihydroflavonol reductase down-regulates tannin accumulation in "hairy root" cultures. *Theor. Appl. Genet.* **1994**, *87*, 1006–1015.

Charlwood, B. V.; Rhodes, M. J. C. *Secondary Products from Plant Tissue Culture*. Claredon Press-Oxford: Osford, **1990**; pp. 1–288.

Chávez-Moctezuma, M. P.; Lozoya-Gloria, E. Biosynthesis of the sesquiterpenic phytoalexin capsidiol in elicited root cultures of chili pepper (*Capsicum annuum*). *Plant Cell Rep.* **1996**, *15*, 360–366.

Cheetham, R.; Weathers, P.; DiIorio, A.; Glubiak, M.; Hofling, D. *In vitro* growth and development of roots using nutrient mist culture. *In Vitro* **1989**, *25*, Pt.2 59A(Abstract)

Chilton, M. D.; Drummond, M. H.; Merlo, D. J.; Sciaky, D.; Montoya, A. L.; Gordon, M. P.; Nester, E. W. Stable incorporation of plasmid DNA into higher plant cells: the molecular basis of crown gall tumorigenesis. *Cell* **1977**, *11*, 263–271.

Chilton, M. D.; Tepfer, D. A.; Petit, A.; David, C.; Casse-Delbart, F.; Tempé, J. *Agrobacterium rhizogenes* inserts T-DNA into the genomes of the host plant root cells. *Nature* **1982**, *295*, 432–434.

Christen, P.; Roberts, M. F.; Phillipson, J. D.; Evans, W. C. High-yield production of tropane alkaloids by hairy-root cultures of a *Datura candida* hybrid. *Plant Cell Rep.* **1989a**, *8*, 75–77.

Christen, P.; Roberts, M. F.; Phillipson, J. D.; Evans, W. C. GC/MS examination of hairy roots of a *Datura candida* hybrid for alkaloids. *Planta Med.* **1989b**, *55*, 595 (Abstract)

Christen, P.; Roberts, M. F.; Phillipson, J. D.; Evans, W. C. Alkaloids of hairy root cultures of a *Datura candida* hybrid. *Plant Cell Rep.* **1990**, *9*, 101–104.

Christen, P.; Aoki, T.; Shimomura, K. Characteristics of growth and tropane alkaloid production in *Hyoscyamus albus* hairy roots transformed with *Agrobacterium rhizogenes* A4. *Plant Cell Rep.* **1992**, *11*, 597–600.

Chumsri, P.; Vajrabhaya, T.; Bodhipadma, K.; Kamada, H.; Vacumpal, O. Phytohormonal effects on alkaloid contents of in vitro tissue and transformed root cultures of Atropa belladonna. *Planta Med.* **1990**, *56*, 633–634. (Abstract)

Ciau-Uitz, R.; Miranda-Ham, M. L.; Coello-Coello, J.; Chí, B.; Pacheco, L. M.; Loyola-Vargas, V. M. Indole alkaloid production by transformed and non-transformed root cultures of *Catharanthus roseus*. *In Vitro Cell. Dev. Biol.* **1994**, *30P*, 84–88.

Comey, N.; Hook, I.; Sheridan, H. Enhancement of anthocyanin production in cell cultures and hairy roots of *Leontopodium alpinum*. *Planta Med.* **1992**, *58*, A605–6. (Abstract)

Constabel, C. P.; Towers, G. H. N. Thiarubrine accumulation in hairy root cultures of *Chaenactis douglasii*. *J. Plant Physiol.* **1988**, *133*, 67–72.

Constabel, C. P.; Towers, G. H. N. Incorporation of ^{35}S into dithiacyclohexadiene and thiophene polyines in hairy root cultures of *Chaenectis douglasii*. *Phytochem.* **1989**, *28*, 93–95.

Cordell, G. A.; Farnsworth, N. R. *Catharanthus* alkaloids XXXII: Isolation of alkaloids from *Catharanthus trichophyllus* roots and structure elucidation of cathaphylline. *J. Pharmaceut. Sci.* **1976**, *65*, 366–369.

Corio-Costet, M.-F.; Chapuis, L.; Delbecque, J.-P. XXIII *Serratula tinctoria* L. (dyer's savory): *in vitro* culture and the production of ecdysteroids and other secondary metabolites. In Biotechnology in agriculture and forestry. Vol. 37. Medicinal and aromatic plants IX; Bajaj, Y. P. S. Ed.; Springer-Verlag: Berlin, **1996**; pp. 384–401.

Corry, J. P.; Reed, W. L.; Curtis, W. R. Enhanced recovery of solavetivone from *Agrobacterium* transformed root cultures of *Hyoscyamus muticus* using integrated product extraction. *Biotechnol. Bioeng.* **1993**, *42*, 503–508.

Croes, A. F.; Van den Berg, A. J. R.; Bosveld, M.; Breteler, H.; Wullems, G. J. Thiophene accumulation in relation to morphology in roots of *Tagetes patula*. Effects of auxin and transformation by *Agrobacterium*. *Planta* **1989**, *179*, 43–50.

Croes, A. F.; Jacobs, J. J. M. R.; Arroo, R. R. J.; Wullems, G. J. Thiophene biosynthesis in *Tagetes* roots: Molecular versus metabolic regulation. *Plant Cell Tiss. Org. Cult.* **1994**, *38*, 159–165.
Curtis, W. R.; Wang, P.; Humphrey, A. Role of calcium and differentiation in enhanced sesquiterpene elicitation from calcium alginate-immobilized plant tissue. *Enzyme Microb. Technol.* **1995**, *17*, 554–557.
David, C.; Chilton, M. D.; Tempé, J. Conservation of T-DNA in plants renegerated from hairy root cultures. *BioTechnol.* **1984**, *2*, 73–76.
Davioud, E.; Kan, C.; Hamon, J.; Tempé, J.; Husson, H. P. Production of indole alkaloids by *in vitro* root cultures from *Catharanthus trichophyllus*. *Phytochem.* **1989a**, *28*, 2675–2680.
Davioud, E.; Kan, C.; Quirion, J. C.; Das, B. C.; Husson, H. P. Epiallo-yohimbine derivatives isolated from *in vitro* hairy-root cultures of *Catharanthus trichophyllus*. *Phytochem.* **1989b**, *28*, 1383–1387.
De los Santos-Briones, C.; Muñoz-Sánchez, J. A.; Chin-Vera, J.; Loyola-Vargas, V. M.; Hernández-Sotomayor, S. M. T. Phosphatidylinositol 4, 5-biphosphate- phospholipase C activity during the growing phase of *Catharanthus roseus* transformed roots. *J. Plant Physiol.* **1997**, *150*, 707–713.
Deno, H.; Yamagata, H.; Emoto, T.; Yoshioka, T.; Yamada, Y.; Fujita, Y. Scopolamine production by root cultures of *Duboisia myoporoides*. II. Establishment of a hairy root culture by infection with *Agrobacterium rhizogenes*. *J. Plant Physiol.* **1987**, *131*, 315–323.
Di Toppi, L. S.; Gorini, P.; Properzi, G.; Barbieri, L.; Spanò, L. Production of ribosome-inactivating protein from hairy root cultures of *Luffa cylindrica* (L) Roem. *Plant Cell Rep.* **1996**, *15*, 910–913.
Dilorio, A. A.; Weathers, P. J.; Cheetham, R. D. Non-lethal secondary product release from transformed root cultures o *Beta vulgaris*. *Appl. Microbiol. Biotechnol.* **1993**, *39*, 174–180.
Diouf, D.; Gherbi, H.; Prin, Y.; Franche, C.; Duhoux, E.; Bogusz, D. Hairy root nodulation of *Casuarina glauca*: A system for the study of symbiotic gene expression in an actinorhizal tree. *Mol. Plant Microbe Interact.* **1995**, *8*, 532–537.
Díaz, C. L.; Hosselet, M.; Logman, G. J. J.; Van Driessche, E.; Lugtenberg, B. J. J.; Kijne, J. M. Distribution of glucose/manose-specific isolectins in pea (*Pisum sativum* L.) seedlings. *Planta* **1990**, *181*, 451–461.
Díaz, C. L.; Logman, T. J. J.; Stam, H. C.; Kijne, J. W. Sugar-binding activity of pea lectin expressed in white clover hairy roots. *Plant Physiol.* **1995**, *109*, 1167–1177.
Doerk, K.; Jonkova, I.; Witte, L.; Alfermann, A. W. Synthesis of tropane alkaloids in hairy root cultures of *Datura* and *Hyoscyamus* species. *Planta Med.* **1989**, *55*, 688–689. (Abstract)
Doerk, K.; Witte, L.; Alfermann, A. W. Identification of tropane alkaloids in hairy root cultures of Hyoscyamus albus. *Z. Naturforsch. C.* **1991**, *46*, 519–521.
Doerk-Schmitz, K.; Witte, L.; Alfermann, A. W. Tropane alkaloid patterns in plants and hairy roots of *Hyoscyamus albus*. *Phytochem.* **1994**, *35*, 107–110.
Doolittle, R. F. Protein sequence comparisons: Searching databases and aligning sequences. *Curr. Opin. Biotechnol.* **1994**, *5*, 24–28.
Downs, C. G.; Christey, M. C.; Davies, K. M.; King, G. A.; Seelye, J. F.; Sinclair, B. K.; Stevenson, D. G. Hairy roots of *Brassica napus*: II. Glutamine synthetase overexpression alters ammonia assimilation and the response to phosphinothricin. *Plant Cell Rep.* **1994a**, *14*, 41–46.
Downs, C. G.; Christey, M. C.; Maddocks, D.; Seelye, J. F.; Stevenson, D. G. Hairy roots of *Brassica napus*: I. Applied glutamine overcomes the effect of phosphinothricin treatment. *Plant Cell Rep.* **1994b**, *14*, 37–40.
Dörnenburg, H.; Knorr, D. Strategies for the improvement of secondary metabolite production in plant cell cultures. *Enzyme Microb. Technol.* **1995**, *17*, 674–684.
Dräger, B.; Portsteffen, A.; Schaal, A.; McCabe, P. H.; Peerless, A. C. J.; Robins, R. J. Levels of tropinone-reductase activities influence the spectrum of tropane esters found in transformed root cultures of *Datura stramonium* L. *Planta* **1992**, *188*, 581–586.
Drewes, F. E.; Van Staden, J. Initiation of and solasodine production in hairy root cultures of *Solanum mauritianum* Scop. *Plant Growth Reg.* **1995**, *17*, 27–31.
Dunlop, D. S.; Curtis, W. R. Synergistic response of plant hairy-root cultures to phosphate limitation and fungal elicitation. *Biotechnol. Prog.* **1991**, *7*, 434–438.
Dupraz, J.; Christen, P.; Kapetanidis, I. Tropane alkaloids in transformed roots of *Datura quercifolia*. *Planta Med.* **1994**, *60*, 158–162.
Dupraz, J. M.; Christen, P.; Kapetanidis, I. Tropane alkaloid production in *Datura quercifolia* hairy roots. *Planta Med.* **1993**, *59*, A659(Abstract)
Ebel, J.; Cosio, E. G.; Grab, D.; Habereder, H. Stimulation of phytoalexin accumulation in fungus-infected roots and elicitor-treated cell cultures of soybean (*Glycine max* L.). In Primary and secondary metabolism of plant cell cultures II; Kurz, W. G. W. Ed.; Springer Verlag: Berlin, **1989**; pp. 229–236.
Ehlers, U.; Commandeur, U.; Frank, R.; Landsmann, J.; Koenig, R.; Burgermeister, W. Cloning of the coat protein gene from beet necrotic yellow vein virus and its expression in sugar beet hairy roots. *Theor. Appl. Genet.* **1991**, *81*, 777–782.

Eilert, U.; Bohlmann, J. Effect of elicitors on enzymes of the phenylpropanoid pathway in different *in vitro* systems of *Ruta graveolens*. *Planta Med.* **1989**, *55*, 685 (Abstract)

Ermayanti, T. M.; McComb, J. A.; O'Brien, P. A. Stimulation of synthesis and release of swainosonine from transformed roots of *Swainsona galegifolia*. *Phytochem.* **1994a**, *36*, 313–317.

Ermayanti, T. M.; McComb, J. A.; O'Brien, P. A. Growth and swainsonine production of *Swainsona galegifolia* (Andr.) R. Br. untransformed and transformed root cultures. *J. Exp. Bot.* **1994b**, *45*, 633–639.

Falk, L. R.; Doran, P. M. Influence of inoculum morphology on growth of *Atropa belladonna* hairy roots and production of tropane alkaloids. *Biotechnol. Lett.* **1996**, *18*, 1099–1104.

Falkenhagen, H.; Stöckigt, J.; Kuzovkina, I. N.; Alterman, I. E.; Kolshorn, H. Indole alkaloids from "hairy roots" of *Rauwolfia serpentina*. *Can. J. Chem.* **1993**, *71*, 2201–2203.

Fecker, L. F.; Hillebrandt, S.; Rugenhagen, C.; Herminghaus, S.; Landsmann, J.; Berlin, J. Metabolic effects of a bacterial lysine decarboxylase gene expressed in hairy root cultures of *Nicotiana glauca*. *Biotechnol. Lett.* **1992**, *14*, 1035–1040.

Fecker, L. F.; Rügenhagen, C.; Berlin, J. Increased production of cadaverine and anabasine in hairy root cultures of *Nicotiana tabacum* expressing a bacterial lysine decarboxylase gene. *Plant Mol. Biol.* **1993**, *23*, 11–21.

Fitter, A. H. Characteristics and functions of root systems. In Plant Roots. The hidden half; Waisel, Y., Eshel, A., Kafkati, U. Eds.; Marcel Dekker, Inc. New York, Basel, Hong Kong, **1991**; pp. 3–24.

Flint-Wandel, J.; Hjortso, M. A flow cell reactor for the study of growth kinetics of single hairy roots. *BioTechniques* **1993**, *1*, 447–452.

Flores, H. E.; Protacio, C. M.; Signs, M. W. Primary and secondary metabolism of polyamines in plants. In Plant Nitrogen Metabolism. Recent Advances in Phytochemistry. Vol. 23; Poulton, J. E., Romeo, J. T., Conn, E. E. Eds.; Plenum Press: New York and London, **1989**; pp. 329–391.

Flores, H. E. Plant roots as chemical factories. *Chemistry & Industry* **1992**, 374–377.

Flores, H. E.; Yao-rem, D.; Cuello, J. L.; Maldonado-Mendoza, I. E.; Loyola-Vargas, V. M. Green roots: photosynthesis and photoautotrophy in an underground plant organ. *Plant Physiol.* **1993**, *101*, 363–371.

Flores, H. E.; Filner, P. Hairy roots of Solanaceae as a source of alkaloids. *Plant Physiol.* **1985a**, *77*, 12s

Flores, H. E.; Filner, P. Metabolic relationships of putrescine, GABA and alkaloids in cell and root cultures of Solanaceae. In Primary and Secondary Metabolism in Plant Cell Cultures; Neumann, K. H., Barz, W., Reinhard, E. Eds.; Springer-Verlag: Heidelberg, **1985b**; pp. 174–186.

Flores, H. E.; Filner, P. 'Hairy roots' of *Solanaceae* as a source of alkaloids. *In Vitro* **1985c**, *21*, Pt.2 53A(Abstract)

Flores, H. E.; Medina-Bolívar, F. Root culture and plant natural products: unearthing the hidden half of plant metabolism. *Plant Tiss. Cult. Biotechnol.* **1995**, *1*, 59–74.

Follansbee, E.; Cheetham, R.; Weathers, P. Transformation of *Euphorbia lathyris* by *Agrobacterium rhizogenes*. *In Vitro* **1995**, *31*, Pt.2 72A(Abstract)

Ford, Y. Y.; Ratcliffe, R. G.; Robins, R. J. *In vivo* NMR analysis of tropane alkaloid metabolism in transformed root and de-differentiated cultures of *Datura stramonium*. *Phytochem.* **1996**, *43*, 115–120.

Foster, R. C.; Rovira, A. D.; Cock, T. W. The soil, root, and rhizosphere. In Ultrastructure of the root-soil interface; American Phytopathological Society: St. Paul, Minnesota, U.S.A. **1983**; pp. 5–11.

Friesen, J. B.; Burkhouse, P. C.; Biesboer, D. D.; Leete, E. Influence of alkaloid precursors on the alkaloid content of *Nicotiana alata* root cultures. *Phytochem.* **1992**, *31*, 3059–3063.

Fujimoto, Y.; Nakagawa, T.; Yamada, J.; Morisaki, M. Biosynthetic origin of C-26 and C-27 of the phytoecdysteroids cyasterone and 29-norcyasterone in *Ajuga* hairy roots. *J. Chem. Soc. Chem. Commun.* **1996**, *1996*, 2063–2064.

Fujimoto, Y.; Kushiro, T.; Nakamura, K. Biosynthesis of 20-hydroxyecdysone in *Ajuga* hairy roots: Hydrogen migration from C-6 to C-5 during *cis*-A/B ring formation. *Tetrahedron Lett.* **1997**, *38*, 2697–2700.

Furuya, T.; Ushiyama, M.; Asada, Y.; Yoshikawa, T. Biotransformation of 2-phenylpropionic acid in root culture of *Panax ginseng*. *Phytochem.* **1989**, *28*, 483–487.

Furze, J. M.; Hamill, J. D.; Parr, A. J.; Robins, R. J.; Rhodes, M. J. C. Variations in morphology and nicotine alkaloid accumulation in protoplasts derived hairy root cultures of *N. rustica*. *J. Plant Physiol.* **1987**, *131*, 237–246.

Furze, J. M.; Rhodes, M. J. C.; Parr, A. J.; Robins, R. J.; Withehead, I. M.; Threlfall, D. R. Abiotic factors elicit sesquiterpenoid phytoalexin production but not alkaloid production in transformed root cultures of *Datura stramonium*. *Plant Cell Rep.* **1991**, *10*, 111–114.

Galneder, E.; Zenk, M. H. Enzymology of alkaloid production in plant cell cultures. In Progress in plant cellular and molecular biology; Nijkamp, H. J. J., Plas, L. H. W.,van der, Aartrijk, J. v. Eds.; Kluver Academic Publishers: Dordrecht, **1990**; pp. 754–762.

Gamborg, O. L.; Miller, R. A.; Ojima, K. Nutrient requirements of suspension cultures of soybean root cells. *Exp. Cell Res.* **1968**, *50*, 151–158.

Giulietti, A. M.; Parr, A. J.; Rhodes, M. J. C. Tropane alkaloid production in transformed root cultures of *Brugmansia candida*. *Planta Med.* **1993**, *59*, 428–431.

Glinski, J.; Lipiec, J. Roots and theur functions. Morphological and anatomical characteristics of root systems. In Soil physical conditions and plant roots; Glinski, J., Lipiec, J. Eds.; CRC Press Inc. Boca Raton, Florida, **1990**; pp. 33–55.

Godoy-Hernández, G.; Loyola-Vargas, V. M. Effect of acetylsalicylic acid on secondary metabolism of *Catharanthus roseus* tumor suspension culture. *Plant Cell Rep.* **1997**, *16*, 287–290.

Gomez-Barrios, M. L.; Parodi, F. J.; Vargas, D.; Quijano, L.; Hjortso, M. A.; Flores, H. E.; Fischer, N. H. Studies on the biosynthesis of thiarubrine A in hairy root cultures of *Ambrosia artemisiifolia* using ^{13}C-labelled acetates. *Phytochem.* **1992**, *31*, 2703–2707.

Goodwin, T. W.; Britton, G. Distribution and analysis of carotenoids. In Plant pigments; Anonymous Academic Press: London, **1988**; pp. 61–132.

Gränicher, F.; Christen, P.; Kapetanidis, I. High-yield production of valepotriates by hairy root cultures of *Valeriana officinalis* L. var. *sambucifolia* Mikan. *Plant Cell Rep.* **1992**, *11*, 339–342.

Gränicher, F.; Christen, P.; Kamalaprija, P.; Burger, U. An iridoid diester from *Valeriana officinalis* var. *sambucifolia* hairy roots. *Phytochem.* **1995a**, *38*, 103–105.

Gränicher, F.; Christen, P.; Kapetanidis, I. Essential oils from normal and hairy roots of *Valeriana officinalis* var *sambucifolia*. *Phytochem.* **1995b**, *40*, 1421–1424.

Gränicher, F.; Christen, P.; Kapétanidis, I. Production of valepotriates by hairy root cultures of *Centranthus ruber* DC. *Plant Cell Rep.* **1995c**, *14*, 294–298.

Green, K. D.; Thomas, N. H.; Callow, J. A. Product enhancement and recovery from transformed root cultures of *Nicotiana glauca*. *Biotechnol. Bioeng.* **1992**, *39*, 195–202.

Gussman, C. D.; Porter, J. R. Podophyllotoxin production by tissue and organ cultures of *Podophyllum peltatum*. *Plant Physiol.* **1989**, *89*, 11 (Abstract)

Hagimori, M.; Matsumoto, T.; Obi, Y. Studies on the production of *Digitalis* cardenolids by plant tissue culture. III. Effects of nutrients on digitoxin formation by shoot-forming culltures of *Digitalis purpurea* L. grown in liquid media. *Plant Cell Physiol.* **1982**, *23*, 1205–1211.

Hamill, J. D.; Parr, A. J.; Robins, R. J.; Rhodes, M. J. C. Secondary product formation by cultures of *Beta vulgaris* and *Nicotiana rustica* transformed with *Agrobacterium rhizogenes*. *Plant Cell Rep.* **1986**, *5*, 111–114.

Hamill, J. D.; Robins, R. J.; Rhodes, M. J. C. Alkaloid production by transformed root cultures of *Cinchona ledgeriana*. *Planta Med.* **1989**, *55*, 354–357.

Han, K.-H.; Keathley, D. E.; Davis, J. M.; Gordon, M. P. Regeneration of a transgenic woody legume (*Robinia pseudoacacia* L., black locust) and morphological alterations induced by *Agrobacterium rhizogenes*-mediated transformation. *Plant Sci.* **1993**, *88*, 149–157.

Hartmann, T.; Witte, L.; Oprach, F.; Toppel, G. Reinvestigation of the alkaloid composition of *Atropa belladonna* plants, root cultures, and cell suspension cultures. *Planta Med.* **1986**, *52*, 390–395.

Hashimoto, T.; Yukimune, Y.; Yamada, Y. Tropane alkaloid production in *Hyoscyamus* root culture. *J. Plant Physiol.* **1986**, *124*, 61–75.

Hashimoto, T.; Yukimune, Y.; Yamada, Y. Putrescine and putrescine N-methyltransferase in the biosynthesis of tropane alkaloids in cultured roots of *Hyoscyamus albus* II. Incorporation of labeled precursors. *Planta* **1989a**, *178*, 131–137.

Hashimoto, T.; Yukimune, Y.; Yamada, Y. Putrescine and putrescine N-methyltransferase in the biosynthesis of tropane alkaloids in cultured roots of *Hyoscyamus albus*. *Planta* **1989b**, *178*, 123–130.

Hashimoto, T.; Mitani, A.; Yamada, Y. Diamine oxidase from cultured roots of *Hyoscyamus niger*. *Plant Physiol.* **1990**, *93*, 216–221.

Hashimoto, T.; Hayashi, A.; Amano, Y.; Kohno, J.; Iwanari, H.; Usuda, S.; Yamada, Y. Hyoscyamine 6ß-hydroxylase, an enzyme involved in tropane alkaloid biosynthesis, is localized at the pericycle of the root. *J. Biol. Chem.* **1991**, *266*, 4648–4653.

Hashimoto, T.; Nakajima, K.; Ongena, G.; Yamada, Y. Two tropinone reductases with distinct stereospecificities from cultured roots of *Hyoscyamus niger*. *Plant Physiol.* **1992**, *100*, 836–845.

Hashimoto, T.; Yun, D.-J.; Yamada, Y. Production of tropane alkaloids in genetically engineered root cultures. *Phytochem.* **1993**, *32*, 713–718.

Hayashi, T.; Gotoh, K.; Ohnishi, K.; Okamura, K.; Asamizu, T. 6-methoxy-2-benzoxazolinone in *Scoparia dulcis* and its production by cultured tissues. *Phytochem.* **1994**, *37*, 1611–1614.

Hänisch, C. H.; Sree Ramulu, K.; Dijkhuis, P.; De Groot, B. Genetic stability of cultured hairy roots induced by *Agrobacterium rhizogenes* on tuber discs of potato cv. bintje. *Plant Sci.* **1987**, *49*, 217–222.

Heimer, Y. M. Nitrite-induced development of the nitrate uptake system in plant cells. *Plant Sci. Lett.* **1975**, *4*, 137–139.

Hibi, N.; Fujita, T.; Hatano, M.; Hashimoto, T.; Yamada, Y. Putrescine *N*-methyltransferase in cultured roots of *Hyoscyamus albus*. *n*-Butylamine as a potent inhibitor of the transferase both in vitro and in vivo. *Plant Physiol.* **1992**, *100*, 826–835.

Hibi, N.; Higashiguchi, S.; Hashimoto, T.; Yamada, Y. Gene expression in tobacco low-nicotine mutants. *Plant Cell* **1994**, *6*, 723–735.

Hilton, M. G.; Rhodes, M. J. C. Growth and hyoscyamine production of *Hairy root* cultures of *Datura stramonium* in a modified stirred tank reactor. *Appl. Microbiol. Biotechnol.* **1990**, *33*, 132–138.

Hilton, M. G.; Rhodes, M. J. C. Factors affecting the growth and hyoscyamine production during batch culture of transformed roots of *Datura stramonium*. *Planta Med.* **1993**, *59*, 340–344.

Hilton, M. G.; Rhodes, M. J. C. The effect of varying levels of Gamborg's B5 salts and temperature on the accumulation of starch and hyoscyamine in batch cultures of transformed roots of *Datura stramonium*. *Plant Cell Tiss. Org. Cult.* **1994**, *38*, 45–51.

Hilton, M. G.; Wilson, D. G. Growth and the uptake of sucrose and mineral ions by transformed root cultures of *Datura staramonium, Datura candida* x *aurea*, *Datura wrightii*, *Hyoscyamus muticus* and *Atropa belladonna*. *Planta Med.* **1995**, *61*, 345–350.

Hirakura, K.; Morita, M.; Nakajima, K.; Ikeya, Y.; Mitsuhashi, H. Three acetylated polyacetylenes from the roots of *Panax ginseng*. *Phytochem.* **1991**, *30*, 4053–4055.

Hirotani, M.; Zhou, Y.; Lui, H.; Furuya, T. Astragalosides from hairy root cultures of *Astragalus membranaceus*. *Phytochem.* **1994a**, *36*, 665–670.

Hirotani, M.; Zhou, Y.; Rui, H.; Furuya, T. Cycloartane triterpene glycosides from the hairy root cultures of *Astragalus membranaceus*. *Phytochem.* **1994b**, *37*, 1403–1407.

Hjortso, M.; Mukundan, U. Genetic transformation in *Tagetes* species (Marigolds) for thiophene contents. In Biotechnology in agriculture and forestry Vol. 29. Plant protoplasts and genetic engineering V; Bajaj, Y. P. S. Ed.; Springer-Verlag: Berlin, **1994**; pp. 365–382.

Ho, C.-H.; Shanks, J. V. Effects of initial medium pH on growth and metabolism of *Catharanthus roseus* hairy root cultures. A study with ^{31}P and ^{13}C NMR spectroscopy. *Biotechnol. Lett.* **1992**, *14*, 959–964.

Hoekstra, S. S.; Harkes, P. A. A.; Verpoorte, R.; Libbenga, K. R. Effect of auxin on cytodifferentiation and production of quinoline alkaloids in compact globular structures of *Cinchona ledgeriana*. *Plant Cell Rep.* **1990**, *8*, 571–574.

Holden, M. Prospects for the genetic manipulation of metabolic pathways leading to secondary products. In The impact of biotechnology in agriculture; Sangwan, R. S., Sangwan-Norreel, B. S. Eds.; Kluwer Academic Publ. The Netherlands, **1991**; pp. 103–117.

Homeyer, B. C.; Fry, C. Z.; Roberts, M. F. Alkaloid variations in hairy root clones of a *Datura candida* hybrid. *J. Pharm. Pharmacol.* **1991**, *43*, 21(Abstract)

Hook, I. Secondary metabolites in hairy root cultures of *Leontopodium alpinum* Cass. (Edelweiss). *Plant Cell Tiss. Org. Cult.* **1994**, *38*, 321–326.

Horz, K. H.; Reichling, J. Allylphenol biosynthesis in a transformed root culture of *Coreopsis tinctoria*: Side-chain formation. *Phytochem.* **1993**, *33*, 349–351.

Hu, Z. B.; Alfermann, A. W. Diterpenoid production in hairy root cultures of *Salvia miltiorrhiza*. *Phytochem.* **1993**, *32*, 699–703.

Ikenaga, T.; Oyama, T.; Muranaka, T. Growth and steroidal saponin production in hairy root cultures of *Solanum aculeatissimum*. *Plant Cell Rep.* **1995**, *14*, 413–417.

Ionkova, I.; Witte, L.; Alfermann, A. W. Production of alkaloids by transformed root cultures of *Datura innoxia*. *Planta Med.* **1989**, *55*, 229 (Abstract)

Ionkova, I.; Ninov, S.; Kolev, D.; Alfermann, A. W. Secondary constituents of *in vitro* cell and transformed root cultures of *Althaea officinalis* var. Russalka. *Planta Med.* **1991**, *57*, A41-A42. (Abstract)

Ionkova, I.; Hu, Z. B.; Alfermann, A. W. Natural products from hairy root cultures of *Astragalus species*. *In Vitro* **1992**, *28*, Pt.2 118A (Abstract)

Ionkova, I.; Witte, L.; Alfermann, A. W. Spectrum of tropane alkaloids in transformed roots of *Datura innoxia* and *Hyoscyamus x györffyi* cultivated *in vitro*. *Planta Med.* **1994**, *60*, 382–384.

Ionkova, I.; Alfermann, A. W. Transformation of *Astragalus* species by *Agrobacterium rhizogenes* and their saponin production. *Planta Med.* **1990**, *56*, 634–635. (Abstract)

Ishimaru, K.; Hirose, M.; Takahashi, K.; Koyama, K.; Shimomura, K. Tannin production in root culture of *Sanguisorba officinalis*. *Phytochem.* **1990a**, *29*, 3827–3830.

Ishimaru, K.; Sudo, H.; Satake, M.; Matsunaga, Y.; Hasegawa, Y.; Takemoto, S.; Shimomura, K. Amarogentin, amaroswerin and four xanthones from hairy root cultures of *Swertia japonica*. *Phytochem.* **1990b**, *29*, 1563–1565.

Ishimaru, K.; Sudo, H.; Satake, M.; Shimomura, K. Phenyl glucosides from a hairy root culture of *Swertia japonica*. *Phytochem.* **1990c**, *29*, 3823–3825.

Ishimaru, K.; Yonemitsu, H.; Shimomura, K. Lobetyolin and lobetyol from hairy root culture of *Lobelia inflata*. *Phytochem.* **1991**, *30*, 2255–2257.

Ishimaru, K.; Sadoshima, S.; Neera, S.; Koyama, K.; Takahashi, K.; Shimomura, K. A polyacetylene gentiobioside from hairy roots of *Lobelia inflata*. *Phytochem.* **1992a**, *31*, 1577–1579.

Ishimaru, K.; Yoshimatsu, K.; Yamakawa, T.; Kamada, H.; Shimomura, K. Phenolic constituents in tissue cultures of Phyllanthus niruri. *Phytochem.* **1992b**, *31*, 2015–2018.

Ishimaru, K.; Arakawa, H.; Yamanaka, M.; Shimomura, K. Polyacetylenes in *Lobelia sessilifolia* hairy roots. *Phytochem.* **1994**, *35*, 365–369.

Ishimaru, K.; Shimomura, K. 7-beta-hydroxyhyoscyamine from Duboisia myoporoides-D. leichhardtii hybrid and Hyoscyamus albus. *Phytochem.* **1989**, *28*, 3507–3509.

Ishimaru, K.; Shimomura, K. Tannin production in hairy root culture of *Geranium thunbergii*. *Phytochem.* **1991**, *30*, 825–828.

Islas, I.; Loyola-Vargas, V. M.; Miranda-Ham, M. L. Tryptophan decarboxylase activity in transformed roots from Catharanthus roseus and its relationship to tryptamine, ajmalicine, and catharanthine accumulation during the culture cycle. *In Vitro Cell. Dev. Biol.* **1994**, *30P*, 81–83.

Itai, C.; Birnbaum, H. Synthesis of plant growth regulators by roots. In Plant Roots. The hidden half; Waisel, Y., Eshel, A., Kafkati, U. Eds.; Marcel Dekker, Inc. New York, Basel, Hong Kong, **1991**; pp. 163–177.

Jacobs, J. J.; Arroo, R. R.; De Koning, E. A.; Klunder, A. J.; Croes, A. F.; Wullems, G. J. Isolation and characterization of mutants of thiophene synthesis in *Tagetes erecta*. *Plant Physiol.* **1995**, *107*, 807–814.

Jacobs, J. J. M. R.; Kluijtmans, L.; Kolen, Z. M. H.; Erckens, K.; Arroo, R. R. J.; Croes, A. F.; Wullems, G. J. Characterization of *Tagetes patula* cDNA clones obtained by differential screening in relation to thiophene biosynthesis. *J. Plant Physiol.* **1995**, *146*, 418–422.

Jaziri, M.; Legros, M.; Homes, J.; VanHaelen, M. Tropine alkaloids production by hairy root cultures of *Datura stramonium* and *Hyoscyamus niger*. *Phytochem.* **1988**, *27*, 419–420.

Jaziri, M.; Homès, J.; Shimomura, K. An unusual root tip formation in hairy root culture of *Hyoscyamus muticus*. *Plant Cell Rep.* **1994**, *13*, 349–352.

Jaziri, M.; Shimomura, K.; Yoshimatsu, K.; Fauconnier, M.-L.; Marlier, M.; Homès, J. Establishment of normal and transformed root cultures of *Artemisia annua* L. for artemisinin production. *J. Plant Physiol.* **1995**, *145*, 175–177.

Jha, S.; Sahu, N. P.; Sen, J.; Jha, T. B.; Mahato, S. B. Production of emetine and cephaeline from cell suspension and excised root cultures of *Cephaelis ipecacuanha*. *Phytochem.* **1991**, *30*, 3999–4003.

Joao, K. H. L.; Brown, T. A. Long-term stability of root cultures of tomato transformed with *Agrobacterium rhizogenes* R1601. *J. Exp. Bot.* **1994**, *45*, 641–647.

José-Yacamán, M.; Rendón, L.; Arenas, J.; Puche, M. C. S. Maya blue paint: An ancient nanostructured material. *Science* **1996**, *273*, 223–225.

Jovanovic, V.; Grubisic, D.; Giba, Z.; Menkovic, N.; Ristic, M. Alkaloids in hairy root cultures of *Anisodus luridus*. *Planta Med.* **1991**, *57*, A102(Abstract)

Jung, G.; Tepfer, D. Use of genetic transformation by the Ri T-DNA of *Agrobacterium rhizogenes* to stimulate biomass and tropane alkaloid production in *Atropa belladonna* and *Calystegia sepium* roots grown *in vitro*. *Plant Sci.* **1987**, *50*, 145–151.

Jung, K.; Kwak, S.; Choi, C.; Liu, J. R. Development of two stage culture process by optimization of inorganic salts for improving catharanthine production in hairy root cultures of *Catharanthus roseus*. *J. Ferment. Bioeng.* **1994**, *77*, 57–61.

Jung, K. H.; Kwak, S. S.; Kim, S. W.; Lee, H.; Choi, C. Y.; Liu, J. R. Improvement of catharanthine productivity in hairy root cultures of *C. roseus* by using monosaccharides as carbon source. *Biotechnol. Lett.* **1992**, *14*, 695–700.

Jung, K. H.; Kwak, S. S.; Choi, C. Y.; Liu, J. R. An interchangeable system of hairy root and cell suspension cultures of *Catharanthus roseus* for indole alkaloid production. *Plant Cell Rep.* **1995**, *15*, 51–54.

Kamada, H.; Okamura, N.; Satake, M.; Harada, H.; Shimomura, K. Alkaloid production by hairy root cultures in *Atropa belladonna*. *Plant Cell Rep.* **1986**, *5*, 239–242.

Kanokwaree, K.; Doran, P. M. The extent to which external oxygen transfer limits growth in shake flask culture of hairy roots. *Biotechnol. Bioeng.* **1997**, *55*, 520–526.

Katavic, V.; Jelaska, S.; Bakran-Petricioli, T.; David, C. Host-tissue differences in transformation of pumpkin (*Cucurbita pepo* L.) by *Agrobacterium rhizogenes*. *Plant Cell Tiss. Org. Cult.* **1991**, *24*, 35–42.

Kennedy, A. I.; Deans, S. G.; Svoboda, K. P.; Gray, A. I.; Waterman, P. G. Volatile oils from normal and transformed root of *Artemisia absinthium*. *Phytochem.* **1993**, *32*, 1449–1451.

Kephalas, T. A.; McLennan, C. J.; Evans, W. C.; Phillipson, J. D.; Roberts, M. F. Production of hyoscyamine and scopolamine in transformed roots of *Datura candida* hybrid. *J. Pharm. Pharmacol.* **1989**, *41*, 68 (Abstract)

Kim, C. H.; Park, J. H.; Chung, I. S.; Kim, S. R.; Lee, S. W. Enhanced anthocyanin production in hairy root culture of *Daucus carota* by fungal elicitor. *Hortscience.* **1992**, *27*, 694 (Abstract)

Kim, J. H.; Lee, Y. W.; Yoo, Y. J. Biosynthesis of SOD by carrot hairy roots. *Biotechnol. Lett.* **1994**, *16*, 581–586.

Kim, S.; Hopper, E.; Hjortso, M. Hairy root growth models: Effect of different branching patterns. *Biotechnol. Prog.* **1995**, *11*, 178–186.

Kim, Y. H.; Yoo, Y. J. Peroxidase production from carrot hairy root cell culture. *Enzyme Microb. Technol.* **1996**, *18*, 531–535.

Kinnersley, A. M.; Dougall, D. K. Correlation between the nicotine content of tobacco plants and callus cultures. *Planta* **1980**, *149*, 205–206.

Kisiel, W.; Stojakowska, A.; Malarz, J.; Kohlmünzer, S. Sesquiterpene lactones in *Agrobacterium rhizogenes* - Transformed hairy root culture of *Lactuca virosa*. *Phytochem.* **1995**, *40*, 1139–1140.

Kitamura, Y.; Taura, A.; Kajiya, Y.; Miura, H. Conversion of phenylalanine and tropic acid into tropane alkaloids by *Duboisia leichhardtii* root cultures. *J. Plant Physiol.* **1992**, *140*, 141–146.

Kitamura, Y.; Nishimi, S.; Miura, H.; Kinoshita, T. Phenyllactic acid in *Duboisia leichhardtii* root cultures by feeding of phenyl[1–^{14}C]alanine. *Phytochem.* **1993**, *34*, 425–427.

Knobloch, K.-H.; Berlin, J. Effects of media constitute on the formation of secondary products in cell suspension cultures of *Catharanthus roseus*. *Adv. Biotechnol.* **1981**, *1*, 129–133.

Knopp, E.; Strauss, A.; Wehrli, W. Root induction on several Solanaceae species by *Agrobacterium rhizogenes* and the determination of root tropane alkaloid content. *Plant Cell Rep.* **1988**, *7*, 590–593.

Ko, K. S.; Ebizuka, Y.; Noguchi, H.; Sankawa, U. Production of polyketide pigments in hairy root cultures of *Cassia* plants. *Chem. Pharm. Bull. (Tokyo)* **1995**, *43*, 274–278.

Kochian, L. V. Cellular mechanisms of aluminum toxicity and resistance in plants. *Annu. Rev. Plant Physiol. Plant Mol. Biol.* **1995**, *46*, 237–260.

Kramer P. J. Root growth and funtions in Water relations of plants; Kramer P. J. ed.; Academic Press: Orlando, Florida, **1983**, pp. 120–145.

Kumar, V.; Laouar, L.; Davey, M. R.; Mulligan, B. J.; Lowe, K. C. Effects of Pluronic F-68 on growth of transformed roots of Solanum dulcamara. *Biotechnol. Lett.* **1990**, *12*, 937–940.

Kumar, V.; Laouar, L.; Davey, M. R.; Mulligan, B. J.; Lowe, K. C. Effects of Pluronic F-68 on callus growth and protoplast plating efficiency of Solanum dulcamara. *Plant Cell Rep.* **1991a**, *10*, 52–54.

Kumar, V.; Laouar, L.; Davey, M. R.; Mulligan, B. J.; Lowe, K. C. Surfactant effects on the growth of transformed roots. *J. Exp. Bot.* **1991b**, *42*, 50 (Abstract)

Kumar, V.; Laouar, L.; Davey, M. R.; Mulligan, B. J.; Lowe, K. C. Pluronic F-68 stimulates growth of Solanum dulcamara in culture. *J. Exp. Bot.* **1992**, *43*, 487–493.

Kurz, W. G. W.; Chatson, K. B.; Constabel, F. Biosynthesis and accumulation of indole alkaloids in cell suspension cultures of *Catharanthus roseus* cultivars. In Primary and secondary metabolism of plant cell cultures; Neumann, K.-H., Barz, W., Reinhard, E. Eds.; Springer-Verlag: Heidelberg, **1985**; pp. 143–153.

Kutschera, L. Short review of the present state of root research. In Plant roots and their environment; McMichael, B. L., Persson, H. Eds.; Elsevier Science Publishers B. V. Amsterdam, **1991**; pp. 1–8.

Kwok, K. H.; Doran, P. M. Kinetic and stoichiometric analysis of hairy roots in a segmented bubble column reactor. *Biotechnol. Prog.* **1995**, *11*, 429–435.

Kyo, M.; Miyauchi, Y.; Fujimoto, T.; Mayama, S. Production of nematocidal compounds by hairy root cultures of *Tagetes patula* L. *Plant Cell Rep.* **1990**, *9*, 393–397.

Larkin, P. J.; Scowcroft, W. R. Somaclonal variation -a novel source of variability from cell cultures for plant improvement. *Theor. Appl. Genet.* **1981**, *60*, 197–214.

Lee, M. Y.; Choi, Y.; Kim, S. S. Purification and immunological relationships of six radish isoperoxidases. *Plant Physiol. Biochem.* **1994**, *32*, 259–265.

Leete, E.; Endo, T.; Yamada, Y. Biosynthesis of nicotine and scopolamine in a root culture of *Duboisia leichhardtii*. *Phytochem.* **1990**, *29*, 1847–1851.

Lloyd, C. Plant Morphogenesis: Life on a different plane. *Curr. Biol.* **1995**, *5*, 1085–1087.

Lodhi, A. H.; Bongaerts, R. J. M.; Verpoorte, P.; Coomber, S. A.; Charlwood, B. V. Expression of bacterial isochorismate synthase (EC 5.4.99.6) in transgenic root cultures of *Rubia peregrina*. *Plant Cell Rep.* **1996**, *16*, 54–57.

Lodhi, A. H.; Charlwood, B. V. *Agrobacterium rhizogenes*-mediated transformation of *Rubia peregrina* L: *In vitro* accumulation of anthraquinones. *Plant Cell Tiss. Org. Cult.* **1996**, *46*, 103–108.

Loyola-Vargas, V. M.; Monforte-González, M.; Méndez, M. Serpentine synthesis during greening in normal and tumor tissues of *Catharanthus roseus*. In Abstracts VIIth International Congress on Plant Tissue and Cell Culture; Nijkamp, H. J. J., Plas, L. H. W.,van der, Aartrijk, J. v. Eds.; International Association for Plant Tissue Culture: Amsterdam, **1990**; pp. 362

Loyola-Vargas, V. M.; Hernández-Sotomayor, S. M. T. Plant roots in agriculture and medicine: the uses of radical biology. In Flores, H. E. Ed.; **1997**; In press

Loyola-Vargas, V. M.; Miranda-Ham, M. L. Root culture as a source of secondary metabolites of economic importance. In Recent Advances in Phytochemistry. Volume 29. Phytochemistry of Medicinal Plants; Arnason, J. T., Mata, R., Romeo, J. T. Eds.; Plenum Press: New York, **1995**; pp. 217–248.

Lu, T.; Parodi, F. J.; Vargas, D.; Quijano, L.; Mertooetomo, E. R.; Hjortso, M. A.; Fischer, N. H. Sesquiterpenes and thiarubrines from *Ambrosia trifida* and its transformed roots. *Phytochem.* **1993**, *33*, 113–116.

Lui, J. H. C.; Staba, E. J. Effects of precursors on serially propagated *Digitalis lanata* leaf and root cultures. *Phytochem.* **1979**, *18*, 1913–1916.

Macek, T.; Kotrba, P.; Suchova, M.; Skacel, F.; Demnerova, K.; Ruml, T. Accumulation of cadmium by hairy-root cultures of *Solanum nigrum*. *Biotechnol. Lett.* **1994**, *16*, 621–624.

Maheshari, S. C. Tissue culture, molecular biology and plant biotechnology - a historical review. In The impact of biotechnology in agriculture; Sangwan, R. S., Sangwan-Norreel, B. S. Eds.; Kluwer Academic Pub. The Netherlands, **1990**; pp. 1–12.

Maitani, T.; Kubota, H.; Sato, K.; Takeda, M.; Yoshihira, K. Induction of phytochelatin (class III metallothionein) and incorporation of copper in transformed hairy roots of *Rubia tinctorum* exposed to cadmium. *J. Plant Physiol.* **1996**, *147*, 743–748.

Maldonado-Mendoza, I. E.; Ayora-Talavera, T.; Loyola-Vargas, V. M. Establishment of hairy root cultures of *Datura stramonium*. Characterization and stability of tropane alkaloid production during long periods of subculturing. *Plant Cell Tiss. Org. Cult.* **1993**, *33*, 321–329.

Mano, Y.; Nabeshima, S.; Matsui, C.; Ohkawa, H. Production of tropane alkaloids by hairy root cultures of *Scopolia japonica*. *Agric. Biol. Chem.* **1986**, *50*, 2715–2722.

Mano, Y.; Ohkawa, H.; Yamada, Y. Production of tropane alkaloids by hairy root cultures of *Duboisia leichhardtii* transformed by *Agrobacterium rhizogenes*. *Plant Sci.* **1989**, *59*, 191–201.

Matsuda, J.; Okabe, S.; Hashimoto, T.; Yamada, Y. Molecular cloning of hyoscyamine 6ß-hydroxylase, a 2-oxoglutarate-dependent dioxygenase, from cultured roots of *Hyoscyamus niger*. *J. Biol. Chem.* **1991**, *266*, 9460–9464.

Matsumoto, M.; Shoyama, Y.; Nishioka, I.; Irino, N. Constituents of regenerated shoot and cultured root tissue of *Rehmannia glutinosa*. *Phytochem.* **1989**, *28*, 2331–2332.

Matsumoto, T.; Tanaka, N. Production of phytoecdysteroids by hairy root cultures of *Ajuga reptans* Var Atropurpurea. *Agric. Biol. Chem.* **1991**, *55*, 1019–1025.

McAfee, B. J.; White, E. E.; Pelcher, L. E.; Lapp, M. S. Root induction in pine (*Pinus*) and larch (*Larix*) spp. using *Agrobacterium rhizogenes*. *Plant Cell Tiss. Org. Cult.* **1993**, *34*, 53–62.

Medina-Bolívar, F.; Flores, H. E. Selection for hyoscyamine and cinnamoyl putrescine overproduction in cell and root cultures of *Hyoscyamus muticus*. *Plant Physiol.* **1995**, *108*, 1553–1560.

Merkli, A.; Christen, P.; Kapetanidis, I. Production of diosgenin by hairy root cultures of *Trigonella foenum-graecum* L. *Plant Cell Rep.* **1997**, *16*, 632–636.

Miranda-Ham, M. L. Biosynthesis of secondary metabolites in plant cell and tissue cultures. In Production of Secondary Metabolites from Plant Tissue Cultures and its Biotechnological Perspectives; Loyola-Vargas, V. M. Ed.; CICY: Merida, Yucatan, **1990**; pp. 80–90.

Motomori, Y.; Shimomura, K.; Mori, K.; Kunitake, H.; Nakashima, T.; Tanaka, M.; Miyazaki, S.; Ishimaru, K. Polyphenol production in hairy root cultures of *Fragaria x ananassa*. *Phytochem.* **1995**, *40*, 1425–1428.

Mugnier, J.; Mosse, B. Vesicular-arbuscular mycorrhizal infection in transformed root inducing T-DNA roots grown axenically. *Phytophatol.* **1987**, *77*, 1045–1050.

Mukerji, K. D.; Jagpal, R.; Bali, M.; Rani, R. The importance of mycorrhiza for roots. In Plant roots and their environment; McMichael, B. L., Persson, H. Eds.; Elsevier Science Pub. Amsterdam, London, New York, Tokyo, **1991**; pp. 290–308.

Mukherjee, S.; Ghosh, B.; Jha, T. B.; Jha, S. Genetic transformation of artemisia annua by *Agrobacterium rhizogenes*. *Indian J. Exp. Bot.* **1995**, *33*, 868–871.

Mukundan, U.; Hjortso, M. A. Thiophene accumulation in hairy roots of *Tagetes patula* in response to fungal elicitors. *Biotechnoi. Lett.* **1990a**, *12*, 609–614.

Mukundan, U.; Hjortso, M. A. Effect of fungal elicitor on thiophene production in hairy root cultures of *Tagetes patula*. *Appl. Microbiol. Biotechnol.* **1990b**, *33*, 145–147.

Mukundan, U.; Hjortso, M. A. Thiophene content in normal and transformed root cultures of *Tagetes erecta*: A comparison with thiophene content in roots of intact plants. *J. Exp. Bot.* **1990c**, *41*, 1497–1501.

Mukundan, U.; Hjortso, M. A. Effect of light on growth and thiophene accumulation in transformed roots of *Tagetes patula*. *J. Plant Physiol.* **1991a**, *138*, 252–255.

Mukundan, U.; Hjortso, M. A. Growth and thiophene accumulation by hairy root cultures of *Tagetes patula* in media of varying initial pH. *Plant Cell Rep.* **1991b**, *9*, 627–630.

Muranaka, T.; Ohkawa, H.; Yamada, Y. Scopolamine release into media by *Duboisia leichhardtii* hairy root clones. *Appl. Microbiol. Biotechnol.* **1992**, *37*, 554–559.

Murashige, T.; Skoog, F. A revised medium for rapid growth and bioassays with tobacco tissue cultures. *Physiol. Plant.* **1962**, *15*, 473–497.

Nabeshima, S.; Mano, Y.; Ohkawa, H. Production of tropane alkaloids by *Hairy root* cultures of *Scopolia japonica*. *Symbiosis* **1986**, *2*, 11–18.

Nakajima, K.; Hashimoto, T.; Yamada, Y. Opposite tropinone reductases with different stereospecificities are short-chain dehydrogenases evolved from a common ancestor. *Proc. Natl. Acad. Sci. USA* **1994**, *90*, 9591–9595.

Neill, S. J.; Lenton, J. R.; Wibberley, M. S. Differential effects of elicitors on secondary metabolism in hairy root cultures of tobacco (*Nicotiana tabacum*). *Biochem. Soc. Trans.* **1994**, *22*, 383–387.

Nessler, C. L. Induction of the synthesis of secondary metabolites in plant cell tissue cultures. In Production of Secondary Metabolites from Plant Tissue Cultures and its Biotechnological Perspectives; Loyola-Vargas, V. M. Ed.; CICY: Merida, Yucatan, **1990**; pp. 110–121.

Nguyen, C.; Bourgaud, F.; Forlot, P.; Guckert, A. Establishment of hairy root cultures of *Psoralea* species. *Plant Cell Rep.* **1992**, *11*, 424–427.

Norton, R. A.; Towers, G. H. N. Factors affecting synthesis of polyacetylenes in root cultures of *Bidens alba*. *J. Plant Physiol.* **1986**, *122*, 41–53.

Nuutila, A. M.; Vestberg, M.; Kauppinen, V. Infection of hairy roots of strawberry (*Fragaria x Ananassa* Duch.) with arbuscular mycorrhizal fungus. *Plant Cell Rep.* **1995**, *14*, 505–509.

Nyman, S. Incorporation of arginine, ornithine and phenylalanine into tropane alkaloids in suspension-cultured cells and aseptic roots of intact plants of *Atropa belladonna*. *J. Exp. Bot.* **1994**, *45*, 979–986.

O'Dowd, N. A.; Richardson, D. H. S. Production of tumours and roots by *Ephedra* following *Agrobacterium rhizogenes* infection. *Can. J. Bot.* **1994**, *72*, 203–207.

O'Keefe, B. R.; Mahady, G. B.; Gills, J. J.; Beecher, C. W. W.; Schilling, A. B. Stable vindoline production in transformed cell cultures of *Catharanthus roseus*. *J. Nat. Prod.* **1997**, *60*, 261–264.

Odnevall, A.; Björk, L. Effects of light on growth, morphogenesis and ginsenoside formation in tissue cultures of *Panax ginseng*. *Biochem. Physiol. Pflanz.* **1989**, *185*, 253–259.

Ogasawara, T.; Chiba, K.; Tada, M. Production in high-yield of a naphthoquinone by a hairy root culture of *Sesamum indicum*. *Phytochem.* **1993**, *33*, 1095–1098.

Ohkawa, H.; Kamada, H.; Sudo, H.; Harada, H. Effects of gibberellic acid on hairy root growth in *Datura innoxia*. *J. Plant Physiol.* **1989**, *134*, 633–636.

Oksman-Caldentey, K. M.; Sevón, N.; Vanhala, L.; Hiltunen, R. Effect of nitrogen and sucrose on the primary and secondary metabolism of transformed root cultures of *Hyoscyamus muticus*. *Plant Cell Tiss. Org. Cult.* **1994**, *38*, 263–272.

Olby, R. DNA before watson-crick. *Nature* **1974**, *248*, 782–785.

Olszowska, O.; Alfermann, A. W.; Furmanowa, M. Eugenol from normal and transformed root cultures of *Coluria geoides*. *Plant Cell Tiss. Org. Cult.* **1996**, *45*, 273–276.

Oostdam, A.; Mol, J. N. M.; Van der Plas, L. H. W. Establishment of hairy root cultures of *Linum flavum* producing the lignan 5-methoxypodophyllotoxin. *Plant Cell Rep.* **1993**, *12*, 474–477.

Palazón, J.; Cusidó, R. M.; Roig, C.; Piñol, M. T. Effect of *rol* genes from *Agrobacterium rhizogenes* TL-DNA on nicotine production in tobacco root cultures. *Plant Physiol. Biochem.* **1997**, *35*, 155–162.

Paniego, N. B.; Giulietti, A. M. Artemisinin production by *Artemisia annua* L-transformed organ cultures. *Enzyme Microb. Technol.* **1996**, *18*, 526–530.

Parkinson, M.; Cotter, T.; Dix, P. J. Peroxidase production by cell suspension and hairy root cultures of horseradish (*Armoracia rusticana*). *Plant Sci.* **1990**, *66*, 271–277.

Parodi, F. J.; Fischer, N. H.; Flores, H. E. Benzofuran and bithiophenes from root cultures of *Tagetes patula*. *J. Nat. Prod.* **1988**, *51*, 594–595.

Parr, A. J.; Peerless, A. C. J.; Hamill, J. D.; Walton, N. J.; Robins, R. J.; Rhodes, M. J. C. Alkaloid production by transformed root cultures of *Catharanthus roseus*. *Plant Cell Rep.* **1988**, *7*, 309–312.

Parr, A. J.; Payne, J.; Eagles, J.; Chapman, B. T.; Robins, R. J.; Rhodes, M. J. C. Variation in tropane alkaloid accumulation within the solanaceae and strategies for its exploitation. *Phytochem.* **1990**, *29*, 2545–2550.

Parr, A. J. Alternative metabolic fates of hygrine in transformed root cultures of *Nicandra physaloides*. *Plant Cell Rep.* **1992**, *11*, 270–273.

Parr, A. J.; Hamill, J. D. Relationship between *Agrobacterium rhizogenes* transformed hairy roots and intact, uninfected *Nicotiana* plants. *Phytochem.* **1987**, *26*, 3241–3245.

Payne, J.; Hamill, J. D.; Robins, R. J.; Rhodes, M. J. C. Production of hyoscyamine by "hairy root" cultures of *Datura stramonium*. *Planta Med.* **1987**, *53*, 474–478.

Pereda-Miranda, R.; Mata, R.; Anaya, A. L.; Mahinda, W. D. B.; Pezzuto, J. M.; Douglas, K. A. Tricolorin a, major phytogrowth inhibitor from *Ipomoea tricolor*. *J. Nat. Prod.* **1993**, *56*, 571–582.

Pita-Alvarez, S. I.; Giulietti, A. M. Advantages and limitations in the use of hairy root cultures for the production of tropane alkaloids: use of auxins in the maintenance of normal root morphology. *In Vitro Cell. Dev. Biol.* **1995**, *31P*, 215–220.

Pitta-Alvarez, S. I.; Giulietti, A. M. Effects of gibberellin GA_7 on kinetics of growth and tropane alkaloid accumulation in hairy roots of *Brugmansia candida*. *In Vitro Cell. Dev. Biol.* **1997**, *33*, 147–153.

Portsteffen, A.; Draeger, B.; Nahrstedt, A. Two tropinone reducing enzymes from *Datura stramonium* transformed root cultures. *Phytochem.* **1992**, *31*, 1135–1138.

Ramakrishnan, D; Curtis, W. R. Fluid dynamic studies of plant root reactor systems. In Advances in Plant Biotechnology. Ryu, D. D. Y.; Furusaki, S. eds; Amsterdam, New York: Elsevier, **1994**, pp. 281–305.

Ramakrishnan, D.; Salim, J.; Curtis, W. R. Inoculation and tissue distribution in pilot-scale plant root culture bioreactors. *Biotechnol. Tech.* **1994**, *8*, 639–644.

Reichling, J.; Thron, U. Accumulation of rare phenylpropanoids in Agrobacterium rhizogenes transformed root cultures of Coreopsis tinctoria. *Planta Med.* **1990**, *56*, 488–490.

Repunte, V. P.; Kino-Oka, M.; Taya, M.; Tone, S. Reversible morphology change of horseradish hairy roots cultivated in phytohormone-containing media. *J. Ferment. Bioeng.* **1993**, *75*, 271–275.

Rhodes, M. J. C.; Hilton, M.; Parr, A. J.; Hamill, J. D.; Robins, R. J. Nicotine production by "hairy root" cultures of *Nicotiana rustica*: fermentation and product recovery. *Biotechnol. Lett.* **1986**, *8*, 415–420.

Rhodes, M. J. C.; Hamill, J. D.; Robins, R. J.; Evans, D. M. Expression of enzymes in secondary metabolic pathways in transformed root cultures of Nicotiana rustica and Datura stramonium. *Eur. Conf. Biotechnol.* **1988**, 28–32.

Rhodes, M. J. C.; Robins, R. J.; Lindsay, E.; Aird, H.; Payne, J.; Parr, A. J.; Walton, N. J. Regulation of secondary metabolism in transformed roots cultures. In Primary and Secondary Metabolism of Plant Cell Cultures II; Kurz, W. G. W. Ed.; Springer-Verlag: Berlin Heidelberg, **1989**; pp. 58–72.

Rhodes, M. J. C.; Parr, A. J.; Giulietti, A.; Aird, E. L. H. Influence of exogenous hormones on the growth and secondary metabolite formation in transformed root cultures. *Plant Cell Tiss. Org. Cult.* **1994**, *38*, 143–151.

Robbins, M. P.; Hartnoll, J.; Morris, P. Phenylpropanoid defence responses in transgenic *Lotus corniculatus*. 1. Glutathione elicitation of isoflavan phytoalexins in transformed root cultures. *Plant Cell Rep.* **1991**, *10*, 59–62.

Robbins, M. P.; Thomas, B.; Morris, P. Phenylpropanoid defence responses in transgenic *Lotus corniculatus*. II. Modelling plant defence responses in transgenic root cultures using thiol and carbohydrate elicitors. *J. Exp. Bot.* **1995**, *46*, 513–524.

Robins, R. J.; Hamill, J. D.; Parr, A. J.; Smith, K.; Walton, N. J.; Rhodes, M. J. C. Potential for use of nicotinic acid as a selective agent for isolation of high nicotine-producing lines of *Nicotiana rustica* hairy root cultures. *Plant Cell Rep.* **1987**, *6*, 122–126.

Robins, R. J.; Parr, A. J.; Payne, J.; Walton, N. J.; Rhodes, M. J. C. Factors regulating tropane-alkaloid production in a transformed root culture of a *Datura candida* x *D. aurea* hybrid. *Planta* **1990**, *181*, 414–422.

Robins, R. J.; Bachmann, P.; Robinson, T.; Rhodes, M. J. C.; Yamada, Y. The formation of 3α- and 3ß-acetoxytropanes by *Datura stramonium* transformed root cultures involves two acetyl-CoA-dependent acyltransferases. *FEBS Lett.* **1991a**, *292*, 293–297.

Robins, R. J.; Bent, E. G.; Rhodes, M. J. C. Studies on the biosynthesis of tropane alkaloids by *Datura stramonium* L. transformed root cultures. 3. The relationship between morphological integrity and alkaloid biosynthesis. *Planta* **1991b**, *185*, 385–390.

Robins, R. J.; Parr, A. J.; Bent, E. G.; Rhodes, M. J. C. Studies on the biosynthesis of tropane alkaloids in *Datura stramonium* L. transformed root cultures. 1. The kinetics of alkaloid production and the influence of feeding intermediate metabolites. *Planta* **1991c**, *183*, 185–195.

Robins, R. J.; Woolley, J. G.; Ansarin, M.; Eagles, J.; Goodfellow, B. J. Phenyllactic acid but not tropic acid is an intermediate in the biosynthesis of tropane alkaloids in *Datura* and *Brugmansia* transformed root cultures. *Planta* **1994**, *194*, 86–94.

Rodríguez-Mendiola, M. A.; Stafford, A.; Cresswell, R.; Arias-Castro, C. Bioreactors for growth of plant roots. *Enzyme Microb. Technol.* **1991**, *13*, 697–702.

Rodríguez-Zapata, L. C.; Hernández-Sotomayor, S. M. T. Evidence of protein-tyrosine kinase activity in *Catharanthus roseus* roots transformed by *Agrobacterium rhizogenes*. *Planta* **1997**,

Rovira, A. D.; Foster, R. C.; Martin, J. K. Note on terminology: origin, nature and nomenclature of the organic materials in the rizhosphere. In The soil-root interface; Harley, J. L., Russell, R. S. Eds.; Academic Press: London, New York, San Francisco, **1979**; pp. 1–4.

Ruiter, H. d.; Kollöffel, C. Properties of ornithine carbamoyltransferase from *Pisum sativum* L. *Plant Physiol.* **1985**, *77*, 695–699.

Saito, K.; Yamazaki, M.; Shimomura, K.; Yoshimatsu, K.; Murakoshi, I. Genetic engineering of foxglove (*Digitalis purpurea*). *Plant Physiol.* **1990a**, *93*, 158 (Abstract)

Saito, K.; Yamazaki, M.; Shimomura, K.; Yoshimatsu, K.; Murakoshi, I. Genetic transformation of foxglove (Digitalis purpurea) by chimeric foreign genes and production of cardioactive glycosides. *Plant Cell Rep.* **1990b**, *9*, 121–124.

Saitou, T.; Kamada, H.; Harada, H. Isoperoxidase in hairy roots and regenerated plants of horseradish (*Armoracia lapathifolia*). *Plant Sci.* **1991**, *75*, 195–201.

Santana, N.; Cortés, S.; Martín, J. V.; Montes, S. Adaptación de vitroplantas de embriones somáticos de cafeto (*C. rabica* L.) variedad Catimor (9722). *Cultivos Tropicales* **1995**, *11*, 429–435.

Sato, K.; Yamazaki, T.; Okuyama, E.; Yoshihira, K.; Shimomura, K. Anthraquinone production by transformed root cultures of *Rubia tinctorum*. Influence of phytohormones and sucrose concentration. *Phytochem.* **1991**, *30*, 1507–1509.

Sauerwein, M.; Ishimaru, K.; Shimomura, K. A piperidone alkaloid from *Hyoscyamus albus* roots transformed with *Agrobacterium rhizogenes*. *Phytochem.* **1991a**, *30*, 2977–2978.

Sauerwein, M.; Ishimaru, K.; Shimomura, K. Indole alkaloids in hairy roots of *Amsonia elliptica*. *Phytochem.* **1991b**, *30*, 1153–1155.

Sauerwein, M.; Yamazaki, T.; Shimomura, K. Hernandulcin in hairy root cultures of *Lippia dulcis*. *Plant Cell Rep.* **1991c**, *9*, 579–581.

Sauerwein, M.; Wink, M.; Shimomura, K. Influence of light and phytohormones on alkaloid production in transformed root cultures of *Hyoscyamus albus*. *J. Plant Physiol.* **1992**, *140*, 147–152.

Sauerwein, M.; Shimomura, K.; Wink, M. Incorporation of 1–^{13}C-acetate into tropane alkaloids by hairy root cultures of *Hyoscyamus albus*. *Phytochem.* **1993**, *32*, 905–909.

Sauerwein, M.; Shimomura, K. Alkaloid production in hairy roots of *Hyoscyamus albus* transformed with *Agrobacterium rhizogenes*. *Phytochem.* **1991**, *30*, 3277–3280.

Sauerwein, M.; Wink, M. On the role of opines in plants transformed with *Agrobacterium rhizogenes*: Tropane alkaloid metabolism, insect-toxicity and allelopathic properties. *J. Plant Physiol.* **1993**, *142*, 446–451.

Savary, B. J.; Flores, H. E.; Hill, J. J. Isolation of a class III chitinase produced in root cultures of *Trichosanthes kirilowii* and assessment of accumulation patterns and antifungal activity. *Plant Physiol. Biochem.* **1997**, *35*, 543–551.

Savary, B. J.; Flores, H. Production of extracellular proteins by root cultures of Chinese medicinal cucumber (*Trichosanthes* spp.). *Plant Physiol.* **1991**, *96*, 31 (Abstract)

Savary, B. J.; Flores, H. E. Characterization of bioactive proteins produced in root cultures of *Trichosanthes kirilowi* var. *Japonica*. *Plant Physiol.* **1993**, *102*, 31s (Abstract)

Savary, B. J.; Flores, H. E. Biosynthesis of defense-related proteins in transformed root cultures of *Trichosanthes kirilowii* Maxim. var *japonicum* (Kitam.). *Plant Physiol.* **1994**, *106*, 1195–1204.

Sáenz-Carbonell, L.; Maldonado-Mendoza, I. E.; Loyola-Vargas, V. M. Effect of the nitrogen-source in tropane alkaloid production in *Datura stramonium* hairy root cultures. *Plant Physiol.* **1991**, *96*, 96 (Abstract)

Sáenz-Carbonell, L.; Loyola-Vargas, V. M. *Datura stramonium* hairy roots tropane alkaloid content as a response to changes in Gamborg's B_5 medium. *Appl. Biochem. Biotechnol.* **1996**, *61*, 321–337.

Sen, J.; Sharma, A. K.; Sahu, N. P.; Mahato, S. B. Forskolin production in untransformed root culture of *Coleus forskohlii*. *Phytochem.* **1993**, *34*, 1309–1312.

Senior, I.; Holford, P.; Cooley, R. N.; Newbury, H. J. Transformation of *Antirrhinum majus using Agrobacterium rhizogenes*. *J. Exp. Bot.* **1995**, *46*, 1233–1239.

Sevón, N.; Dräger, B.; Hiltunen, R.; Oksman-Caldentey, K. M. Characterization of transgenic plants derived from hairy roots of *Hyoscyamus muticus*. *Plant Cell Rep.* **1997**, *16*, 605–611.

Sharp, J. M.; Doran, P. M. Characteristics of growth and tropane alkaloid synthesis in *Atropa belladonna* roots transformed by *Agrobacterium rhizogenes*. *J. Biotechnol.* **1990**, *16*, 171–186.

Shen, W.; Williamson, M. S.; Forde, B. G. Functional analysis of the promoter region of a nodule-enhanced glutamine synthetase gene from *Phaseolus vulgaris* L. *Plant Mol. Biol.* **1992**, *19*, 837–846.

Shen, W. H.; Davioud, E.; David, C.; Barbier-Brygoo, H.; Tempé, J.; Guern, J. High sensitivity to auxin is a common feature of hairy root. *Plant Physiol.* **1990**, *94*, 554–560.

Sheng, J.; Citovsky, V. Agrobacterium-Plant Cell DNA transport: have virulence proteins, will travel. Plant Cell **1996**, 8, 1699–1710.

Sherwood, J. E.; Truchet, G. L.; Dazzo, F. B. Effect of nitrate supply on the *in vivo* synthesis and distribution of trifoliin A, a *Rhizobium trifolii*-binding lectin in *Trifolium repens* seedlings. *Planta* **1984**, *162*, 540–547.

Shimomura, K.; Sauerwein, M.; Ishimaru, K. Tropane alkaloids in the adventitious and hairy root cultures of solanaceous plants. *Phytochem.* **1991a**, *30*, 2275–2278.

Shimomura, K.; Sudo, H.; Saga, H.; Kamada, H. Shikonin production and secretion by hairy root cultures of *Lithospermum erythrorhizon*. *Plant Cell Rep.* **1991b**, *10*, 282–285.

Siebertz, R.; Proksch, P.; Wray, V.; Witte, L. Accumulation and biosynthesis of benzofurans in root cultures of *Eupatorium cannabinum*. *Phytochem.* **1989**, *28*, 789–793.

Sikuli, N. N.; Demeyer, K. Influence of the ion-composition of the medium on alkaloid production by "hairy roots" of *Datura stramonium*. *Plant Cell Tiss. Org. Cult.* **1997**, *47*, 261–267.

Sim, S. J.; Chang, H. N. Increased shikonin production by hairy roots of *Lithospermum erythrorhizon* in two phase bubble column reactor. *Biotechnol. Lett.* **1993**, *15*, 145–150.

Singh, G.; Reddy, G. R.; Curtis, W. R. Use of binding measurements to predict elicitor dosage requirements for secondary metabolite production from root cultures. *Biotechnol. Prog.* **1994**, *10*, 365–371.

Slater, G. W.; Guo, H. L. An exactly solvable Ogston model of gel electrophoresis .2. Sieving through periodic gels. *Electrophoresis* **1996**, *17*, 1407–1415.

Smith, R. G.; Caswell, D.; Carriere, A.; Zielke, B. Variation in the ginsenoside content of American ginseng, *Panax quinquefolius* L, roots. *Can. J. Bot.* **1996**, *74*, 1616–1620.

Smith, T. C.; Weathers, P. J.; Cheetham, R. D. Effects of gibberellic acid on hairy root cultures of *Artemisia annua*: growth and artemisinin production. *In Vitro Cell. Dev. Biol.* **1997**, *33*, 75–79.

Song, Q.; Gomez-Barrios, M. L.; Hopper, E. L.; Hjortso, M. A.; Fischer, N. H. Biosynthetic studies of lactucin derivatives in hairy root cultures of *Lactuca floridana*. *Phytochem.* **1995**, *40*, 1659–1665.

Stepan-Sarkissian, G. Biotransformation by plant cell cultures. In Plant cell and tissue culture; Stafford, A., Warren, G. Eds.; Open University Press: Great Britain, Melkshan, Wiltshire, **1991**; pp. 163–204.

Steward, F. C.; Mapes, M. O.; Smith, J. Growth and organized development of cultured cells. I. Growth and division of freely suspended cells. *Am. J. Bot.* **1958**, *45*, 693–703.

Stojakowska, A.; Malarz, J.; Kisiel, W.; Kohlmünzer, S. Callus and hairy root cultures of *Lactuca virosa*. *Planta Med.* **1993**, *59*, 658

Strauss, A.; Spengel, S. M.; Schaffner, W. Saponins from root cultures of *Phytolacca acinosa*. *Phytochem.* **1995**, *38*, 861–865.

Subroto, M. A.; Kwok, K. H.; Hamill, J. D.; Doran, P. M. Coculture of genetically transformed roots and shoots for synthesis, translocation, and biotransformation of secondary metabolites. *Biotechnol. Bioeng.* **1996**, *49*, 481–494.

Subroto, M. A.; Doran, P. M. Production of steroidal alkaloids by hairy roots of *Solanum aviculare* and the effect of gibberellic acid. *Plant Cell Tiss. Org. Cult.* **1994**, *38*, 93–102.

Sugimoto, Y.; Sugimura, Y.; Yamada, Y. Production of bisbenzylisoquinoline alkaloids in cultured roots of *Stephania cepharantha*. *Phytochem.* **1988**, *27*, 1379–1381.

Sugimoto, Y.; Yamada, Y.; Sugimura, Y. Comparison of bisbenzylisoquinoline alkaloids in root cultures of *Stephania cepharantha* with those in the parent plant. *J. Nat. Prod.* **1989**, *52*, 199–202.

Sugimoto, Y.; Sugimura, Y.; Yamada, Y. Biosynthesis of bisbenzylisoquinoline alkaloids in cultured roots of *Stephania cepharantha*. *FEBS Lett.* **1990**, *273*, 82–86.

Sugimoto, Y.; Yoshida, A.; Uchida, S.; Inanaga, S.; Yamada, Y. Dauricine production in cultured roots of *Menispermum dauricum*. *Phytochem.* **1994**, *36*, 679–683.

Sugimoto, Y.; Uchida, S.; Inanaga, S.; Kimura, Y.; Hashimoto, M.; Isogai, A. Early steps of dauricine biosynthesis in cultured roots of *Menispermum dauricum*. *Biosci. Biotech. Biochem.* **1996**, *60*, 503–505.

Sugimoto, Y.; Yamada, Y. Involvement of tyramine in bisbenzylisoquinoline biosynthesis in cultured roots of *Stephania cepharantha*. *J. Nat. Prod.* **1993**, *56*, 1229–1233.

Tada, H.; Shimomura, K.; Ishimaru, K. Polyacetylenes in hairy root cultures of *Lobelia chinensis* Lour. *J. Plant Physiol.* **1995a**, *146*, 199–202.

Tada, H.; Shimomura, K.; Ishimaru, K. Polyacetylenes in *Platycodon grandiflorum* hairy root and campanulaceous plants. *J. Plant Physiol.* **1995b**, *145*, 7–10.

Tada, H.; Murakami, Y.; Omoto, T.; Shimomura, K.; Ishimaru, K. Rosmarinic acid and related phenolics in hairy root cultures of *Ocimum basilicum*. *Phytochem.* **1996a**, *42*, 431–434.

Tada, H.; Nakashima, T.; Kunitake, H.; Mori, K.; Tanaka, M.; Ishimaru, K. Polyacetylenes in hairy root cultures of *Campanula medium* L. *J. Plant Physiol.* **1996b**, *147*, 617–619.

Taiz, L.; Zeiger, E. *Plant Physiology*. The Benjamin/Cummings Publishing: California, **1991**; pp. 1–559.

Talou, J. R.; Cascone, O.; Giulietti, A. M. Content of thiophenes in transformed root cultures of Argentinian species of Tagetes. *Planta Med.* **1994**, *60*, 260–262.

Talou, J. R.; Giulietti, A. M. In vitro thiophene production by transformed root cultures of Tagetes laxa (Cabrera). *Biotechnol. Lett.* **1995**, *17*, 1337–1342.

Tanaka, N.; Takao, M.; Matsumoto, T. *Agrobacterium rhizogenes*-mediated transformation and regeneration of *Vinca minor* L. *Plant Tissue Culture Letters* **1994**, *11*, 191–198.

Tanaka, N.; Takao, M.; Matsumoto, T. Vincamine production in multiple shoot culture derived from hairy roots of *Vinca minor*. *Plant Cell Tiss. Org. Cult.* **1995**, *41*, 61–64.

Tanaka, N.; Matsumoto, T. Characterization of *Ajuga* plants regenerated from hairy roots. *Plant Tissue Culture Letters* **1993**, *10*, 78–83.

Tanford, C. Mechanism of free energy coupling in active transport. *Annu. Rev. Biochem.* **1983**, *52*, 379–409.

Taya, M.; Yoyama, A.; Kondo, O.; Kobayashi, T.; Matsui, C. Growth characteristics of plant hairy roots and their cultures in bioreactors. *J. Chem. Eng. (Japan)* **1989**, *22*, 84–89.

Taya, M.; Sato, H.; Kino-Oka, M.; Tone, S. Characterization of pak-bung green hairy roots cultivated under light irradiation. *J. Ferment. Bioeng.* **1994**, *78*, 42–48.

Teoh, K. H.; Weathers, P. J.; Cheetham, R. D.; Walcerz, D. B. Cryopreservation of transformed (hairy) roots of *Artemisia annua*. *Cryobiology* **1996**, *33*, 106–117.

Tepfer, D. Ri t-DNA from *Agrobacterium rhizogenes*: a source of genes having applications in rhizosphere biology and and plant development, ecology and evolution. In Plant-Microbe Interactions; Kosuge, T., Nester, E. Eds.; McGraw Hill: New York, **1989**; pp. 294–342.

Tepfer, D.; Metzger, L.; Prost, R. Use of roots transformed by *Agrobacterium rhizogenes* in rhizosphere research: Applications in studies of cadmium assimilation from sewage sludges. *Plant Mol. Biol.* **1989**, *13*, 295–302.

Thorpe, T. A. The current status of plant tissue culture. In Developments in crop science 19. Plant tissue culture: applications and limitations; Bhojwani, S. S. Ed., Elsevier: Amsterdam, **1990**; pp. 1–33.

Thorup, J. E.; McDonald, K. A.; Jackman, A. P.; Bhatia, N.; Dandekar, A. M. Ribosome inactivating protein production from Trichosanthes kirilowii plant cell cultures. *Biotechnol. Prog.* **1994**, *10*, 345–352.

Toivonen, L.; Balsevich, J.; Kurz, W. G. W. Indole alkaloid production by hairy root cultures of *Catharanthus roseus*. *Plant Cell Tiss. Org. Cult.* **1989**, *18*, 79–93.

Toivonen, L.; Ojala, M.; Kauppinen, V. Indole alkaloid production by hairy root cultures of *Catharanthus roseus*: growth kinetics and fermentation. *Biotechnol. Lett.* **1990**, *12*, 519–524.

Toivonen, L.; Ojala, M.; Kauppinen, V. Studies on the optimization of growth and indole alkaloid production by hairy root cultures of *Catharanthus roseus*. *Biotechnol. Bioeng.* **1991**, *37*, 673–680.

Toivonen, L.; Laakso, S.; Rosenqvist, H. The effect of temperature on growth, indole alkaloid accumulation and lipid composition of *Catharanthus roseus* cell suspension cultures. *Plant Cell Rep.* **1992**, *11*, 390–394.

Toivonen, L.; Rosenqvist, H. Establishment and growth characteristics of *Glycyrrhiza glabra* hairy root cultures. *Plant Cell Tiss. Org. Cult.* **1995**, *41*, 249–258.

Trotin, F.; Moumou, Y.; Vasseur, J. Flavanol production by *Fagopyrum esculentum* hairy and normal root cultures. *Phytochem.* **1993**, *32*, 929–931.

Trypsteen, M.; Vanlijsebettens, M.; Vanseveren, R.; Vanmontagu, M. *Agrobacterium-rhizogenes*-mediated transformation of *Echinacea purpurea*. *Plant Cell Rep.* **1991**, *10*, 85–89.

Uesato, S.; Ogawa, Y.; Inouya, H.; Saiki, K.; Zenk, M. H. Synthesis of iridodial by cell free extracts from *Rauwolfia serpentina* cell suspension cultures. *Tetrahedrom Lett.* **1986**, *13*, 2893–2896.

Uozumi, N.; Kohketsu, K.; Kondo, O.; Honda, H.; Kobayashi, T. Fed-batch culture of hairy root using fructores as a carbon source. *J. Ferment. Bioeng.* **1991**, *72*, 457–460.

Uozumi, N.; Kato, Y.; Nakashimada, Y.; Kobayashi, T. Excretion of peroxidase from horseradish hairy root in combination with ion supplementation. *Appl. Microbiol. Biotechnol.* **1992**, *37*, 560–565.

Uozumi, N.; Kohketsu, K.; Kobayashi, T. Growth and kinetic parameters of *Ajuga* hairy root in fed-batch culture on monosaccharide medium. *J. Chem. Tech. Biotechnol.* **1993**, *57*, 155–161.

Ushiyama, M.; Asada, Y.; Yoshikawa, T.; Furuya, T. Biotransformation of aromatic carboxylic acids by root culture of *Panax ginseng*. *Phytochem.* **1989**, *28*, 1859–1869.

Ushiyama, M.; Furuya, T. Glycosylation of phenolic compounds by root culture of *Panax ginseng*. *Phytochem.* **1989**, *28*, 3009–3013.

Van Der Heijden, R.; Verpoorte, R.; Hoekstra, S. S.; Hoge, C. J. H. Nordamnacanthal, a major anthraquinone from an *Agrobacterium rhizogenes* induced root culture of *Rubia tinctorum*. *Plant Physiol. Biochem.* **1994**, *32*, 399–404.

Van Eijsden, R. R.; Díaz, C. L.; De Pater, B. S.; Kijne, J. W. Sugar-binding activity of pea (*Pisum sativum*) lectin is essential for heterologous infection of transgenic white clover hairy roots by *Rhizobium leguminosarum* biovar viciae. *Plant Mol. Biol.* **1995**, *29*, 431–439.

Van Noordwijk, M.; De Willigen, P. Root functions in agricultural systems. In Plant Roots. The hidden half; Waisel, Y., Eshel, A., Kafkati, U. Eds.; Marcel Dekker, Inc. New York, Basel, Hong Kong, **1991**; pp. 381–395.

Van Uden, W.; Pras, N.; Homan, B.; Malingré, T. M. Improvement of the production of 5-methoxypodophyllotoxin using a new selected root culture of *Linum flavum* L. *Plant Cell Tiss. Org. Cult.* **1991**, *27*, 115–121.

Vanhala, L.; Hiltunen, R.; Oksman-Caldentey, K. M. Virulence of different *Agrobacterium* strains on hairy root formation of *Hyoscyamus muticus*. *Plant Cell Rep.* **1995**, *14*, 236–240.

Vázquez-Flota, F.; Coello, J.; Loyola-Vargas, V. M. Growth kinetics and alkaloid production in hairy root cultures of *Catharanthus roseus*. *Plant Physiol.* **1992**, *99*, 49s (Abstract)

Vázquez-Flota, F.; Moreno-Valenzuela, O.; Miranda-Ham, M. L.; Coello-Coello, J.; Loyola-Vargas, V. M. Catharanthine and ajmalicine synthesis in *Catharanthus roseus* hairy root cultures. Medium optimization and elicitation. *Plant Cell Tiss. Org. Cult.* **1994**, *38*, 273–279.

Vogt, K. A.; Vogt, D. J.; Bloonfield, J. Input of organic matter to the soil by tree roots. In Plant roots and their environment; McMichael, B. L., Persson, H. Eds.; Elsevier Science Pub. Amsterdam, London, New York, Tokyo, **1991**; pp. 171–190.

Wagner, L. J.; Flores, H. W. Patterns of protein and secondary metabolite production in root cultures of *Momordica charantia*. *Plant Physiol.* **1992**, *99*, 48 (Abstract)

Walton, N. J.; Robins, R. J.; Rhodes, M. J. C. Perturbation of alkaloid production by cadaverine in hairy root cultures of *Nicotiana rustica*. *Plant Sci.* **1988**, *54*, 125–131.

Walton, N. J.; Robins, R. J.; Peerless, A. C. J. Enzymes of N-methylputrescine biosynthesis in relation to hyoscyamine formation in transformed root cultures of *Datura stramonium* and *Atropa belladonna*. *Planta* **1990**, *182*, 136–141.

Walton, N. J.; Peerless, A. C. J.; Robins, R. J.; Rhodes, M. J. C.; Boswell, H. D.; Robins, D. J. Purification and properties of putrescine *N*-methyltransferase from transformed roots of *Datura stramonium* L. *Planta* **1994**, *193*, 9–15.

Walton, N. J.; Belshaw, N. J. The effect of cadaverine on the formation of anabasine from lysine in hairy root cultures of *Nicotiana hesperis*. *Plant Cell Rep.* **1988**, *7*, 115–118.

Walton, N. J.; McLauchlan, W. R. Diamine oxidation and alkaloid production in transformed root cultures of *Nicotiana tabacum*. *Phytochem.* **1990**, *29*, 1455–1457.

Wang, P.; Flores, H. E.; Humphrey, A. E. Production of expansin from light/dark growing *Trichosanthes kirilowii* var. *Japonicum* root cultures. *Biotechnol. Lett.* **1994**, *16*, 955–958.

Weathers, P. J.; Cheetham, R. D.; Follansbee, E.; Teoh, K. Artemisinin production by transformed roots of *Artemisia annua*. *Biotechnol. Lett.* **1994**, *16*, 1281–1286.

Westcott, R. J. Thiophene production from "hairy roots" of *Tagetes*. In Manipulating Secondary Metabolism in Culture; Robins, R. J., Rhodes, M. J. C. Eds.; Cambridge University Press: Cambridge, **1988**; pp. 233–237.

Wibberley, M. S.; Lenton, J. R.; Neill, S. J. Sesquiterpenoid phytoalexins produced by hairy roots of *Nicotiana tabacum*. *Phytochem.* **1994**, *37*, 349–351.

Wickremesinhe, E. R.; Arteca, R. N. Establishment of root cultures from *Cephalotaxus harringtonia* callus. *Hortscience.* **1991**, *26*, 717 (Abstract)

Wilcox, H. E. Mycorrhizae. The rhizosphere (part V). In Plant Roots. The hidden half; Waisel, Y., Eshel, A., Kafkati, U. Eds.; Marcel Dekker, Inc. New York, Basel, Hong Kong, **1991**; pp. 731–765.

Williams, R. D.; Ellis, B. E. Patterns of benzylisoquinoline alkaloids in *Papaver somniferum* plants and *Agrobacterium rhizogenes* transformed tissues. *Phytochem.Soc.Newsletter* **1988**, *28*, 12

Williams, R. D.; Ellis, B. E. Alkaloids from *Agrobacterium rhizogenes*-transformed *Papaver somniferum* cultures. *Phytochem.* **1993**, *32*, 719–723.

Wilson, P. D. G.; Hilton, M. G.; Robins, R. J.; Rhodes, M. J. C. Fermentation studies of transformed root cultures. In Bioreactors and Biotransformations; Moody, G. W., Baker, P. B. Eds.; Elsevier Applied Sci. Pub. New York, **1989**; pp. 38–51.

Wink, M.; Witte, L. Alkaloids in stem roots of *Nicotiana tabacum* and *Spartium junceum* transformed by *Agrobacterium rhizogenes*. *Z. Naturforsch.* **1987**, *42c*, 69–72.

Witty, M. Thaumatin II: A simple marker gene for use in plants. *Nucleic Acids Res.* **1989**, *17*, 3312

Witty, M. Thaumatin II - a palatability protein. *Trends. Biotechnol.* **1990a**, *8*, 113–116.

Witty, M. Preprothaumatin II is processed to biological activity in *Solanum tuberosum*. *Biotechnol. Lett.* **1990b**, *12*, 131–136.

Wongsamuth, R.; Doran, P. M. Production of monoclonal antibodies by tobacco hairy roots. *Biotechnol. Bioeng.* **1997**, *54*, 401–415.

Xu, Z.; Jia, J.; Hu, Z. Enhancement by osmotic treatment of hairy root transformation of alfalfa suspension cultures, and chromosomal variation in the transformed tissues. *Aust. J. Plant Physiol.* **1997**, *24*, 345–351.

Xu, Z. Q.; Jia, J. F. The reduction of chromosome number and the loss of regeneration ability during subculture of hairy root cultures of *Onobrychis viciaefolia* transformed by *Agrobacterium rhizogenes* A_4. *Plant Sci.* **1996**, *120*, 107–112.

Yagi, T.; Morisaki, M.; Kushiro, T.; Yoshida, H.; Fujimoto, Y. Biosynthesis of 24ß-alkyl-Delta^{25}-sterols in hairy roots of *Ajuga reptans* var *atropurpurea*. *Phytochem.* **1996**, *41*, 1057–1064.

Yamada, Y.; Hashimoto, T. Biosynthesis of tropane alkaloids. In Applications of Plant Cell and Tissue Culture; Bock, G., Marsh, J. Eds.; John Wiley & Sons: New York, **1988**; pp. 199–212.

Yamamoto, Y.; Kinoshita, Y.; Watanabe, S.; Yamada, Y. Anthocyanin production in suspension cultures of high-producing cells of *Euphorbia millii*. *Agric. Biol. Chem.* **1989**, *53*, 417–423.

Yamanaka, M.; Shimomura, K.; Sasaki, K.; Yoshihira, K.; Ishimaru, K. Glucosylation of phenolics by hairy root cultures of *Lobelia sessilifolia*. *Phytochem.* **1995**, *40*, 1149–1150.

Yamazaki, M.; Lin, S.; Hayashi, T.; Morita, N.; Asamizu, T.; Mourakoshi, I.; Saito, K. Transgenic fertile *Scoparia dulcis* L, a folk medicinal plant, conferred with a herbicide-resistant trait using an Ri binary vector. *Plant Cell Rep.* **1996**, *15*, 317–321.

Yamazaki, T.; Flores, H. E.; Shimomura, K.; Yoshihira, K. Examination of steviol glucosides production by hairy root and shoot cultures of *Stevia rebaudiana*. *J. Nat. Prod.* **1991**, *54*, 986–992.

Yeoman, M. M.; Yeoman, C. L. Manipulating secondary metabolism in cultured plant cells. *New Phytol.* **1996**, *134*, 553–569.

Yonemitsu, H.; Shimomura, K.; Satake, M.; Mochida, S.; Tanaka, M.; Endo, T.; Kaji, A. Lobeline production by hairy root culture of *Lobelia inflata* L. *Plant Cell Rep.* **1990**, *9*, 307–310.

Yoshikawa, T.; Asada, Y.; Furuya, T. Continuous production of glycosides by a bioreactor using ginseng hairy root culture. *Appl. Microbiol. Biotechnol.* **1993**, *39*, 460–464.

Yoshikawa, T.; Furuya, T. Saponin production by cultures of *Panax ginseng* with *Agrobacterium rhizogenes*. *Plant Cell Rep.* **1987**, *6*, 449–453.

Yoshimatsu, K.; Yamaguchi, H.; Shimomura, K. Traits of *Panax ginseng* hairy roots after cold storage and cryopreservation. *Plant Cell Rep.* **1996**, *15*, 555–560.

Yoshimatsu, K.; Shimomura, K. Emetic alkaloid formation in root culture of *Cephaelis ipecacuanha*. *Phytochem.* **1991**, *30*, 505–507.

Yu, S. X.; Kwok, K. H.; Doran, P. M. Effect of sucrose, exogenous product concentration, and other culture conditions on growth and steroidal alkaloid production by *Solanum aviculare* hairy roots. *Enzyme Microb. Technol.* **1996**, *18*, 238–243.

Yukimune, Y.; Hara, Y.; Yamada, Y. Tropane alkaloid production in root cultures of *Duboisia myoporoides* obtained by repeated selection. *Biosci. Biotechnol. Biochem.* **1994a**, *58*, 1443–1446.

Yukimune, Y.; Tabata, H.; Hara, Y.; Yamada, Y. Increase of scopolamine production by high density culture of *Duboisia myoporoides* roots. *Biosci. Biotechnol. Biochem.* **1994b**, *58*, 1447–1450.

Yun, D.-J.; Hashimoto, T.; Yamada, Y. Metabolic engineering of medicinal plants: transgenic *Atropa belladonna* with an improved alkaloid composition. *Proc. Natl. Acad. Sci. USA* **1992**, *89*, 11799–11803.

Zenk, M. H.; El-Shagi, H.; Arens, H.; Stöckigt, J.; Weiler, E. W.; Deus, B. Formation of the indole alkaloids serpentine and ajmalicine in cell suspension cultures of *Catharanthus roseus*. In Plant Tissue Culture and its Bio-technological Application; Barz, W., Reinhard, E., Zenk, M. H. Eds.; Springer-Verlag: Berlin, **1977**; pp. 27–43.

Zhou, Y.; Hirotani, M.; Rui, H.; Furuya, T. Two triglycosidic triterpene astragalosides from hairy root cultures of *Astragalus membranaceus*. *Phytochem.* **1995**, *38*, 1407–1410.

Zhou, Y.; Hirotani, M.; Yoshikawa, T.; Furuya, T. Flavonoids and phenylethanoids from hairy root cultures of *Scutellaria baicalensis*. *Phytochem.* **1997**, *44*, 83–87.

Zocher, R.; Weckwerth, W.; Hacker, C.; Kammer, B.; Hornbogen, T.; Ewald, D. Biosynthesis of taxol: enzymatic acetylation of 10-deacetylbaccatin-III to baccatin-III in crude extracts from roots of *Taxus baccata*. *Biochem. Biophys. Res. Commun.* **1996**, *229*, 16–20.

INDEX